Nanomaterials for Energy and Sensor Applications

Editors

Vidya Nand Singh

CSIR-National Physical Laboratory
New Delhi, India

Sunil Singh Kushvaha

CSIR-National Physical Laboratory
New Delhi, India

CRC Press

Taylor & Francis Group
Boca Raton London New York

CRC Press is an imprint of the
Taylor & Francis Group, an **informa** business

A SCIENCE PUBLISHERS BOOK

Cover credit: Image reproduced by kind courtesy of the authors of Chapter 4.

First edition published 2024
by CRC Press
2385 NW Executive Center Drive, Suite 320, Boca Raton FL 33431

and by CRC Press
4 Park Square, Milton Park, Abingdon, Oxon, OX14 4RN

CRC Press is an imprint of Taylor & Francis Group, LLC

Library of Congress Cataloging-in-Publication Data (applied for)

ISBN: 978-1-032-39700-9 (hbk)
ISBN: 978-1-032-39702-3 (pbk)
ISBN: 978-1-003-35096-5 (ebk)

DOI: 10.1201/9781003350965

Typeset in Palatino
by Shubham Creation

Preface

The book, "Nanomaterials for Energy and Sensor Applications," edited by Dr. Vidya Nand Singh and Dr. Sunil Singh Kushvaha and published by CRC Press, Taylor & Francis Group, LLC, is the result of joint efforts of some well-known researchers in the field. Nowadays, most of the research work focuses on nanomaterials in which one of the dimensions falls in the range of 1–100 nm. These nanomaterials can be in the form of thin films, quantum dots, nanowires, nanopyramids, and nanoclusters. Thus, nanomaterials are touching almost all aspects of materials for various applications in the field of emerging energy and sensor devices. The book comprises ten chapters and discusses various aspects of nanomaterials for their application in the field of energy, solar cells, water splitting and sensors, etc. The book aims to cater to the need of budding researchers in the field of synthesis and application of nanomaterials and post-graduate students.

Chapter 1 provides a brief introduction to nanomaterials used for energy storage devices such as supercapacitors, pseudo-capacitors, etc. This chapter summarized various carbon nanofibers based nanostructured, e.g., hollow, porous, porous-hollow structures to enhance the electrochemical performance of supercapacitors. Moreover, it discusses the role of mesopore volume and total pore volume in determining the electrochemical performance of supercapacitors.

The need for renewable energy sources for the sustainable growth of human society is one of the key parameters. The various nanocomposite development tools for energy production have been discussed in Chapter 2. Furthermore, it presents several methods of harvesting energy such as hydropower resources, ocean and tide energy, wind energy, solar power, etc., and summarized the several applications of energy harvesting devices. Additionally, this chapter also provides insight into futuristic flexible electronics by using flexible polymer-based materials.

Chapter 3 explores silicon (Si) based nanostructures for solar energy conversion applications and also discusses the electronic and optical properties of Si nanostructures. This chapter provides wide synthesis procedures to prepare Si nanostructure such as the vapor-liquid-solid (VLS) method, chemical vapor deposition (CVD), electrochemical etching, electropolishing, etc. It explores the merit and demerits of Si nanostructures and provides a possible solution to overcome their limitations.

The hydrogen production by photoelectrochemical (PEC) water splitting technique is one of the ideal techniques to utilize the solar spectrum. Chapter 4 discusses the need for highly efficient photocatalysts to hunt the PEC technology in a commercial platform. Further, it explores the role of binary chalcogenides material in PEC application. The prepared film is characterized by X-ray diffraction (XRD), Raman spectroscopy, field emission scanning electron microscopy (FESEM), and X-ray photoelectron spectroscopy (XPS). A few methods to enhance the PEC performance of Bi_2Se_3 material by designing various heterostructures, doping of co-catalyst, Z-scheme structure, and p-n junction etch have been also discussed. Moreover, flexible photoelectrode can be beneficial to large-scale roll-to-roll PEC device fabrication and other applications such as flexible solar cells, flexible thermoelectric generators, etc.

Chapter 5 provides a review of the quantum-cutting phosphors phenomenon and their application in various solar thermal devices, such as CRT tubing, plasma display, light bulbs, and X-ray conversion screens, etc. and also in various biomedical applications (such as tumor monitoring, diagnosing ischemia, etc.) This chapter highlights various key point which affects the thermal stability of phosphors and gives appropriate solution to overcome these limitations.

Chapter 6 is a review of the recent development of a flexible sensor for their practical use in wearable electronics, artificial intelligence, etc. Further, the author introduces sensor parameters and the basic working mechanism of various types of sensors. This chapter also discusses many fabrication methods and various kinds of flexible substrates polyimide, PET, PEN, etc. to fabricate flexible sensors to development of futuristic flexible electronics.

Chapter 7 presents the recent insight into Pb and Pb-free organic-inorganic metal halide (OIMH) perovskite materials. In this chapter, the authors present the pros or cons of Pb and Pb free (OIMH) perovskite materials and give appropriate solutions to tackle their limitations, such as stability and efficiency, etc. Furthermore, this chapter provides different methods for improving the stability and quantum efficiency of Pb-free as compared to Pb-based perovskite inks synthesized by different solution-processed colloidal methods.

The advantage of dye sensitization in solar cell (DSSCs) application to increase the solar conversion efficiency of photoelectrodes is one of the innovative approaches. Chapter 8 presents the construction of DSSCs and details of working electrodes, including conductive substrates, compact layers, mesoporous active layers, etc. The preparation of TiO_2 powders, doping, composite structures, junction formations, and unique and mixed morphological influences on the DSSC device performance was also reviewed. At last, the authors ended this chapter by considering how DSSC devices integrated into modules from small scale to large scale for real-world applications.

Chapter 9 is mainly focused on the need for renewable energy sources and their importance for the sustainable development of human society. It provides a brief introduction to hydrogen production via photocatalytic (PC) and its working mechanism. Various semiconductors, e.g., metal oxides, like titanates, niobates, etc., have been discussed for PC applications. To increase the efficiency of photocatalysts in the PC process, various engineering processes such as doping, forming heterojunctions, and co-catalysts have been widely discussed in the literature.

Chapter 10 deals with a fundamental understanding of the correlations development of nanofluids for heat transfer systems and energy applications. It discusses various nanomaterials for widespread applications of the nanofluidic, the mechanism of heat transfer in nanofluids, and also discusses key parameters that affect the heat transfer behavior in nanofluids. Moreover, the challenges and sustainability aspects of nanofluids in heat transfer applications have been discussed in detail for large-scale implementation in commercial applications.

We would like to thank all those who kindly contributed chapters to this book. We are also indebted to the editorial office and the publishing and production teams at CRC Press for their assistance in preparing and publishing this book. Finally, we hope that the efforts of various authors will be helpful to the research community.

Vidya Nand Singh
Sunil Singh Kushvaha

Acknowledgments

First and foremost, we humbly express our gratitude to the Omnipotent God for showering us with his blessings throughout this venture. Without His grace, this book could not have become a reality.

We would like to take this opportunity to thank everyone who has been instrumental in the successful completion of this book. We are especially grateful to all the authors who have made significant contributions to this book. Their timely inputs and willingness to respond to every query aided us in undertaking and accomplishing this challenging task. By presenting their work on the latest areas of nanomaterials and their applications in energy and sensor devices, all the authors have justified their contribution to this book, for which we are truly indebted.

We would like to extend our sincere thanks to Prof. Venugopal Achanta, Director, CSIR-National Physical Laboratory, New Delhi (India), for his constant support and encouragement in creating an academic and research environment that fosters higher education. We would also like to thank Dr. Nahar Singh and Sh. J.C. Biswas for their kind support whenever we needed it.

We are equally thankful to the editorial team of CRC Press, Taylor & Francis, for their efforts in bringing the book to its present form and ensuring its timely publication.

Vidya Nand Singh
Sunil Singh Kushvaha

Contents

Porous and Hollow Carbon Nanofibrous Electrode Materials from Electrospinning for Supercapacitor Energy Storage

Kingsford Asare[1], Md Faruque Hasan[1],
Abolghasem Shahbazi[2]* and Lifeng Zhang[1]*

[1]Department of Nanoengineering, Joint School of Nanoscience and Nanoengineering, North Carolina A&T State University, 2907 E Gate City Blvd, Greensboro, NC 27401, USA.

[2]Department of Natural Resources and Environmental Design, College of Agriculture and Environmental Sciences, North Carolina A&T State University, 1601 E Market St, Greensboro, NC 27411, USA.

1.1 INTRODUCTION TO SUPERCAPACITOR

Electrical energy storage is essential in modern society and spans uses from consumer electronics to motorized vehicles. Among all electrical energy storage devices, supercapacitors have demonstrated their great promises in the energy storage market owing to their high power

*For Correspondence: Lifeng Zhang (lzhang@ncat.edu), Abolghasem Shahbazi (ash@ncat.edu)

density, faster charge and discharge rates, and cycle stability [1]. They are best employed for devices that require high current in a short time. According to the mechanism of charge storage, supercapacitors can be classified into three categories: electric double-layer capacitors (EDLC), pseudocapacitors, and hybrid capacitors. EDLC stores electrical charges at the interface between the electrode and electrolyte upon ion adsorption when a voltage is applied. The capacitance of EDLC is proportional to the accessible surface area of its electrode. EDLC can provide ultrahigh power and excellent cycle stability due to the fast and non-destructive process between electrode and electrolyte [2, 3]. Pseudocapacitor stores electrical charges by making use of fast and reversible redox reactions at the interface between the electrode surface and electrolyte. Charges that are associated with the redox reaction transfer across the abovementioned interface and show pseudocapacitance [4]. Pseudocapacitors can have greater specific capacitance and energy density than EDLC, but they have lower power performance and rate capability. The hybrid capacitors combine both EDLC capacitance and pseudocapacitance with improved electrochemical performance.

1.2   ELECTROSPUN CARBON NANOFIBROUS MATERIALS FOR SUPERCAPACITOR ELECTRODE

The electrochemical performance of supercapacitors relies on quite a few factors like electrode materials, electrolytes, and voltage, among which the characteristics of electrode material are the most critical. Carbon has been a choice of supercapacitor electrode material due to its excellent electrical conductivity, chemical and thermal stability, and low cost [5]. A variety of carbon-based electrode materials have been investigated so far including activated carbon, carbon nanotube, carbon cloth, carbon aerogel, carbide-derived carbon, etc.

In recent years, carbon nanofibers from electrospinning have been extensively studied as promising electrode materials for supercapacitors [6–8]. Unlike conventional fiber spinning techniques, like wet spinning and dry spinning, electrospinning utilizes electrical driving force and has a unique thinning mechanism (whipping instability) [9]. As a result, fibers with diameters at least one order of magnitude smaller than conventional fibers are obtained from electrospinning. In general, the preparation of electrospun carbon nanofibers (ECNFs) follows three steps: electrospinning, stabilization, and carbonization. Firstly, a carbon precursor polymer typically polyacrylonitrile (PAN), the most popular carbon precursor polymer with high carbon yield, is dissolved in N,

N-dimethyl formamide (DMF) to make an electrospinning solution (spin dope). The solution is then electrospun to obtain PAN nanofibers. Secondly, the electrospun PAN nanofibers are heated to 200–300 °C in air to cyclize PAN molecules. The PAN cyclization reaction can facilitate its formation of a ladder molecular structure in carbonization process and thus reduce mass loss and dimension shrinkage therefrom. Thirdly, the stabilized PAN electrospun nanofibers are further heated to a temperature over 800 °C in an inert atmosphere to acquire ECNFs upon H and N extraction. The final ECNFs hold an advantage of conductive network structure for charge transfer and can be used as a stand-alone and binder-free electrode material, which warrants better electrochemical performance of the resultant supercapacitor electrode.

1.3 POROUS/HOLLOW CARBON NANOFIBROUS MATERIALS FOR SUPERCAPACITOR ELECTRODE

It is noted that the ECNFs from PAN alone have a solid structure, possess relatively low specific surface area, and correspondingly present a low EDLC capacitance as supercapacitor electrode material [8]. In order to improve the EDLC capacitance of ECNFs, a lot of efforts have been carried out to improve the specific surface area of ECNFs by creating porous or hollow structures [10]. For example, porous ECNFs were prepared by *in-situ* activation of ECNFs with chemical agents (such as H_3PO_4 [11]), integration of nanoscale templates (such as $CaCO_3$ nanoparticles in ECNFs followed by selective removal [12]), inclusion of sacrificial component under high temperature (such as poly(methyl methacrylate) (PMMA) with PAN in electrospinning [13]), or use of PAN-based block copolymers (such as poly(acrylonitrile-block-methyl methacrylate) (PAN-b-PMMA) for electrospinning and subsequent carbonization [14]). At the same time, electrode materials with 1D hollow nanostructures have demonstrated their advantages in supercapacitor applications by serving as "ion-buffering reservoirs" and reducing the diffusion path of electrons and ions [15, 16]. Co-axial electrospinning is a common and convenient electrospinning technique to make core-shell or hollow carbon nanofibers [17]. In co-axial electrospinning, precursor solutions for core and shell components are fed separately and simultaneously to the corresponding inner and outer nozzle, which are concentrically aligned as co-axial spinneret, and result in core-shell electrospun nanofibers. By selectively extracting the core component, hollow nanofibers are attained. Hollow ECNFs, for example, can be acquired by co-axial electrospinning with PAN solution as a shell solution and sacrificial PMMA solution as a core solution [18]. In recent years, hollow ECNFs as well as porous hollow ECNFs have been developed through co-axial electrospinning and explored as supercapacitor electrode materials [18–20]. Compared to solid ECNFs, these porous and/or hollow ECNFs exhibited improvement in electrochemical performance as electrode materials for supercapacitor applications.

1.4 COMPARATIVE STUDY OF POROUS AND HOLLOW CARBON NANOFIBROUS ELECTRODE MATERIALS FOR SUPERCAPACITOR APPLICATION

It is noteworthy that all other research about the electrochemical performance of porous and/or hollow ECNFs only reported one type of carbon nanofibrous structure, i.e. porous, hollow, or hollow porous, at a time and the electrochemical data from these reports are not directly comparable due to different electrospinning conditions and various electrochemical stations for electrochemical measurements. It is unknown if a certain 1D carbon nanofibrous structure outperforms another in electrode material for supercapacitors. For example, hollow ECNFs might outperform porous ECNFs as supercapacitor electrode material due to the "ion-buffering reservoir" effect while hollow porous ECNFs might outperform hollow ECNFs or porous ECNFs as electrode material for supercapacitor due to their chance to take advantages of both hollow and porous nanostructures. However, there is no such conclusion or evidence yet. To verify this hypothesis, we prepared four types of supercapacitor electrode materials with different carbon nanofibrous structures, i.e. solid, porous, hollow, and hollow porous ECNFs, respectively, and conducted a side-by-side comparison for their electrochemical performance as electrode materials for EDLC supercapacitor. The findings herein are expected to provide an in-depth understanding of the relationship between 1D carbon nanofibrous structures from electrospinning including solid, porous, hollow, and hollow porous nanostructures and their corresponding electrochemical performance as supercapacitor electrode materials, which can benefit the advances of electrode materials for high-performance supercapacitors.

1.4.1 Preparation of Electrode Materials

Based on the fact that PAN and PMMA have phase separation in their bicomponent nanofibers and PMMA can decompose completely under heat and generate only volatile products as a sacrificial component

[21], four types of carbon nanofibrous materials including solid carbon nanofibers (ECNFs), porous carbon nanofibers (P-ECNFs), hollow carbon nanofibers (H-ECNFs), and hollow porous carbon nanofibers (HP-ECNFs) were prepared in this research through electrospinning and co-axial electrospinning followed by stabilization and carbonization [22]. Specifically, ECNFs were prepared from electrospinning a 12 wt.% PAN (Mw = 150,000) DMF solution. P-ECNFs were prepared from electrospinning a 12 wt.% PAN/PMMA (Mw = 120,000) DMF solution with PAN/PMMA compositions at 90/10, 70/30, and 50/50 and denoted as P-ECNF-90-10, P-ECNF-70-30, and P-ECNF-50-50, respectively. In both cases, the electrospinning was conducted at 15 kV with a solution feed rate of 1 mL/h. To make H-ECNFs and HP-ECNFs, a co-axial electrospinning setup was used with certain polymer solutions for shell and core components, respectively. In the co-electrospinning for H-ECNFs, the shell solution was fixed with 12 wt.% PAN DMF solution while the core solutions were PMMA DMF solutions with varied PMMA concentrations at 10 wt.%, 20 wt.%, and 30 wt.% and labeled as H-ECNF-10, H-ECNF-20, and H-ECNF-30, respectively. In the co-axial electrospinning for HP-ECNFs, the core solution was fixed with 20 wt.% PMMA DMF solution while the shell solutions were 12 wt.% PAN/PMMA DMF solutions but with varied PAN/PMMA compositions at 90/10, 70/30, and 50/50 and labeled as HP-ECNF-90-10, HP-ECNF-70-30, and HP-ECNF-50-50, respectively. All the co-axial electrospinning was conducted at 15 kV with a feeding rate of 1.5 mL/h for shell solution and 1 mL/h for core solution. Stabilization of as-spun nanofibrous mats was carried out in air at 280 °C for 6 hours with a heating rate of 1 °C/min from room temperature. The stabilized nanofibrous mats were cooled down to room temperature and then re-heated in a nitrogen atmosphere to 900 °C at a heating rate of 5 °C/min. The nanofibrous mats were further maintained at 900 °C for 1 hour to be fully carbonized before cooling down to room temperature. Herein, PMMA was used as a sacrificial component and it completely decomposed in the process of carbonization [21] for hollow and porous nanostructures as described.

1.4.2   Electrode Materials Characterization

1.4.2.1   *Morphology*

The respective morphology of the carbon nanofibrous electrode materials was examined using a Zeiss Auriga field emission scanning electron microscope (FESEM). ECNFs were solid fibers with an average diameter of 1034 ± 95 nm (Figure 1.1A). P-ECNFs showed porous structure both on the surface and inside (Figure 1.1B–1.1D). At PAN/PMMA = 50/50, elongated pores and/or short channels were observed. With the increase of PMMA content from 10 wt.% to 50 wt.% in PAN/PMMA bicomponent fibers, the average fiber diameter reduced from 998 ± 91 nm at 10 wt.% PMMA to 679 ± 72 nm at 30 wt.% PMMA and to 565 ± 46 nm at 50 wt.% PMMA. H-ECNFs presented a hollow structure as expected. With the increase of PMMA concentration from 10 wt.% to 30 wt.% in the core solution of co-axial electrospinning for H-ECNFs, the average fiber diameter increased from 513 ± 48 nm at 10 wt.% PMMA to 921 ± 89 nm at 20 wt.% PMMA and to 1679 ± 94 nm at 30 wt.% PMMA, respectively (Figure 1.1E–1.1G). The wall thickness of these hollow fibers was similar, i.e. 150–180 nm. HP-ECNFs showed a combined hollow and porous morphology, i.e. hollow fibers with porous walls (Figure 1.1H–1.1J). With the increase of PMMA content from 10 wt.% to 50 wt.% in the shell solution of co-axial electrospinning for HP-ECNFs, the average fiber diameter reduced from 1588 ± 85 nm at 10 wt.% PMMA to 782 ± 52 nm at 30 wt.% PMMA and to 503 ± 59 nm at 50 wt.% PMMA with wall thickness reduction from ~200 nm to ~100 nm, correspondingly.

It is well known that PMMA completely degrades and gasifies in a temperature range of 300–400 °C [21]. In hollow carbon nanofibers, the fiber size, porous structure, hollow channel size, and wall thickness are determined by both an inflation effect due to the discharge of a relatively large volume of volatile PMMA degradation/gasification products and a volume shrinking effect due to PMMA removal and PAN carbonization in the process of carbonization. Polymer-polymer phase separation in the PAN/PMMA bicomponent as-electrospun nanofibers occurs and domains of PMMA in the bicomponent nanofibers can completely decompose during carbonization. A larger proportion of PMMA would lead to a larger volume shrinking in carbonization and reduce average fiber size in the case of P-ECNFs. In the meantime, a larger proportion of PMMA could form more elongated domains under electrical driving force in the electrospinning process and result in a "short channel" structure in the carbonized fibers. In the case of H-ECNFs, a larger amount of PMMA in the core of the core-shell PAN/PMMA carbon precursor nanofibers from higher PMMA concentration in the core solution of the co-axial electrospinning could generate a larger volume of volatile products upon degradation/gasification in the process of carbonization, counteract the fiber contraction in that process, and thus resulted in larger hollow channels. As for HP-ECNFs, the porous wall structure of hollow nanofibers could reduce the thickness as well as mechanical strength of the fiber walls and lead to huge size shrinkage/collapse in the process of carbonization. With PAN/PMMA = 50/50 in the shell solution of the co-axial electrospinning for HP-ECNFs, some long and open slits on the fiber surface were observed.

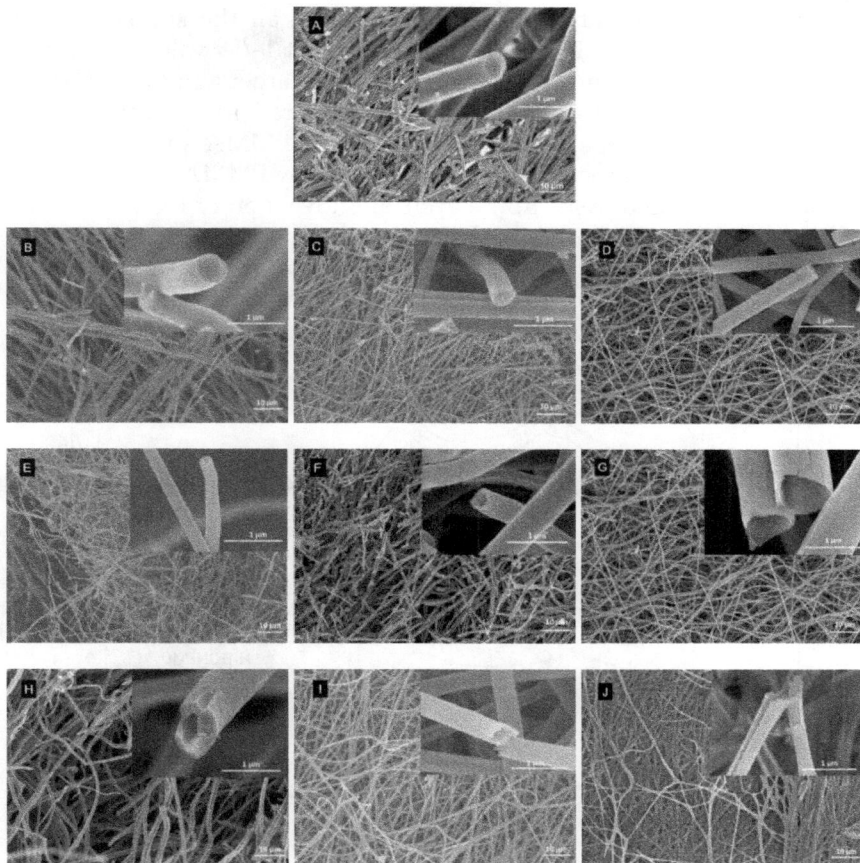

Figure 1.1   Representative SEM images of electrospun carbon nanofibrous materials: (A) ECNFs; (B) P-ECNF-90-10; (C) P-ECNF-70-30; (D) P-ECNF-50-50; (E) H-ECNF–10; (F) H-ECNF-20; (G) H-ECNF-30; (H) HP-ECNF-90-10; (I) HP-ECNF-70-30; (J) HP-ECNF-50-50 [22].

1.4.2.2   *Structure*

Raman spectroscopy was used to characterize carbon structure of the electrospun carbon nanofibrous materials and was done at room temperature using a Horiba Raman Confocal Microscope at an excitation wavelength of 532 nm. In Raman spectra of all the carbon nanofibrous materials (Figure 1.2), the "D-band" between 1,297 and 1,355 cm^{-1} corresponded to the sp^3 hybridized disordered carbonaceous structures while the "G-band" between 1,548 and 1,591 cm^{-1} indicated the sp^2 hybridized graphitic phase of carbon [23]. The intensity ratio of G-band to D-band (I_G/I_D) was used to characterize the carbon structure of these carbon nanofibrous materials. Among all the studied carbon nanofibrous materials, ECNFs showed the least I_G/I_D value, indicating the least ordered carbon structure. The use of a larger amount of PMMA for P-ECNFs, H-ECNFs, and HP-ECNFs resulted in a more ordered carbon structure. Apparently, the addition of PMMA promoted PAN carbonization. This promotion might be caused by PMMA's hydroxyl end groups, which could assist in the cyclization reaction of PAN molecules via an ionic mechanism [24]. Under this circumstance, the activation energy of the PAN cyclization reaction is lowered, the cyclization reaction temperature is reduced, and the enthalpy of the cyclization reaction is increased, consequently leading to the formation of a more ordered carbon structure.

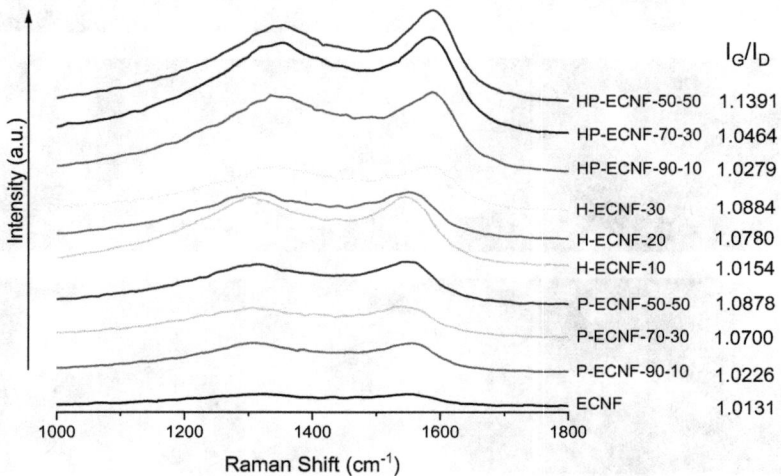

	I_G/I_D
HP-ECNF-50-50	1.1391
HP-ECNF-70-30	1.0464
HP-ECNF-90-10	1.0279
H-ECNF-30	1.0884
H-ECNF-20	1.0780
H-ECNF-10	1.0154
P-ECNF-50-50	1.0878
P-ECNF-70-30	1.0700
P-ECNF-90-10	1.0226
ECNF	1.0131

Figure 1.2   Raman spectra of the electrospun carbon nanofibrous materials [22].

BET surface area and porosity analyses of the electrospun carbon nanofibrous materials were performed by using a Micromeritics ASAP 2020 Surface Area and Porosity Analyzer. The N_2 adsorption isotherm as well as pore size distribution based on BJH adsorption of the electrospun carbon nanofibrous materials are shown in Figure 1.3. The N_2 adsorption curves of the electrospun carbon nanofibrous materials can be classified as IUPAC Type II. The increase of N_2 adsorption in low-pressure region ($P/P_0 < 0.1$) indicated N_2 adsorption in micropores, while the large increase of N_2 adsorption in high-pressure region ($P/P_0 > 0.9$) indicated N_2 adsorption in mesopores [25]. The N_2 adsorption curves, as well as pore size distribution results, suggested that P-ECNF-90-10, P-ECNF-50-50, H-ECNF-30, and HP-ENCF-90-10 possessed relatively large amounts of micropores while all the porous, hollow, and hollow porous samples had a significant amount of mesopores.

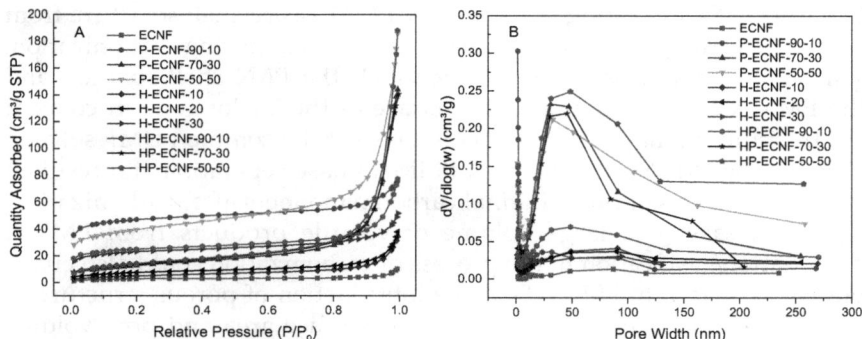

Figure 1.3 N$_2$ adsorption isotherms (A) and pore size distributions (B) of electrospun carbon nanofibrous materials. The pore size distribution was obtained via the BJH adsorption [22].

According to BET-specific surface area and porosity analysis results (Table 1.1), ECNFs showed the lowest specific surface area and the smallest pore volume, indicating a solid fiber structure. P-ECNFs demonstrated the largest specific surface area of 146 m^2/g in the case of PAN/PMMA = 90/10 in electrospinning solution. However, the corresponding pore volume was the smallest among the three P-ECNF samples. This is probably caused by the micropore formation in P-ECNF-90-10. At this low proportion, PMMA formed much smaller domains in PAN/PMMA bicomponent nanofibers and resulted in the largest micropore volume upon degradation/gasification. With the increase of PMMA content in the electrospinning solution, the corresponding micropore volume of P-ECNFs reduced, while its mesopore and macropore volumes increased. Although P-ECNF-50-50 possessed the largest pore volume of 0.2763 cm^3/g, the largest proportion of its pores was mesopores and its specific surface area was still less than that of P-ECNF-90-10. Compared to P-ECNF-90-10, P-ECNF-70-30 contained less micropore volume due to larger PMMA phase separation domains in the PAN/PMMA bicomponent precursor fibers. Compared to P-ECNF-50-50, P-ECNF-70-30 produced less amounts of PMMA volatile products upon degradation/gasification and created fewer pores, resulting in smaller pore volume (micropore, mesopore, macropore, and total pore volumes) and thus smaller specific surface area than that of P-ECNF-50-50. H-ECNFs possessed the least pore volume compared to P-ECNFs and HP-ECNFs, indicating solid walls in these hollow nanofibers. Meanwhile, the hollow space did not contribute much to micro-, meso- or macro-porous structures. The pore volumes of all H-ECNF samples were close and there were some increases in total pore volume with the increase of PMMA concentration in core solution of the co-axial electrospinning for H-ECNFs. The total pore volumes of H-ECNFs, however, were still more than four times that of ECNFs, indicating that the PAN/PMMA core-shell structure from co-axial electrospinning did generate extra pores during carbonization. This could be ascribed to two aspects: (1) the PAN shell solution and the PMMA core solution at the interface of the Taylor cone in co-axial electrospinning may partially mix because of the common DMF solvent. This mixing led to very fine PAN-PMMA phase separation and resulted in different levels of pores in final carbon nanofibers after carbonization; (2) the release of a large volume of volatile products from PMMA degradation/gasification in the process of carbonization could also create pores. Compared to H-ECNF-20, the introduction of porous structure in walls of hollow fibers increased specific surface area and pore volume of HP-ECNFs. With the increase of PMMA proportion in shell solution of the co-axial electrospinning, the micropore volume of HP-ECNFs reduced with increase of mesopore, macropore, and total pore volumes. The reduction of specific surface area of HP-ECNFs with increase of PMMA proportion in shell solution of the co-axial electrospinning may be attributed to shrinkage and collapse of the thinner and more porous fiber walls in the process of carbonization.

Table 1.1   BET-specific surface area and porosity of electrospun carbon nanofibrous materials [22]

Carbon Nanofibrous Materials	S_{BET} (m²/g)	V_{micro} (cm³/g)	V_{meso} (cm³/g)	V_{macro} (cm³/g)	V_{total} (cm³/g)	BJH Average Pore Size (nm)
ECNF	9.4	0.0006	0.0065	0.0053	0.0124	16.877
Pore proportion		4.8%	52.4%	42.7%		
P-ECNF-90-10	146	0.0260	0.0637	0.0094	0.0991	3.582
Pore proportion		26.2%	64.3%	9.5%		
P-ECNF-70-30	45.9	0.0021	0.1706	0.0476	0.2204	19.173
Pore proportion		1.0%	77.4%	21.6%		
P-ECNF-50-50	130	0.01	0.2031	0.0633	0.2763	11.069
Pore proportion		3.6%	73.5%	22.9%		
H-ECNF-10	18.1	0.0014	0.0324	0.0164	0.0503	13.756
Pore proportion		2.8%	64.4%	32.6%		
H-ECNF-20	26.3	0.0023	0.0333	0.0188	0.0544	11.572
Pore proportion		4.2%	61.2%	34.6%		
H-ECNF-30	73.3	0.0068	0.0422	0.0103	0.0592	5.738
Pore proportion		11.5%	71.3%	17.4%		
HP-ECNF-90-10	78.8	0.0062	0.0722	0.0282	0.1066	8.604
Pore proportion		5.8%	67.7%	26.5%		
HP-ECNF-70-30	47.7	0.0015	0.1580	0.0545	0.2141	18.132
Pore proportion		0.7%	73.8%	25.5%		
HP-ECNF-50-50	51.5	0.0017	0.1877	0.0933	0.2827	24.790
Pore proportion		0.7%	66.4%	33%		

S_{BET} – BET specific surface area; V_{micro} – micropore volume; V_{meso} – mesopore volume; V_{macro} – macropore volume; V_{total} – total pore volume

1.4.3   Electrochemical Evaluation

The electrochemical performance of all the electrospun carbon nanofibrous materials was evaluated by cyclic voltammetry (CV) and galvanostatic charge-discharge (CD) tests in 6M KOH aqueous electrolyte. Specifically, a square piece (1 cm × 1 cm) of each as-prepared carbon nanofibrous mat was cut and weighed. The square piece was then firmly attached to an Au working electrode using conductive carbon glue and placed in 6M KOH aqueous solution in combination with a reference electrode of Ag/AgCl and a counter electrode of platinum rod to construct a three-electrode electrochemical system, which was connected to a CHI660E electrochemical workstation for electrochemical measurement. CV test was conducted using a potential range of 0.0 to –0.8 V with a scan rate of 5, 10, 20, 50, and 100 mV/s, respectively. The galvanostatic CD test was performed at current densities of 0.5, 1, and 2 A/g, respectively, in the potential range of 0.0 to –0.8 V.

The specific capacitance (C_{sp}) was calculated based on CV analysis from the following formula:

$$C_{sp} = \frac{\int_{V_1}^{V_2} I(V)dV}{2 \times m \times v \times \Delta V}$$

where I is current, m is mass of electrode material, v is scan rate, and $\Delta V = V_2 - V_1$, which is the sweeping potential window (0.8 V) and CD analysis from the following formula:

$$C_{sp} = \frac{I \times \Delta t}{m \times \Delta V}$$

where I is discharge current, m is mass of electrode material, Δt is discharge time, and ΔV is potential window (0.8 V).

1.4.3.1   Specific Capacitance

CV profiles of all the carbon nanofibrous electrode materials at a scan rate of 5 mV/s as well as their CD profiles at the current density of 0.5 A/g are shown in Figure 1.4. The specific capacitances of these carbon nanofibrous electrode materials from both CV and CD tests were compared in Figure 1.5 and these values generally matched with each other, i.e. specific capacitance from CV (Figure 1.5A) vs. specific capacitance from CD (Figure 1.5B). Compared to ECNFs, all the carbon nanofibrous electrode materials with porous and/or hollow nanostructures demonstrated better electrochemical performance. In each type of electrospun carbon nanofibrous material, i.e. P-ECNFs,

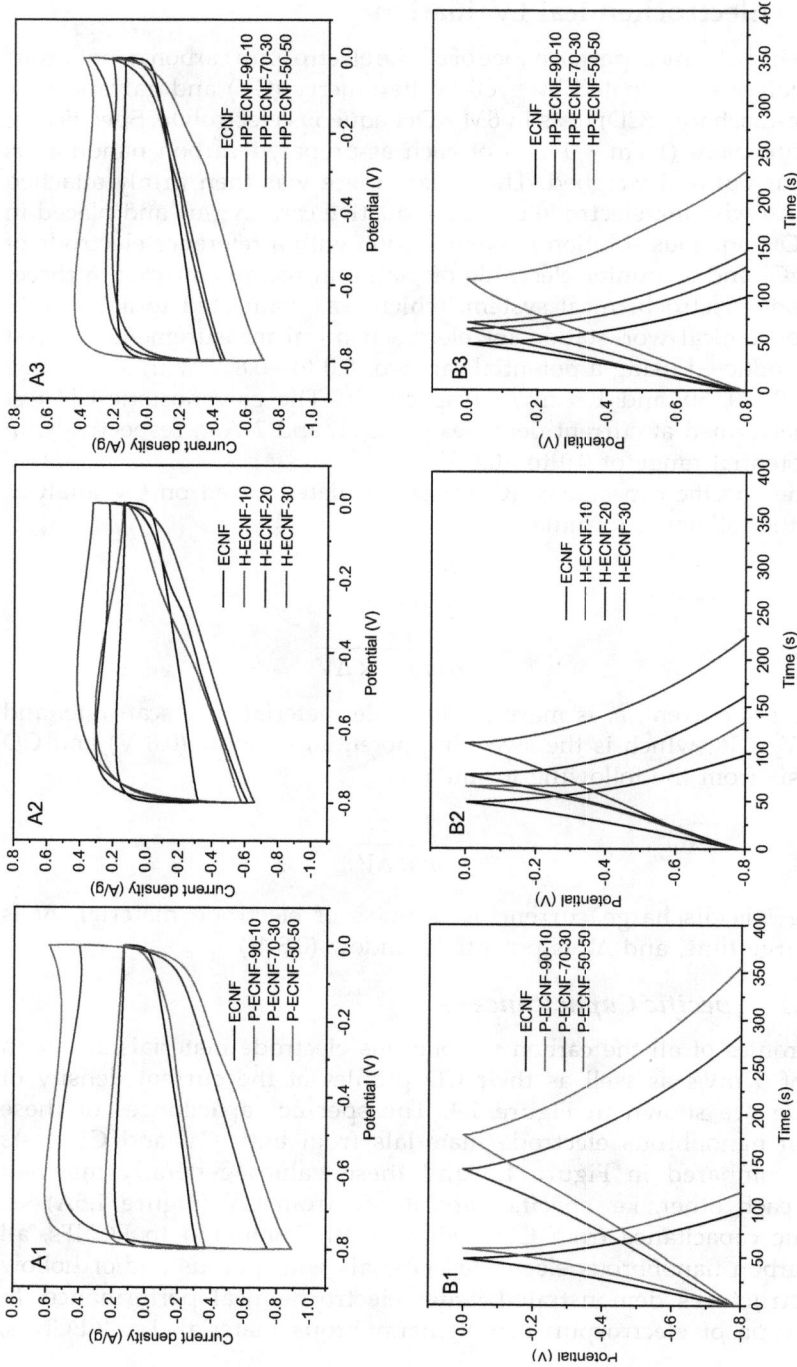

Figure 1.4 CV (A) and galvanostatic CD (B) profiles of P-ECNFs (labeled as 1), H-ECNFs (labeled as 2), and HP-ECNFs (labeled as 3). ECNFs are used as a reference in all plots [22].

H-ECNFs, and HP-ECNFs, electrochemical performance increased with the content of PMMA in electrospinning and co-axial electrospinning solutions. With the largest PMMA content in electrospinning or co-axial electrospinning solution for each type of carbon nanofibrous material, the sequence of electrical performance is P-ECNF-50-50 > HP-ECNF-50-50 > H-ECNF-30. P-ECNF-50-50 was further scanned at varied scan rates of 5–100 mV/s for CV test and examined with varied current densities at 0.5, 1, and 2 A/g, respectively, for CD test (Figure 1.6). With the increase of voltage sweep rate, the CV curves retained their quasi-rectangular shape. All the CD curves were almost linear and exhibited isosceles triangles. All these results indicated excellent supercapacitor behavior of P-ECNF-50-50.

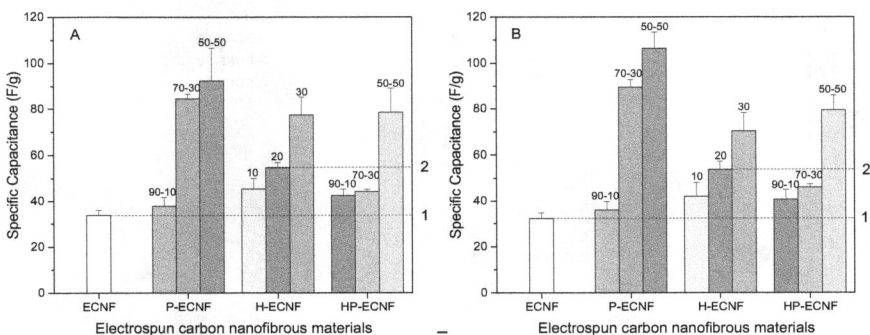

Figure 1.5 Specific capacitances of electrospun carbon nanofibrous materials including ECNFs, P-ECNFs, H-ECNFs, and HP-ECNFs from CV at 5 mV/s (A) and galvanostatic CD at 0.5 A/g (B). ECNFs is used as a control of all the electrospun carbon nanofibrous materials and marked with line 1 while H-ECNF-20 is used as a control of all the HP-ECNF samples with a fixed core solution of 20 wt.% PMMA in DMF and marked with line 2 [22].

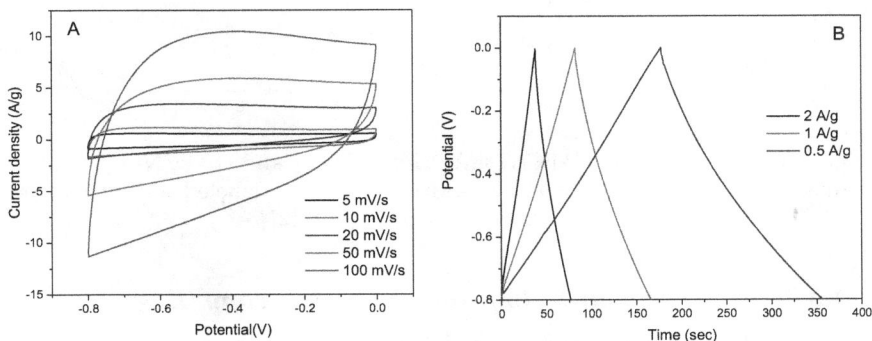

Figure 1.6 CV (A) and galvanostatic CD (B) profiles of P-ECNF-50-50 at different scan rates (A) and current densities (B) [22].

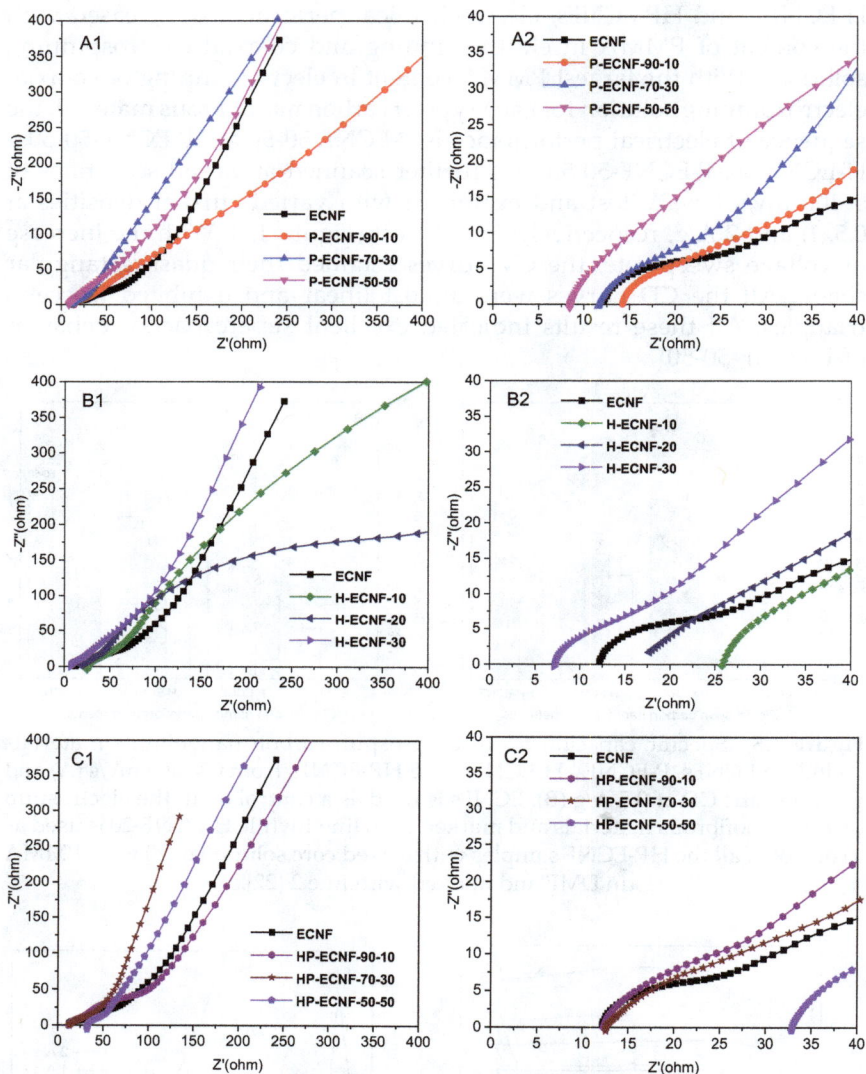

Figure 1.7 EIS profiles of P-ECNFs (A), H-ECNFs (B), and HP-ECNFs (C) in full frequency range (labeled as 1) and high-frequency range (labeled as 2). ECNFs are used as a reference in all plots [22].

1.4.3.2 *Electrochemical Impedance Spectroscopy (EIS)*

EIS has been extensively used to characterize electrode materials for supercapacitors [26]. EIS test herein was performed within the frequency range of 100 kHz to 0.1 Hz. Nyquist plots of all the electrospun carbon nanofibrous electrodes showed no or much-depressed semi-circles at high frequency and nearly straight lines at low frequency (Figure 1.7). For P-ECNFs, the sample from the electrospinning solution with the largest PMMA content (P-ECNF-50-50) exhibited the smallest electrode resistance (the intercept at the axis of the real part of complex impedance) and the smallest electrolyte resistance (the size of the semi-circle at high frequency), while the sample from the electrospinning solution with the least PMMA content (P-ECNF-90-10) showed the largest electrode resistance, the largest electrolyte resistance, and the least capacitive behavior (the angle between the straight line and the axis of the real part of complex impedance). For H-ECNFs, the sample from the core solution of co-axial electrospinning with the largest PMMA content (H-ECNF-30) exhibited the lowest electrode resistance, the smallest electrolyte resistance, and the best capacitive behavior. For HP-ECNFs, the sample from the shell solution of co-axial electrospinning with the largest PMMA content (HP-ECNF-50-50) demonstrated the smallest electrolyte resistance but the largest electrode resistance.

1.4.3.3 Cycling Stability

The cycling stability of the electrode materials including P-ECNF-50-50, H-ECNF-30, and HP-ECNF-50-50 was evaluated by monitoring the variation of specific capacitance with 3,000 cycles of CD test at 5 A/g within the potential window of 0.0 to −0.8 V (Figure 1.8). It is observed that there was almost no loss in specific capacitance for these nanofibrous electrode materials after 3,000 cycles of charging/discharging, indicating their long-term durability for supercapacitor use.

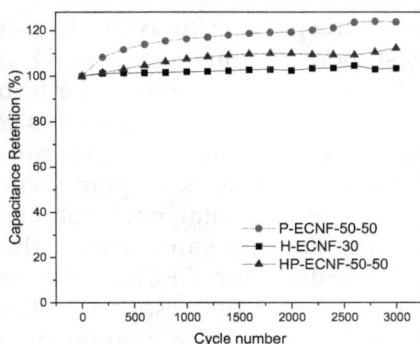

Figure 1.8 Cycling stability of electrospun carbon nanofibrous electrode materials at current density of 5 A/g [22].

Particularly P-ECNF-50-50 and HP-ECNF-50-50 even showed some increases in specific capacitance during the cycling test. This could be attributed to the wettability improvement and activation of the electrode caused by continuous diffusion of electrolyte ions into previously inaccessible pores and graphitic layers, which led to an increase of active charge storage sites of the electrode. P-ECNF-50-50 possessed a relatively large amount of micropores among the three P-ECNF samples and these micropores might gradually become accessible for electrolyte ions with charging/discharging cycles. Therefore P-ECNF-50-50 exhibited the most increase in specific capacitance during the cycling test.

1.4.3.4 *Discussion*

The electrochemical performance of these electrospun carbon nanofibrous materials showed a sequence of porous structure > hollow porous structure > hollow structure > solid structure. Compared to the porous structure, the hollow structure did not demonstrate the previously mentioned "ion-buffering reservoir" effect toward the electrode's specific capacitance. Meanwhile, the electrochemical performance of these electrospun carbon nanofibrous materials did not correlate linearly with their specific surface area, indicating that multiple factors instead of just specific surface area determine the electrochemical performance (specific capacitance) of these electrode materials. These factors may include overall accessible surface area, electrolyte resistance, electrode resistance, etc. It is observed that the samples with a larger proportion of micropore volume such as P-ECNF-90-10 and HP-ECNF-90-10 possessed higher specific areas but lower electrochemical performance in their respective material groups. They also exhibited the largest electrolyte resistance. These results indicated that electrolyte ions had more resistance to access micropores. In the meantime, the samples in each group, i.e. P-ECNFs, H-ECNFs, and HP-ECNFs, with the largest PMMA content in electrospinning/co-axial electrospinning solution exhibited the largest pore volume particularly mesopore volume and concurrently the best electrochemical performance. When the specific capacitance of samples from each group is compared with the corresponding sample's pore volume including micropore, mesopore, and macropore volumes as well as total pore volume, it is observed that the mesopore volume and total pore volume follow the same sequence as that of specific capacitance for all the samples in each group (Figure 1.9). Particularly for P-ECNFs, the specific capacitance is proportional to the amount of mesopore volume (Coefficient of Determination, R^2 = 0.99998) and the amount of total pore volume (R^2 = 0.99293) instead of the proportion of mesopore. These results indicated that mesopore volume and total pore volume are primary in determining the electrochemical performance of these electrospun carbon nanofibrous electrode materials. Total pore volume determines the total number of available ion adsorption sites. Mesopore volume not only contributes to ion adsorption sites but also determines the transportation

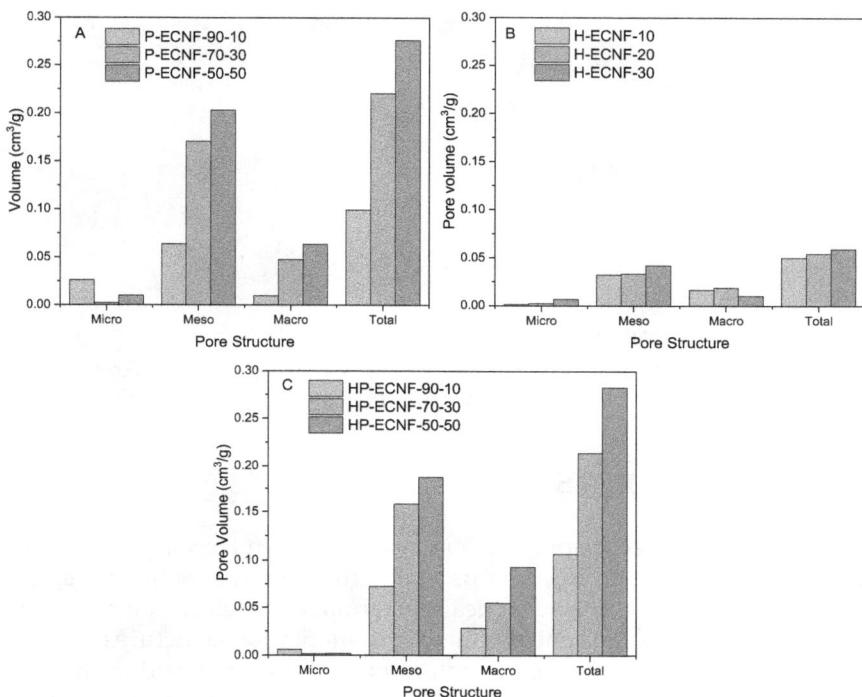

Figure 1.9   Comparison of pore volumes of P-ECNFs (A), H-ECNFs (B), and HP-ECNFs(C) [22].

difficulty of electrolyte ions to available adsorption sites. The mesopores may serve as connection paths and temporary reservoirs for electrolyte ions and assist electrolyte ions in access of micropores for charge storage (Figure 1.10). The mesopore volume sequence of electrospun carbon nanofibrous electrode materials matched their electrolyte resistance sequence from EIS very well (Figure 1.7). P-ECNF-50-50 possessed the highest mesopore volume and thus exhibited the highest electrochemical performance. When combining pore volume information with EIS results, it is also observed that micropores hindered electrolyte ion transport and increased electrolyte resistance, while hollow structures facilitated electrolyte transport and reduced electrolyte resistance. Electrode resistance also played a role in the final electrochemical performance. A larger amount of PMMA in the electrospinning/co-axial electrospinning solution led to a more ordered carbon structure and could result in lower electrode resistance. Nevertheless, the electrode resistance of HP-ECNF-50-50 increased significantly. This might be due to the open long slits on HP-ECNF-50-50 nanofibers as observed from SEM (Figure 1.1J) as well as the largest average pore size, which reduced short moving paths of electrons.

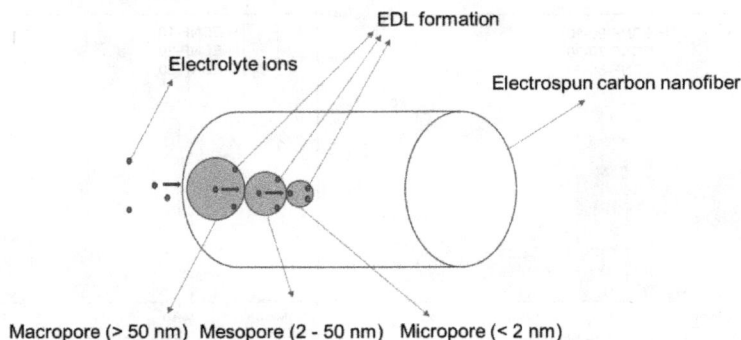

Figure 1.10   Schematic diagram of electrolyte ion transfer in electrospun carbon nanofibers [22].

1.5   CONCLUSIONS

In this chapter, electrospun carbon nanofibrous materials with solid, porous, hollow, and hollow porous nanostructures are comprehensively compared for their electrochemical performance as electrode materials for supercapacitors. In contrast to solid nanofibrous structures, all the carbon nanofibrous electrode materials with porous and/or hollow nanostructures demonstrate better electrochemical performance. Despite the variation of morphology, i.e. porous, hollow, and hollow porous nanostructures, mesopore volume and total pore volume are primary in determining the electrochemical performance (specific capacitance) of the nanofibrous electrode materials. Total pore volume determines the total number of available charge storage sites while mesopore volume controls electrolyte resistance. Mesopores can serve as connection paths and temporary reservoirs for electrolyte ions and assist electrolyte ions in access of micropores for more charge storage. The hollow nanofibrous structure can benefit electrolyte transport and reduce electrolyte resistance of the supercapacitor electrode but is not the decisive factor for its electrochemical performance. Electrode resistance of these electrospun carbon nanofibrous materials also plays a role in the final electrochemical performance through corresponding electrical resistance, which is dependent on carbon internal structure as well as overall fiber morphology.

ACKNOWLEDGMENTS

This work was performed in whole/part at the Joint School of Nanoscience and Nanoengineering, a member of the National Nanotechnology

Coordinated Infrastructure (NNCI), which is supported by the National Science Foundation (Grant ECCS-2025462).

REFERENCES

[1]   Kim, B.K., S. Sy, A. Yu and J. Zhang. 2015. Electrochemical Supercapacitors for Energy Storage and Conversion. pp. 1–25. *In:* J. Yan (ed.). Handbook of Clean Energy Systems. John Wiley & Sons.

[2]   Simon, P. and Y. Gogotsi. 2008. Materials for electrochemical capacitors. Nat. Mater. 7: 845–854.

[3]   Najib, S. and E. Erdem. 2019. Current progress achieved in novel materials for supercapacitor electrodes: mini review. Nanoscale Adv. 1: 2817–2827.

[4]   Schoetz, T., L.W. Gordon, S. Ivanov, A. Bund, D. Mandler and R.J. Messinger. 2022. Disentangling faradaic, pseudocapacitive, and capacitive charge storage: A tutorial for the characterization of batteries, supercapacitors, and hybrid systems. Electrochim. Acta 412: 140072.

[5]   Zhang, L.L. and X.S. Zhao. 2009. Carbon-based materials as supercapacitor electrodes. Chem. Soc. Rev. 38: 2520–2531.

[6]   Zhang, L., A. Aboagye, A. Kelkar, C. Lai and H. Fong. 2014. A review: carbon nanofibers from electrospun polyacrylonitrile and their applications. J. Mater. Sci. 49: 463–480.

[7]   Lai, C., Z. Zhou, L. Zhang, X. Wang, Q. Zhou, Y. Zhao, et al. 2014. Free-standing and mechanically flexible mats consisting of electrospun carbon nanofibers made from a natural product of alkali lignin as binder-free electrodes for high-performance supercapacitors. J. Power Sources 247: 134–141.

[8]   Aboagye, A., Y. Liu, J.G. Ryan, J. Wei and L. Zhang. 2018. Hierarchical carbon composite nanofibrous electrode material for high-performance aqueous supercapacitors. Mater. Chem. Phys. 214: 557–563.

[9]   Shin, Y.M., M.M. Hohman, M.P. Brenner and G.C. Rutledge. 2001. Experimental characterization of electrospinning: the electrically forced jet and instabilities. Polymer 42: 9955–9967.

[10] Zhang, B., F. Kang, J.-M. Tarascon and J.-K. Kim. 2016. Recent advances in electrospun carbon nanofibers and their application in electrochemical energy storage. Prog. Mater. Sci. 76: 319–380.

[11] Zhi, M., S. Liu, Z. Hong and N. Wu. 2014. Electrospun activated carbon nanofibers for supercapacitor electrodes. RSC Advances 4: 43619–43623.

[12] Zhang, L., Y. Jiang, L. Wang, C. Zhang and S. Liu. 2016. Hierarchical porous carbon nanofibers as binder-free electrode for high-performance supercapacitor. Electrochim. Acta 196: 189–196.

[13] He, G., Y. Song, S. Chen and L. Wang. 2018. Porous carbon nanofiber mats from electrospun polyacrylonitrile/polymethylmethacrylate composite nanofibers for supercapacitor electrode materials. J. Mater. Sci. 53: 9721–9730.

[14] Zhou, Z., T. Liu, A.U. Khan and G. Liu. 2019. Block copolymer-based porous carbon fibers. Sci. Adv. 5: eaau6852.

[15] Peng, S., L. Li, Y. Hu, M. Srinivasan, F. Cheng, J. Chen, et al. 2015. Fabrication of spinel one-dimensional architectures by single-spinneret electrospinning for energy storage applications. ACS Nano 9: 1945–1954.

[16] Xu, K., S. Li, J. Yang and J. Hu. 2018. Hierarchical hollow MnO_2 nanofibers with enhanced supercapacitor performance. J. Colloid Interface Sci. 513: 448–454.

[17] Han, D. and A.J. Steckl. 2019. Coaxial electrospinning formation of complex polymer fibers and their applications. ChemPlusChem. 84: 1453–1497.

[18] Shilpa and A. Sharma. 2016. Free standing hollow carbon nanofiber mats for supercapacitor electrodes. RSC Adv. 6: 78528–78537.

[19] Kim, J.-G., H.-C. Kim, N.D. Kim and M.-S. Khil. 2020. N-doped hierarchical porous hollow carbon nanofibers based on PAN/PVP@ SAN structure for high performance supercapacitor. Composites Part B 186: 107825.

[20] Le, T.H., Y. Yang, L. Yu, T. Gao, Z. Huang and F. Kang. 2016. Polyimide-based porous hollow carbon nanofibers for supercapacitor electrode. J. Appl. Polym. Sci. 133: 43397.

[21] Zhang, L. and Y.-L. Hsieh. 2009. Carbon nanofibers with nanoporosity and hollow channels from binary polyacrylonitrile systems. Eur. Polym. J. 45: 47–56.

[22] Asare, K., M.F. Hasan, A. Shahbazi and L. Zhang. 2021. A comparative study of porous and hollow carbon nanofibrous structures from electrospinning for supercapacitor electrode material development. Surf. Interfaces 26: 101386.

[23] Zhou, Z., C. Lai, L. Zhang, Y. Qian, H. Hou, D.H. Reneker, et al. 2009. Development of carbon nanofibers from aligned electrospun polyacrylonitrile nanofiber bundles and characterization of their microstructural, electrical, and mechanical properties. Polymer 50: 2999–3006.

[24] Zhang, L. and Y.-L. 2006. Hsieh. Nanoporous ultrahigh specific surface polyacrylonitrile fibres. Nanotechnology 17: 4416–4423.

[25] Lee, H.-M., H.-R. Kang, K.-H. An, H.-G. Kim and B.-J. Kim. 2013. Comparative studies of porous carbon nanofibers by various activation methods. Carbon Lett. 14: 180–185.

[26] Mei, B.-A., O. Munteshari, J. Lau, B. Dunn and L. Pilon. 2018. Physical interpretations of Nyquist plots for EDLC electrodes and devices. J. Phys. Chem. C. 122: 194–206.

Chapter 2

Energy and Sensor Applications of Polymer Nanocomposites

Ankit Kumar Srivastava[1], Swasti Saxena[2]* and
Surendra K. Yadav[3]

[1]Department of Physics, Indrashil University, Mehsana 382740, India.
E-mail: pushpankit@gmail.com

[2]Department of Physics, Sardar Vallabhbhai National Institute of Technology,
Surat 395007, India.
E-mail: swastisaxenaa@gmail.com

[3]Department of Physics, Sri Venkateswara College, University of Delhi,
Benito Juarez Marg, 110021, New Delhi.
E-mail: surendraky@gmail.com

2.1 INTRODUCTION

The development of energy sources that can meet the world's rising energy demand while being ecologically benign is a fundamental challenge the world faces at the beginning of the twenty-first century. The world's energy requirements are constantly rising due to population growth and quickening economic growth. Approximately 60% of the electrical energy is lost as heat losses during the production process [1]. Also, 8–15% is lost as heat during the transmission and transformation of electricity [2]. Because of this, only 35% of the energy produced in a power plant makes it to our houses. Another illustration is transportation efficiency. Overall,

*For Correspondence: E-mail: swastisaxenaa@gmail.com

40% of the energy produced in a car is lost as heat. Another 30% is used to cool the engine. Thus, 70% of energy is wasted, even without considering the CO_2 emissions caused by the additional 70% of the fuel that must be used. To put it another way, much low-quality thermal energy must be produced to sustain daily life, and industrial activity, which is regrettably wasted [3]. Renewable thermoelectric functional materials can generate heat and electricity by using the mobility of solid internal carriers, even at very low-temperature variations from room temperature.

In contrast to conventional new energy technologies, thermal energy (TE) devices provide several distinctive qualities, including the absence of moving parts and noise, a long operational lifetime, and the potential to replace existing energy materials [4–6]. Thermo electrical materials are frequently used in military, aerospace, and other high-tech fields, as well as in microsensors, medical thermostats, and other non-military applications [7–8]. Thermoelectric generators, or devices that generate electricity from waste heat, can contribute to a more sustainable environment by using their ability to convert temperature changes into energy [9–10]. Burning fossil fuels releases significant atmospheric CO_2, a known greenhouse gas that accelerates climate change. To prevent a catastrophe due to climate change, it is necessary to stabilise CO_2 levels at appropriate goal levels, which can be achieved through the large-scale development of carbon-free renewable sources.

2.2   ENERGY HARVESTING

Developers had no reason to worry about energy harvesting before the advent of ultra-low-energy MCUs. But as portable gadgets became more and more common and battery innovation lagged, real attention began to be paid to energy harvesting. Without ultra-low-energy MCUs, wireless sensor networks, for instance, would not be feasible. Micropower harvesting systems also support these MCUs [11–12]. The most prevalent energy-harvesting systems use solar, thermal, RF, and piezoelectric energy sources.

1.   Photovoltaic (PV) panels or solar cells convert light energy into electricity. Photovoltaic cells are among the energy-collection technologies with the highest power density and output.

2.   Thermoelectric energy harvesters turn heat into electricity. When the temperature difference between their bi-metal junctions is measured, their thermocouple (TC) arrays—which are what they are made of—produce a voltage (effect of Seebeck effect). On the contrary, when power is applied to a thermocouple (TC) junction, one junction heats up while the other cools, which is how heat pumps function due to the Peltier Effect.

Figure 2.1 shows the Phonon scattering mechanism which is the best example of thermocouple see beck effect where two ends of thermocouple maintain the flow of electricity, this is also called Peltier effect.

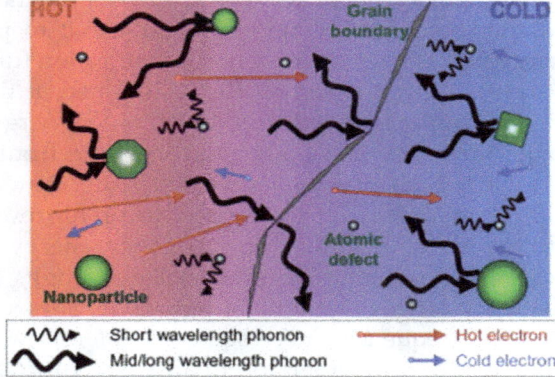

Figure 2.1 Schematic diagram illustrating phonon scattering mechanisms via atomic defects, nanoparticles, and grain boundaries.

Thermoelectric materials convert thermal gradients and electric fields for power generation and refrigeration, respectively. Thermoelectric devices currently find only limited applications because of their poor efficiency, which is benchmarked by the so-called thermoelectric figure of merit:

$$ZT = \frac{S^2 \sigma}{k} T \qquad \text{(Dimensionless)} \qquad (1)$$

S is the Seebeck coefficient, the electrical conductivity σ, average temperature T, and thermal conductivity κ. The latter contains both electronic and phononic contributions. Maximising ZT is challenging because optimising one physical parameter often adversely affects another. In metals, electrons contribute equally to the electrical and thermal transport (Wiedemann–Franz law); in insulators, only phonons give a non-negligible contribution to thermal conductivity; in semiconductors, both subsystems strongly contribute to the thermal transport, while the electrical conductivity can be dramatically changed via doping. For materials having similar thermal conductivities, the term power factor (PF) is used to define the performance of thermoelectric,

$$PF = S^2 \sigma \qquad (2)$$

The Arrhenius equation can explain the temperature dependence of conductivity for conducting substances; accordingly, the activation energy value is low for the highest conductivity.

$$\sigma = \sigma_0 e^{(-E_a/k_b T)} \tag{3}$$

Where σ is the conductivity, σ_0 is the pre-exponential factor, E_a is the activation energy and κ is the Boltzmann constant.

3. Radiofrequency harvesters gather environmental radiofrequency radiation, rectify it, amplify it, and use it to power ultra-low-energy devices. Radiofrequency identification functions similarly by responding to a strong RF field aimed at the sensor rather than absorbing ambient RF. Protecting RF energy from the radio environment and using it in low-voltage electronic devices are the two main goals of RF energy harvesting. Antenna-like patches with ultra-wideband characteristics or narrow-band antennas are needed to detect radio frequency energy emitted by the radio environment. The other's use, though, depends on the frequency ranges that will be picked up. For example, detecting GSM-900 frequencies necessitate an antenna with narrow-band characteristics. In order to enhance the functional qualities of wireless communication systems, multiple-input multiple-output systems are also used.

4. Piezoelectric transducers are used to transform pressure or tension into electricity. Roadbed vibrations caused by motors, airfoils, and piezoelectric energy harvesters are frequently reported as anomalies. Although other energy-collection systems are being looked into, the redisplay is now in the lead.

Due to the expansion of battery-powered portable consumer, commercial, and diagnostic supplies, these four energy-harvesting businesses will continue to expand quickly for many years.

2.3 ENERGY-HARVESTING SOURCES

Resources that can provide enough energy entirely or partially to power sensor networks in smart environments are known as energy-harvesting resources. Based on their properties, energy-collection sources can be categorised into two classes: natural resources, such as sun, wind, and geothermal energy, which are readily available from the environment, and synthetic sources, which are generated by human or system activity.

They are not a part of the ecosystem naturally. Examples include human movement, pressure from jogging or walking on floor slab inserts, and system vibrations during operation. Table 2.1 lists several energy sources for energy harvesting and information about the source type and typical harvesting power. System designers must consider the source of the energy-efficient solution for two reasons. Natural factors, like weather, temperature, and season, impact natural sources, while the schedules and impacts of human and mechanical systems impact artificial sources. For instance, the prediction methods of each generating source will be impacted by physical conditions.

Second, there is no need for more energy to generate natural resources. Our research on microscale energy-harvesting systems does not address the potential environmental effects of large-scale resource extraction. On the other side, artificial resources demand energy from human/machine systems to produce atmospheric harvestable energy. It should not be considered a cost if energy is mainly used for other purposes, such as lighting a room or powering a computer system. Consequently, the energy harvested through such a process is merely a by-product. However, it is considered a cost if energy production is primarily used to produce harvestable energy. It can occur if a light is left on for a few more hours to charge a sensor with solar energy or if a radio spectrum is created to charge an RFID sensor. Renewable energy thermoelectric (TE) materials can convert heat to electricity directly.

They could be utilised in TE generators for energy collection and local cooling. Wasted fuel and exceptionally tiny heat losses could be advantageous as well. Due to synergistic effects that highlight the advantages of carbon nanoparticles and polymers, organic conducting polymers carbon nanocomposites, which are employed as TE composites, have recently attracted much attention. To find alternative energy sources, thermoelectric generators are being used to help increase the effectiveness of the real energy system by recovering heat that has been lost today. TE nanostructure features, ranging from so-called 3D nano bulk materials to the inclusion of 0D quantum dots in TE structures [61], are studied.

Table 2.1 Comparison in the study of various forms of energy consumption

Serial No.	Sources of Energy	Energy Amount (TW)
1.	Hydropower resources	≤ 0.50
2.	Ocean and tide energy	≤ 2.00
3.	Wind energy	2.00 – 4.00
4.	Solar power	120,000

According to our extensive literature review, achieving favourable thermal properties for CNT nanofluids over time requires maintaining a homogenous dispersion and long-term stability. The goal of preserving the criteria mentioned above is challenging, given that CNTs are hydrophobic to most fluids and have a strong van der Waals interaction with one another. Nevertheless, many academics have made various attempts to satisfy these requirements. However, several issues need to be identified and fixed for various CNT nanofluid applications, especially for solar systems. The commercialisation of this technology is hindered due to major issues: stability and production expenses. Because of this, most collector designs must be modified to accommodate the practical needs of water heating systems used in home and industrial settings.

If these challenges are overcome, nanofluids are predicted to have a substantial impact not only on companies and technological areas but also on raising the standard of living for people. Recent studies [13–15] have shown the viability of using hybrid/composite materials and magnetic nanofluids to increase heat transmission. The most significant level of effectiveness of nanofluid in solar thermal engineering devices must be ensured, although none have been used in solar energy systems. Researchers have shown that carbon nanotube yarns may be used to gather energy. These researchers led the group from the University of Texas in Dallas, and their specialists have been developing yarns made of carbon nanotubes for over ten years.

Finally, energy-harvesting devices have been created by stretching and twisting carbon nanotube threads. According to the initial research, these nanotubes might be used right away to power tiny sensor nodes for Internet of Things (IoT) applications. Nanotube yarns may produce a significant amount of energy by flexing and stretching in response to the motions of breaking waves, according to scientists. The nanotube appears to utilise the piezoelectric effect.

2.4 ENERGY-HARVESTING STORAGE

1. Even if micropower energy is not intermittent, the output is typically so low that a boost valve is required to control the build-up of energy.
2. Small rechargeable Li-ion batteries are frequently used in portable applications with limited space.
3. Micropower energy devices find it challenging to deal with current spikes, such as when a sensor displays data bursts. In addition to the need for power regulation, energy harvesters frequently require a large supercapacitor or capacitor to dampen rapid surges in output.
4. Electrical double layers capacitors are also known as supercapacitors due to the great proximity of their conducting layers. These supercapacitors have a lower energy density than batteries but a significantly higher power density. They are ideally suited to handle sudden increases in demand since, unlike batteries, they may be completely exhausted in a matter of seconds. They are widely used in combination with ultra-low-

power applications because they have less internal resistance than thin-film batteries, which enables them to drain more slowly over time.

2.5   ENERGY COLLECTION FROM CONDUCTING NANOCOMPOSITES DEVELOPMENT TOOLS

2.5.1   Development Tool for TE Harvesting

Energy harvesting is a broad category of technologies that are constantly evolving. The significance of development tools is growing as design complexity and design cycle lengthen. Designers can evaluate and become familiar with the most recent power harvesting technologies and products using development kits and boards in the industry. The iterative design approach must include development and testing, and circuit modelling. Engineers are frequently compelled to use the Breadboard, which is purportedly inescapable. Development kits offer a simple approach to reaching the development core with little setup time.

In addition to a quicker time to market, development tools can benefit immediately applicable, tested circuits, widely available printed circuit layouts, and a similar level from which to build designs. Energy harvesting is a broad category of technologies that are constantly evolving. The significance of development tools is growing as design complexity and design cycle lengthen. Designers can evaluate and become familiar with the most recent energy-harvesting technologies and products using developer kits and boards in the field.

2.5.2   A Quick Look at Carbon Nanotubes (CNT) and Graphene

Having a wide variety of allotropes, carbon has historically been the most adaptable and all-purpose element, with numerous applications in the world of materials. There are many other forms of carbon, but carbon nanotubes and graphene have emerged as the most exciting ones, offering countless research opportunities in every field of science. After being discovered by the renowned scientist Iijima, the carbon nanotube has long been a household word in science [16]. CNTs are typically in the micrometre range in length. CNTs display three different chirality: armchair, zigzag, and chiral, based on the lattice vectors and chiral angles [17]. With a wide range of applications from industrial to nanotechnology, including energy storage, modular electronics, conductive polymers, and structural composites, CNTs have emerged as the most anticipated substance in society.

Graphene is a different kind of carbon that has recently been researched for various uses. Graphene is a relatively new material but because of its unusual structure, it has already shown excellent mechanical, thermal, and electrical properties. A hexagonal lattice of carbon atoms on a planar surface makes up the two-dimensional allotrope of carbon known as graphene. Graphene is the building block for other carbon nanostructures, including fullerene and carbon nanotubes. For the first time, graphene became a fact in 2004. After that, curiosity about the structure and properties of graphene suddenly soared, and it did not disappoint. The detailed measurements of graphene's properties, including charge transfer and tensile strength, revealed that they are both relatively high. These distinctive qualities allow graphene to be employed in a wide range of applications.

Only a few themes covered [18–23] include thin and flexible displays, solar cells, electronics, medical, pharmaceutical, and industrial activities. In order to enhance electrical and mechanical properties, CNTs and graphene have been extensively investigated in various thermoplastic and thermosetting polymer systems. Research using polyurethane, polystyrene, polycarbonate, ABS, PMMA, polyethylene, epoxy, and phenolic systems has been described by several researchers [24–33]. Aside from being used with polymers to create self-healing nanocomposites, CNT and graphene have also been used for their superior heat conductivity and stability [21]. The following sections provide a brief overview of the self-healing phenomenon achieved with graphene and the polymers that subsequently used CNTs similarly.

2.5.3   Self-Healing Polymer Composites Based on Graphene

Graphene's potential for healing is its capacity to correct ingrained faults. The graphene structure's reconstruction (knitting) produces the healing action [35]. When a vacancy defect develops, carbon atoms from the surrounding areas rush in to fill the gaps [36]. Impurities, such as hydrocarbon contaminants, often cause these extra carbon atoms. When graphene is scraped in the presence of metals, carbon atoms are transferred from the nearby hydrocarbon impurities and fill the hole that is left behind [37]. This unusual reknitting of graphene holes is the outcome of this process. It is because graphene's innate ability to mend itself creates new possibilities for applications that use various techniques, including the e-beam technique and the etching process. A stiff C60 molecule that causes nano-damage in a suspended graphene monolayer is studied using molecular dynamics simulation, and the results show that the correct heat treatment can effectively repair the damage [38].

The self-healing mechanism is explained as a two-step process: (a) the formation of local curvature around the defects brought on by the damage, and (b) defect rebuilding leading to noise filtering of the contour brought on by the destruction, ultimately leading to the destruction being minimised. Two factors dominate the ability of graphene to self-heal. The first one is the damage's extent, and the former is the temperature change. As a result, graphene, being a single atom layer, is worth highlighting, may preserve the power of moving current, and thus damage repair in the same manner. Graphene may readily repair damage in microstructures when combined with polymer composites. The self-healing in polymer composites using CNTs and graphene is the primary emphasis of this chapter. Because of their flat structure and superior thermal contraction, high-performance thermal interface materials are another excellent use of graphene-based polymer composites. Much research has found that adding graphene to a material improves heat conductivity. It will increase the polymer matrix's heat transport capabilities, improving self-healing in response to thermal or other external stimuli. Epoxy and polyurethane are two instances of polymers that incorporate graphene. There have been numerous assessments of the self-healing properties.

2.5.4 Carbon Nanotube-Based as Self-Healing Polymer Nanocomposites

Due to their exceptional mechanical properties, carbon nanotubes (CNTs) have been the focus of material research ever since their discovery. There are several applications for CNT-reinforced materials across numerous sectors. In addition to the abovementioned applications, CNT has been thoroughly explored to fully appreciate the self-healing properties of polymer composites based on CNT.

CNT can be used in self-healing materials in both extrinsic and intrinsic approaches. On the one hand, it has a 1D tubular structure with strong reinforcing qualities; on the other, it has excellent thermal conductivity and heat transfer capabilities.

2.5.5 Extrinsic Self-Healing Polymers with CNTs

Both the fluid overload and capsule-based extrinsic self-healing strategies can be applied. Due to their efficient 1D nanotubular structure, CNTs can be used as a nanoreservoir for the healing agent. Additionally, because it enhances structural integrity, CNTs can be combined with other polymers as a healing agent in fluid overload self-healing. Finally, CNTs can be reinforced in polymers with microcapsules to regain mechanical strength.

2.5.6   Carbon Nanotubes as Nano Reservoirs

In microvascular self-healing techniques, single-walled carbon nanotubes (SWCNTs) can be used as nano reservoirs of healing chemicals [39]. According to the dynamics study, the fluid from a burst SWCNT after the injury is thoroughly studied, with the fluid resembling a healing agent. Using SWCNTs as a self-healing container reinforces the overall mechanical strength of the system. The following are the main factors that affect how much of a healing agent needs to be retained in the SWCNT reservoir to facilitate self-healing.

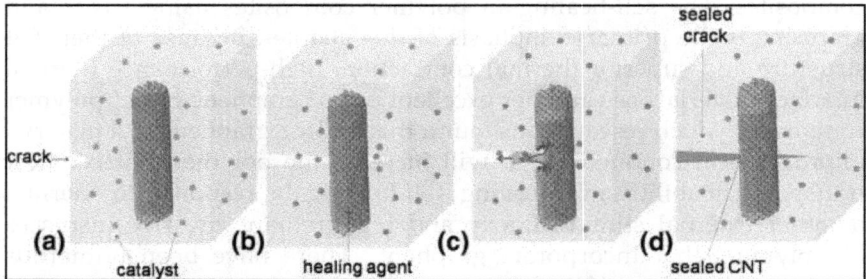

Figure 2.2   Carbon nanotubes act as nano reservoirs in a self-healing mechanism (D. Ponnamma, K.K. Sadasivuni, M. Strankowski, Q. Guo, S. Thomas, Soft Matter 9(43): 10343–10353).

The final applications of the material will dictate these requirements for developing a practical CNT-based self-healing system [39]. Using the self-sustained diffusion method, liquid monomers utilised as therapeutic treatments were intercalated into CNTs [40]. The healing agent (dicyclopentadiene (DCPD) or isophorone diisocyanate) was first coupled with an empty CNT solution that was semisolid and dissolved in benzene. The benzene in the solution evaporated after sonication, allowing the solutes to penetrate the CNTs. Fresh benzene was added after the intercalation and sonicated for 3–4 minutes to sanitise the CNTs' exterior. Mass production of intelligent composites with integrated self-healing agents might start if a more practical method for loading self-healing compounds into CNTs can be created. Different self-healing capsule and vascular-based polymer systems have three main drawbacks:

- It is challenging to embed capsules containing healing agents into polymer systems;
- Capsule and vascular implantation reduce the mechanical strength of the material;
- The mending agent runs out after just one crack.

The vast electrical conductivity and exceptional mechanical properties of CNTs make them preferable to pure polymers as reinforcing conductive fillers. As a result, the strength loss brought on by capsule embedment can be made up for by CNT reinforcement. An epoxy-based self-healing coating with CNT fillers was produced by microencapsulating the healing agent inside the polymer system [41–44]. Even though it is well known that adding microcapsules to a polymer system reduces its mechanical strength, this problem was fixed by integrating SWCNTs. Nanoindentation experiments significantly improved the samples' elastic modulus and hardness after adding SWCNTs to the system. As a result, CNTs may easily enhance the physical properties of composites, thereby expanding their applications and helping to re-establish the mechanical properties lost due to capsule embedment.

2.5.7   Carbon Nanotubes as Effective Healing Agents

In vascular-based self-healing (SH) polymers, CNTs have also been used as healing agents. The vacant channels of an epoxy resin system were filled with a nanocomposite made of SWCNTs and 5 ethylidene-2 norbornene (5E-2N) that was combined with ruthenium Grubb's catalyst [45] and used as a healing agent. An example defines a microvascular SH method.

A mass was applied to raise the impact hole to evaluate the self-healing behaviour. The damaged sample was treated with the 5E-2N/SWNT composite healing agent and thermally repaired for 15 minutes at 60 degrees Celsius. Although it could not measure the epoxy system's mechanical healing effectiveness, it recovered its structure after 30 minutes. Similarly, ethyl-phenylacetate and 2.50% carbon nanotubes were mixed and used as medicinal agents in a capsule-based method [46]. CNTs boost both electrical and mechanical healing capability when paired with healing agents. As a result, the role of carbon nanotubes in extrinsic self-healing materials has been intensively explored. Due to their tubular shape, CNTs can restore lost mechanical properties in micro-capsule-based healable polymer nanocomposite and a nanoreservoir for healing agents. In addition to this study, carbon nanotubes have been used to create carbon nanotube-based polymer composites with built-in healing properties. These composites will be discussed in the following section.

2.5.8   Intrinsic Self-Healing Using CNTs Composites Made of Polymers

Carbon nanotubes are combined with polymers to make healable composites, which can be used in applications requiring enhanced flexibility, durability, and fracture resistance. Some key uses include shear-stiffening materials for body armour and conducive healable polymers for robotics. The following sections go through all these different uses for polymers based on carbon nanotubes with inherent healing properties.

2.5.9 Healable-Conductive Polymer Composites with Multiple Functions

SH multifunctional conductivity can be used to restore circuit conductance, prevent damage, and lengthen the lifespan of electronic appliances. Carbon nanotubes exhibit very high mechanical and electrical conductivities. A conductive elastomer with autonomic healing capacity was made utilising a nanocomposite of poly-2-hydroxyethyl methacrylate and SWNTs connected via interactions [47] by introducing carbon nanotubes into an elastomer system. Both mechanical and electrical healing capacities in the sample were identified. An LED light and a power source were connected to the nanocomposite sample.

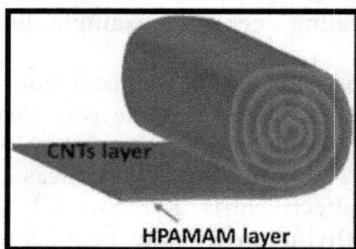

Figure 2.3 Sushi-style architecture displays conductance Schematic representation of the disconnection of HPAMAM/carbon nanotube composite (Copyright 2015). For this reproduction, permission has been obtained from the Royal Society of Chemistry.

The sample was immediately split in half, and the LED went off due to a lack of connectivity. However, when kept close to one another and the LED lamp was turned back on, the test parts repaired themselves in just five minutes. The electrical healing efficiency was calculated for all the various sample types, and it was found to be roughly 95% with different SWNT concentrations. Consequently, using SWCNTs, an elastomer with electrical conductivity and structural stability was made. This elastomer might be used to manufacture complex sensing devices. Another composite with SH characteristics and a 20-minute perfect electric-conductance recovery time was developed using hyperbranched polyamidoamine (HPAMAM) polymers covered in carbon nanotube films (Figure 2.2). The light was on when the sample was first connected to the circuit. It was off when it was detached, but it was turned back on when it was reconnected. The light was off when resistance was measured both before and after separation.

As seen in Figure 2.3, the structural integrity and conductivity were restored by the sushi-like construction of the SH conductive composite. The conduction channels are restored because the surfaces at the site of damage meet thanks to the spiral coiling of the CNT layers.

2.5.10 Self-Healing Polymer Nanocomposites with Shear-Stiffening

Shear-thickening materials fall within the topic of intelligent materials. These materials' viscosity has an uncommon feature that causes them to rise sharply when the applied tension exceeds the critical shear rate. S-ST polymer composites have lately emerged as a significant research area because of their crucial applications in military body armour. Combining S-ST with other functionalities will create the best composites for high-performance body armour. Because of their exceptional mechanical properties, low density, and electrical conductivity, MWCNTs are excellent nanofillers for reinforcing polymer materials. Due to their better dispersity than SWCNTs, MWCNTs can be easily combined with the polymer matrix to produce multifunctional nanocomposites.

Several pieces of research have already demonstrated that incorporating MWCNTs into polymers enhances their mechanical properties. Body armour materials with good shielding and compressing rate-dependent conductivity may be the MWNT-based S-ST polymer composite. An MWNT/S-ST composite based on a poly-boron dimethyl siloxane (PBDMS) derivative has demonstrated electrical self-healing. On the opposite side, an LED light was connected in a circuit using the MWNT/S-ST-polymer composite. The LED glowed brightly with a 9 V power supply when the composite was split into two sections before abruptly going out. The LED shone brightly once more when the broken components were put back together. It highlights the material's capacity for self-healing and how little conductivity was lost in the repaired specimen. Because of this, the MWCNT/S-ST polymer composite has an exceptional capacity for self-healing at room temperature. Due to their very high specific stiffness and strength, CNTs are intriguing candidates for the production of composite materials. CNTs have excellent thermal and electrical conductivities. Due to these characteristics, their high aspect ratio, one-dimensional (1D) honeycomb lattice, and low density, CNTs have been extensively researched for the construction of various composite and smart materials [48].

Graphene, a monolayer of carbon atoms tightly packed into a two-dimensional (2D) flat form, is the fundamental building block of graphite materials. Open-ended CNTs are suitable for the construction of composite materials because the geometrical (2D) and electronic effects of graphene on their field-emission properties affect their mechanical and electrical properties [49]. In addition, graphene has the lowest energy due to the overlap of the 2pz orbitals of carbon atoms, which gives its composite materials an anisotropic quality.

Due to their distinctive properties and low cost, graphene-based composites are potential fillers due to the changes in carbon atom bonding in-plane and out-of-plane, as well as their three-dimensional (3D) geometrical qualities. Carbon nanotube (CNT)/graphene-filled organic composites have much potential for developing less expensive thermoelectric materials for energy-harvesting applications due to their low cost, low density, straightforward preparation pathways, variety of process capabilities, and low thermal conductivity. These characteristics elevate them above other hybrid alloys that have been previously reported. CNT and graphene are currently the most widely used nanofillers because of their unique structures and characteristics, such as superconductivity, low weight, high stiffness, and axial strength.

2.5.11   Carbon Nanotubes with Customised Shapes Produce Energy-Collecting Textile

Numerous pieces of research have examined the viability of extracting energy from textiles. It makes sense because fabric-based harvesting presents a promising area for such exploitation. One material used to accomplish this is carbon nanotubes (CNTs), which have already demonstrated various distinctive and beneficial qualities in this field. Researchers from Rice University and Tokyo Metropolitan University collaborated to create a flexible cotton fabric with improved fibres that uses carbon nanotubes as a thermoelectric (TE) energy source. Using the well-known Seebeck effect, it transforms heat into sufficient energy to illuminate an LED. Using carbon nanotubes has disadvantages. However, given their one-dimensionality and distinctive qualities, such as flexibility and lightweight, they seem like appealing options. However, maintaining the massive energy factor of isolated nanostructures in the macroscopic assembly has proven challenging due to poor sample shape and inadequate Fermi energy tuning [50–54].

2.6   ENERGY-COLLECTING MODES

The many methods of harvesting energy are described below in the following sections.

2.6.1 Energy Harvesting for Fossil Fuel Alternatives

Also, with Paris Agreement, most world leaders have come together to pledge to do more to combat the climate change threat. Each nation must reduce emissions to prevent an increase in world average temperatures of more than 2 degrees Celsius. If this goal is to be accomplished, time is running out, and more fresh, clean energy solutions must be developed. The new economy looks at a few of the fascinating green energy technologies now under development.

2.6.2 Elephant Grass Energy Harvesting

For centuries, biomass energy was the chosen fuel before it was more convenient to access coal, oil, and gas. Today's CO_2 emissions are causing havoc on the environment and are once more a significant player in the world's energy mix. Any biological material derived from plants or animals is called biomass, but wood is the most prevalent. A system using an alternative to wood pellets has been created by Sweden-based clean-tech startup Next Fuel. With elephant grass, Next Fuel provides a direct replacement for fossil fuels that emit no greenhouse gases.

Elephant grass is a unique plant that can reach a height of 4 metres in just 100 days and produces many harvests each year. After the grass is collected, Next Fuel's method uses less energy to transform it into briquettes during manufacturing. Because less CO_2 is released into the environment when the fuel is burned than was taken in from the atmosphere a few months earlier when the grass was growing, the entire carbon cycle shifts to a negative state annually.

2.6.3 Energy-Harvesting Hydrogen Fuel Cells

One of the most prevalent elements in the world, hydrogen, has long been used in the power sector, but recent intriguing developments have reignited interest in it. Clean energy can be produced using hydrogen fuel cells from a variety of sources. Like lithium-ion batteries, they can be used in the transportation sector but do not require charging.

1. Electricity can be produced or transported using hydrogen, a renewable energy source. When hydrogen is consumed, all that is generated is heat and water.
2. We must first isolate hydrogen from a mixture to acquire pure hydrogen.
3. A method for dissolving a molecule of water into its hydrogen and oxygen atoms is electrolysis.
4. An electric current is produced when electrons are forced to flow across a circuit. After completing the circuits, the electrons

combine with oxygen and hydrogen molecules to generate water (H_2O), which is then heated.

Figure 2.4   Schematic diagram of hydrogen fuel cell.

5. Hydrogen fuel cells can be used to help create a zero-emission power system because they do not release any CO_2 or other pollutants.

6. A wide range of electrical devices, including cars, aeroplanes, and buildings, can be powered by fuel cells (Figure 2.4). To generate the enormous amounts of electricity needed to run automobiles and other electrical equipment, many fuel cells can be connected to form a fuel cell stack.

7. The anode electrode allows hydrogen to enter the fuel cell. Hydrogen atoms are divided into protons with a positive charge and electrons (Figure 2.5) with a negative charge (positive charge).

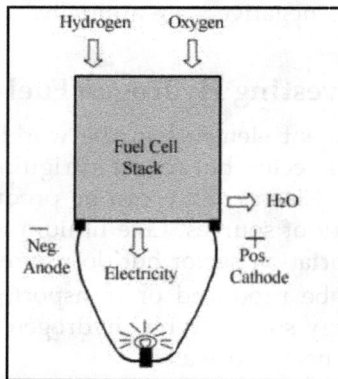

Figure 2.5   Fuel cell diagram.

8. The fuel cell membrane allows the positive charge protons to pass through. Electrons with negative charges cannot pass through the membrane and must travel through a circuit instead.

9. Electrons travelling through a circuit produce electrical current.

Germany was the first nation in the world to operate passenger trains fuelled by hydrogen-based fuel cells due to the high cost of hydrogen technology. European Union has decided to work together to improve the possibilities for hydrogen in the power and transportation industries.

2.6.4 Solar Paint as a Source of Energy

One of the most popular sustainable fossil fuel alternatives is solar panels, but what if one can harness the sun's energy without being concerned about the panels' harmful effects on the environment? After developing a paint that can generate energy, researchers may have found the solution. The substance may capture solar energy and humidity from the environment by fusing synthetic molybdenum-sulphide with titanium oxide, which is present in many wall paints. A concrete structure might be converted into a fuel and energy source by adding a new composition. The process is easy to follow:

The paint is made of synthetic molybdenum-sulphide and titanium oxide, which are two new chemicals (Figure 2.6). The characteristics of that mouthful to wick away moisture are comparable to those of the silica gel package that comes with new pairs of sneakers and other goods.

Figure 2.6 Solar paint applied on the wall of a house (Wikipedia source).

Synthetic molybdenum-sulphide absorbs solar energy before splitting it into hydrogen and oxygen. After that, the hydrogen can be captured and used to power a house, car, truck, boat, or all-terrain vehicle (ATV).

2.6.5 Energy Harvesting from Waves

It seems easy to harness the energy of the ocean's waves. Unfortunately, it is far more challenging in practice. An optimal design has been the focus of years of research. A small-scale "wave snake" experiment operated between 2008 and 2009 off the coast of Portugal, but when the Scottish company that invented the technology went bankrupt, Wave Energy Scotland, a public organisation, received the intellectual property.

Research on wave energy is constantly going on. The most significant engineering company in the world, Lockheed Martin, disclosed plans. The EU is working with Wave Generating Scotland on an initiative to provide open-source software for wave and tidal energy systems to promote private investment (Figure 2.7).

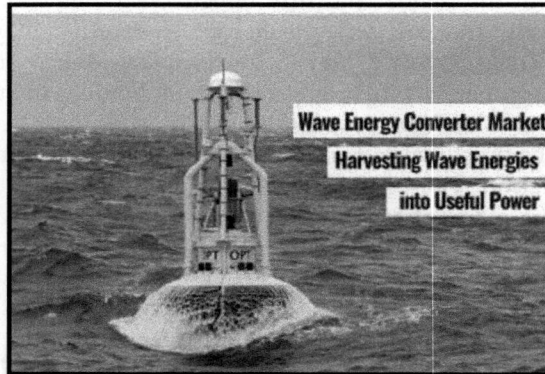

Figure 2.7   Energy harvesting through wave energy.

Wave energy harvesters outperform other ocean energy harvesting technologies, like tidal currents, ocean thermal energy conversion (OTEC), and salinity processes. Waves are better suited for energy gathering than tides because they are present throughout the ocean.

2.6.6   Energy-Harvesting Whisky

Whisky is a source of energy. Scotland has been making whisky for centuries, and distilleries have developed strategies to deal with the trash produced as a by-product, typically selling it to farmers as cow feed. With 4 million tonnes of waste produced annually, the sector is looking for a more inventive solution. The market for whisky waste by-products may be worth £140 million ($184 million), according to a 2015 assessment by the Green Alliance. Certain distilleries use anaerobic digester facilities to produce bio-gas, which is transformed into steam energy that powers their operations. Diageo's Glendullan Distillery produced 6,000 MW hours of TE in its first year of operations, cutting down on the need for fossil fuels by 25% [11]. By-products from whisky may also be used to generate heat and power automobiles.

2.6.7 Vehicle Energy-Harvesting System

High electricity consumption and constant traffic are two unrelated aspects of city living. On the other hand, the Vehicle Energy Harvesting System (VEHS) has developed a cutting-edge link that uses traffic flow to increase access to electricity. The VEHS uses an engineered overlay layer to gather pressure from any new or existing road, which is then used to drive a traffic turbine and generate electricity [12].

The system might be finished and operational in six months, enabling quicker access to power. The VEHS's straightforward design will permit local assembly, a deliberate move to local manufacture, and greater electrical access. We anticipate that this will give local communities the skills and employment opportunities they need and aid in reviving their ailing economies.

2.6.8 Energy Harvesting from a Sustainable Power Supply

The field of wireless gadgets has expanded exponentially since its inception. However, the battery life of these gadgets frequently limits the variety of services they can offer. Therefore, having an independent power source would help us advance and fully utilise the capabilities of such devices. The current technological revolution is driven by miniaturisation, and as devices get smaller, less power is needed. It has led scientists to wonder if the battery may be enhanced by devices that continuously capture the environment's energy that would otherwise be wasted. Converting otherwise wasted renewable sources into a usable form is known as energy harvesting or scavenging. Energy harvesting might even replace the need for unfavourable or challenging-to-reach battery replacement and maintenance when used with a wireless system's long-term power source. Safety monitoring devices, microsensors incorporated in structures, and implants for treating patients are a few examples of energy harvesting benefits of environmentally renewable sources, including waves, heat, light, and water into electricity. It creates a "battery-free" solution.

2.6.9 Harvesting Mechanical Energy

Transportation, fluid flow like air motions, biological locomotion like walking, and internal body motion, like chest and heart, are all mechanical energy sources. They also include vibrations and noise from industrial gear and equipment. The two fundamental techniques for extracting energy from mechanical sources are kinematic and inertial [54–56]. The resistance to an acceleration of a mass absorbs inertial energy. These systems have a single connection point between the base and the spring-mass-damper system. Due to its inertia, most of the base vibrates, and these vibrations can turn into electrical energy. Pendulums, magnetoelastic oscillators, and cantilever beams are examples of harvesters that use this concept of inertia.

The energy-collecting transducer is directly connected to various parts of the source in kinematic energy harvesting. The relative movement between these elements causes the transducer to deform, ultimately transforming into electrical energy. Two instances of energy harvesting are bending a tyre wall to check the pressure and flexing and extending limbs to charge a phone. Electrostatic, electromagnetic, and piezoelectric transduction methods are being used to transform mechanical energy into electrical energy [57–60].

2.7 ADVANCE APPLICATIONS AND TECHNOLOGIES OF ENERGY HARVESTING

There are now several energy-collection devices, and new ones are on the horizon. The most prevalent forms of energy include light, heat, vibration, and radio frequency (RF). An enormous and rapidly expanding power harvesting business has resulted from the invention of ultralow-energy microcontroller devices. Low-power wireless sensors were produced due to energy harvesting and are now widely used. There are many applications of energy harvesting (Figure 2.8) on the other hand, the ripple effect will spread throughout the commercial, scientific, and consumer sectors, leading to new uses, such as creating compact battery-powered devices.

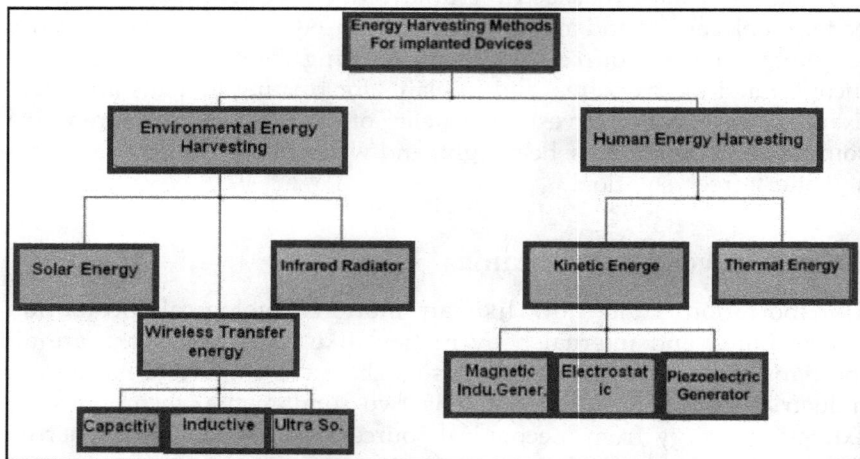

Figure 2.8   Applications of energy harvesting.

2.7.1 Mobile Phone

One can achieve the low-power consumption needed to run a cell phone by collecting energy from the environment. Mobile phones require a good battery life while advancing wireless analogue telephony technologies to portable computers, including website browsing, movies, sports, and email capabilities. Low power has always been the most crucial electrical design goal for at least the previous 10 years. Many industrial, medical, and commercial purposes use wireless sensor networks with minimal power consumption.

2.7.2 Solar Power

Small solar cells are used in consumer and industrial products, such as calculators, smart watches, entertainment, street light controllers, transportable energy sources, and satellites. Because light beams are frequently intermittent, solar energy cells are used on batteries with long lifespans to provide a continuous energy supply.

2.7.3 Thermoelectric

The piezoelectric effect, which occurs when two different metals have a difference in temperature and produce a voltage, is the basis for thermoelectric harvesters. TEGs are constructed from several thermocouples connected to a heat source, such as a heater, an engine, or even a solar panel.

2.7.4 Piezoelectric

Piezoelectric transducers are excellent for accelerometer sensors with energy-collecting modules that identify aeroplane wing vibrations and motor bearing noise because they generate electricity while under strain (Figure 2.9). When the cantilever vibrates, it produces an AC output voltage that can be rectified, regulated, or used as a battery's thin-film or supercapacitor foundation.

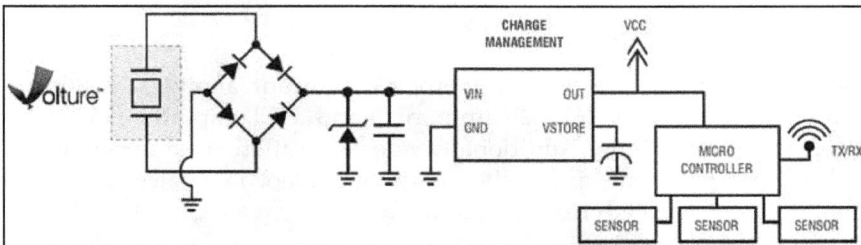

Figure 2.9 Midé Volture™ piezoelectric energy harvester
(Courtesy of Midé et al.).

2.8 INNOVATIVE TECHNIQUES AND TECHNOLOGIES

Some incredibly innovative energy-harvesting labs may change the direction of the industry in the upcoming years.

2.8.1 Medical and Fitness Equipment

A few cutting-edge uses for piezoelectric energy collection are beginning to materialise. This might subsequently be implanted or used to control a pacemaker, perhaps eliminating the need for battery maintenance. The potential for other implanted devices to be driven by body heat, motion, or vibration is being researched.

- The patient utilises a low-frequency RF emitter attached to the chair, which the device picks up, corrects, and stores.
- People who frequent the gym will be delighted to learn that they can somewhat recover the power they waste there. The creation of a piezoelectric power harvesting unit that attaches to the knee and produces electricity while using a treadmill has been a team effort among three British colleges.

2.8.2 Antennas

NEC devices have been effectively prototyped on silicon and HDPE substrates, but additional funding and work will be needed to achieve cost-effective mass manufacturing procedures. The researchers foresee a system that utilises previously untapped infrared energy in addition to conventional PV solar panels.

2.9 SENSOR APPLICATIONS OF POLYMER NANOCOMPOSITES

2.9.1 Polyaniline

Electrochemical sensors for numerous significant analytes, including serotonin, ammonia, [61–62] ammonia, and HCl vapours [63], have integrated polyaniline. Additionally, the fabrication of a sensor using polyaniline, graphene, and carbon nanotube nanocomposites effectively identified several environmental hazards, such as Cs, phenol, and 4-aminophenol [64].

2.9.2 Polypyrrole

The functionalisation of Au nanoparticles on PPy is known to have an enhanced response to ammonia [65], and p-polypyrrole—n-tungsten oxide hybrid nanocomposites were used to detect nitrogen dioxide gas at ambient temperature [66]. A polycrystalline hybrid nanocomposite material based on cobalt hexacyanoferrate/carbon nanofibers/polypyrrole (CoHCF/CNF/PPY) was created to bridge the gap between supercapacitors and batteries. Using a variety of characterisation techniques, the structural and elemental characterisations of the nanocomposite revealed the formation of a PPY layer on the CNF surface to which CoHCF nanoparticles were attached. Therefore, the proposed CoHCF/CNF/PPY ternary hybrid nanocomposite can be a strong candidate for the upcoming energy storage technologies [67].

2.9.3 Graphene and Its Derivatives

2.9.3.1 Graphene

A promising use for hydrogen sensing was demonstrated by advanced electrode materials made from multi-layer graphene nanoribbons functionalised with Pd. The development of a novel method to increase the energy density of flexible solid-state supercapacitors using hierarchically designed graphene nanocomposite (GNC) electrode material and an ionic liquid gel polymer electrolyte also represents a breakthrough in the field of supercapacitors [68]. This method aims to provide the next generation of wearable and portable electronic devices with high energy density, durability, and flexibility of supercapacitors. Additionally, silicon boron carbon nitride (SiBCN) ceramic composites filled with nitrogen sulphur dual-doped graphene (NSG) sheet (SiBCN/NSGs) were designed and synthesised, and their viability to function as lithium-ion battery anodes were tested [69]. These improvements significantly increased the Li-ion loading capacity and gave a higher rate ability.

After flocculation treatment, the MnO_2/graphene nanocomposite was created to remove the remaining tetracycline from pharmaceutical effluent. Due to the nanocomposite's strong water solubility, the tetracycline removal rate was up to 99.4%. Additionally, the addition of MnO_2 nanorods made it possible to prevent the needless loading of prepared graphene sheets during the adsorption process, giving the tetracycline molecules more chances to come into contact with the adsorbents [70]. Additionally, 3D electrochemical GQDs can detect Fe^{3+} ions optically in a sensitive and focused manner [71]. In addition to the aforementioned uses, graphene-based fibre exhibits extremely high sensibility to high tensile deformation and has dazzling bending and torsion-sensitive capabilities due to a variable microstructure under various mechanical stimuli [72].

2.9.3.2   Graphene Oxide

The development of rGO-Pt-NiO nanocomposites for enzyme-free glucose sensing. Additionally, a concave tetrahedral Pd NCs@CuO composite and reduced graphene oxide was used to create an ultra-low detection limit glucose sensor [73]. Additionally, many electrochemical sensors based on reduced graphene oxide have been successfully developed to detect the incorporation of NO_2 gas into ZnO, Fe_2O_3, and SnO_2 nanoparticles [74–75]. Additionally, a TiO_2 nanoparticles adorned reduced graphene oxide nanocomposite can be employed for ultrasensitive electrochemical detection of rifampicin [78], in conjunction with CdS, as a non-enzymatic biosensor for H_2O_2 [76], and also Cu_2O [77].

For the susceptible electrochemical measurement of essential cancer and oxidative stress biomarker 8-hydroxy-2′-deoxyguanosine (8-HDG), multi-layer based graphene oxide coated zinc oxide nanoflower (ZnO NFs @ GOS) was used as an electrode material [79]. Additionally, silica/graphene oxide nanocomposites were employed as solid-phase extraction adsorbents in conjunction with high-performance liquid chromatography to analyse aflatoxins in cereal crops [80].

2.9.3.3   Carbon Nanotubes

In conjunction with metals and metal oxides, carbon nanotubes serve as a substrate for several electrochemical sensors. When CNT is assembled atop graphene, a synergistic effect makes it easier for conductive channels to form, improving electrical conductivity [81]. Additionally, compared to CNT-based and WO_3 nano brick-based sensors used independently, the ammonia gas-sensing characteristics at low temperatures were synergistically improved by carbon nanotube and tungsten oxide nano brick sensors [82]. Additionally, a composite made of multi-walled carbon nanotubes, magnetite nanoparticles, and poly (2-aminopyrimidine) can perform solid-phase extraction of acidic, essential, and amphoteric medicines for the quantitative assessment of these pharmaceuticals in bodily fluids and wastewater. The composite exhibits a greater adsorption capacity and improves medication dispersion in aqueous media [83].

An extremely sensitive electrochemical sensor made of Fe_3O_4-SWCNTs/MOCTlCl/CPE was successfully made to analyse the anticancer agent epirubicin in actual samples [84]. Additionally, the devices that combine electrochemical systems and nanosensors are ideally suited for quick and accurate trace analysis of dangerous substances in the environment, food, or health applications [85]. By building a stereoselective sensor based on multi-walled carbon nanotubes cross-linked with chiral nanocomposites based on Hydroxypropyl-b-cyclodextrin, atorvastatin isomers could be discriminated [86]. PCL-CNT nanocomposites were created using an oil-in-water emulsion solvent evaporation technique. Applications involving medication delivery make use of these nanocomposites [87]. Recently, a promising electrode that was sensitive and selective was modified using MWCNTs/THI/AuNPs nanocomposites to inhibit the antibodies to 17-estradiol, resulting in the creation of an easy-to-make, inexpensive, label-free 17-estradiol immunosensor [88]. Additionally, electrochemical oxidation of BZ at CPE/nMBZBr/NiO-NPs can be used to simultaneously determine benserazide and levodopa by voltammetry in tablet formulation and human urine [89]. Multi-wall carbon nanotubes—poly (p-phenylene terephthalamide) nanoparticles were produced and described to produce a nanocomposite with increased electrical conductivity [90].

2.10 CONCLUSION AND FUTURE SCOPE

In this era of polymer and essential flexible electronics, society is striving for damaged and crack-free materials. In actuality, self-healing material production on an industrial scale will usher in a new technological era, with the aerospace and military sectors undergoing profound change. It will increase flexibility and decrease maintenance costs while extending the usable life of all manufactured materials. The polymer has better healing capacity due to the inclusion of graphene and carbon nanotubes. Although graphene seems to be a viable SH material, incorporating it into polymers is still challenging. Society is working to develop materials impervious to breakage and cracking in this era of polymer and crucial flexible electronics.

The industrial manufacturing of self-healing materials will usher in a new technological era, with the aerospace and defence industries experiencing significant upheaval. It will prolong the useful life of all manufactured materials while increasing flexibility and lowering maintenance expenses. Because graphene and carbon nanotubes are present in the polymer, it has a more vital ability to repair. Although graphene appears to be a potential SH material, adding it to polymers remains difficult. The impact of CNTs and graphene on an intrinsic self-healing behaviour in a polymer composite was investigated. Two other extrinsic strategies are examining how CNT is used in practise as a nanoreservoir and conducting in-depth self-healing investigations.

We are not far from the day manufactured materials will be able to restore their structural integrity in the case of a breakdown, even though self-healing is still a science fiction concept. The term "ohmic overpotential" refers to the voltage loss caused by the electrolyte's resistance, and the term "mass transport overpotential" refers to the loss caused by the time it takes for ions to move through the electrolyte. Conductive polymer composites for energy harvesting are materials that experience volume changes due to electrochemical reactions. If the length between the electrodes were lowered, the resistance and ohmic overpotential would also be decreased. The mass transport overpotential and the amount of time it takes for ions to move between electrodes are decreased by reducing the distance between them.

Therefore, it is necessary to choose the electrode distance to balance the two. Conductive polymer composites are ineffective at converting mechanical energy into electrical energy. It is because a chemical process, rather than a direct conversion, is used to mediate the conversion efficiency from mechanical to electrical. They are useless as transducers in mechanical energy harvesters as a result. Conductive polymers with carbon nanotube fillers are appealing options for creating electrodes for thermogalvanic cells because of their low production costs, chemical and thermal stability, and rapid electron transfer kinetics. All energy conversion processes are inefficient to some extent. Energy is wasted as heat in the following situations: motors, power transistors, car engines, and light bulbs. Radio stations generate megawatts of RF, but the signals they send to antennas are only microwatts. A fraction of this wasted energy is captured, converted to power, and used by energy-harvesting systems. The two most important energy sources nowadays are air turbines and solar power panels. The most common energy-collecting collectors are air turbines and large solar power panels. With proper design, energy-harvesting devices may be capable of replacing batteries in some applications.

The output of an energy-harvesting unit is often intermittent. Therefore, a system containing a controller for a thin-film or charging Li-ion battery, a valve for the MCU or sensors, and a wireless networking module must be properly constructed. The possibility of an embedded system becoming battery-free increases with the degree to which an energy-harvesting device can satisfy all of its requirements. For mass, motion, temperature, humidity, and light sensing applications, a wide range of sensing materials, including metals, metal-organic frameworks, metal oxide, solid electrolytes, sol-gel materials, carbon nanotubes, and graphene nanocomposites, are utilised to make both direct and complex sensors. As a result, we have made an effort to include many detection materials as practical in our research and evaluate the overall data linked with them, along with their potential interests and applications in current sensors.

REFERENCES

[1] Agency Energy and environment (EE) Report. 2008.

[2] Commission IE Efficient Electrical Energy Transmission and Distribution. 2007.

[3] Tritt, T.M., H. Böttner and L. Chen. 2008. Thermoelectrics: direct solar thermal energy conversion. MRS Bull. 33(4): 366–368.

[4] Abhat, A. 1981. Low temperature latent heat thermal energy storage: heat storage materials. Sol. Energy 30(4): 313–32.

[5] Sharma, S.D., D. Buddhi and R.L. Sawhney. 1999. Accelerated thermal cycle test of latent heat-storage materials, Sol. Energy 66: 483–490.

[6] Davis, E.S. and R. Bartern, 1975. Stratification in solar water heater storage tanks. pp. 38–42. In: Proc. Workshop on Solar Energy Storage Subsystems for Heating and Cooling of Buildings. Charlottesville, Virginia.

[7] George, A. 1989. Hand book of thermal design. In: Guyer, C. (ed.). Phase Change Thermal Storage Materials. McGraw Hill Book Co.

[8] Heremans, J.P., C.M. Thrush, D.T. Morelli and M.C. Wu. 2002. Thermoelectric power of bismuth nanocomposites. Phys. Rev. Lett. 88: 4361–5.

[9] He, W., G. Zhang, X. Zhang, J. Ji, G. Li and X. Zhao. 2015. Recent development and application of thermoelectric generator and cooler. Appl. Energy. 143: 1–25.

[10] Yang, J. and T. Caillat. 2006. Thermoelectric materials for space and automotive power generation. MRS Bull. 31(3): 224–229.

[11] Changyoon, J., J. Chanwoo, L. Seonghwan, Q.F. Maria and B.P. Young. 2020. Carbon nanocomposite based mechanical sensing and energy harvesting. Int. J. Precis. Eng. Manuf. Green Technol. 1: 85.

[12] Dey, A., O.P. Bajpai, A.K. Sikder, S. Chattopadhyay and M.A.S. Khan. 2016. Recent advances in CNT/graphene based thermoelectric polymer nanocomposite: A proficient move towards waste energy harvesting. Renewable Sustainable Energy Rev. 53: 653–671.

[13] Thostenson, E.T., Z. Ren and T.W. Chou. 2001. Advances in the science and technology of carbon nanotubes and their composites: A review. Compos. Sci. Technol. 61(13): 1899–1912.

[14] Baughman, R.H., A.A. Zakhidov and W.A. de Heer. 2002. Carbon nanotubes—the route toward applications. Science. 297(5582): 787–792

[15] Prasek, J., J. Drbohlavova, J. Chomoucka, J. Hubalek, O. Jasek, V. Adam, et al. 2011. Methods for carbon nanotubes synthesis—review. J. Mater. Chem. 21(40): 15872–15884.

[16] Iijima, S. 1991. Helical microtubules of graphitic carbon. Nature 354(6348): 56–58.

[17] Farukh, M., R. Dhawan, B.P. Singh and S. Dhawan. 2015. Sandwich composites of polyurethane reinforced with poly(3,4-ethylene dioxythiophene)—coated multi-walled carbon nanotubes with

exceptional electromagnetic interference shielding properties. RSC Adv. 5(92): 75229–75238.

[18] Mathur, R., S. Pande, B. Singh and T. Dhami. 2008. Electrical and mechanical properties of multi-walled carbon nanotubes reinforced PMMA and PS composites. Polym. Compos. 29(7): 717–727.

[19] Graham, A.P., G.S. Duesberg, W. Hoenlein, F. Kreupl, M. Liebau, R. Martin, et al. 2005. How do carbon nanotubes fit into the semiconductor roadmap? Appl. Phys. A 80(6): 1141–1151.

[20] Jindal, P., S. Pande, P. Sharma, V. Mangla, A. Chaudhury, D. Patel, et al. 2013. High strain rate behavior of multi-walled carbon nanotubes-polycarbonate composites. Compos. B Eng. 45(1): 417–422.

[21] Pande, S., B.P. Singh and R.B. Mathur. 2014. Processing and properties of carbon nanotube/polycarbonate composites. pp. 333–364. In: V. Mittal (ed.). Polymer Nanotubes Nanocomposites: Synthesis, Properties and Applications, 2nd Ed. Wiley. New Jersey.

[22] Singh, B.P., P. Saini, T.K. Gupta, P. Garg, G. Kumar, I. Pande, et al. 2011. Designing of multi-walled carbon nanotubes reinforced low density polyethylene nanocomposites for suppression of electromagnetic radiation. J. Nanopart Res. 13(12): 7065–7074.

[23] Fim, F.C., N.R. Basso, AP. Graebin, D.S. Azambuja and G.B. Galland. 2013. Thermal, electrical, and mechanical properties of polyethylene–graphene nanocomposites obtained by in situ polymerization, J. Appl. Polym. Sci. 128(5): 2630–2637.

[24] Li, B. and, J. Zhang. 2015. Polysiloxane/multi-walled carbon nanotubes nanocomposites and their applications as ultrastable, healable and super hydrophobic coatings. Carbon. 93: 648–658.

[25] Wang, Z., Y. Yang, R. Burtovyy, I. Luzinov and M.W. Urban. 2014. UV-induced self-repairing polydimethylsiloxane–polyurethane (PDMS–PUR) and polyethylene glycol–polyurethane (PEG–PUR) Cu-catalyzed networks. J. Mater. Chem. A 2(37): 15527–15534.

[26] Ling, J., M.Z. Rong and M.Q. Zhang. 2012. Photo-stimulated self-healing polyurethane containing dihydroxyl coumarin derivatives. Polymer 53(13): 2691–2698.

[27] Verma, M., P. Verma, S. Dhawan and V. Choudhary. 2015. Tailored graphene based polyurethane composites for efficient electrostatic dissipation and electromagnetic interference shielding applications. RSC Adv. 5(118): 97349–97358.

[28] Saini, P., V. Choudhary, B. Singh, R. Mathur and S. Dhawan. 2011. Enhanced microwave absorption behavior of polyaniline-CNT/polystyrene blend in 12.4–18.0 GHz range. Synth. Met. 161(15): 1522–1526.

[29] Shahzad, F., S. Yu, P. Kumar, J.-W. Lee, Y.-H. Kim, S.M. Hong, et al. 2015. Sulfur doped graphene/polystyrene nanocomposites for electromagnetic interference shielding. Compos. Struct. 133: 1267–1275.

[30] Babal, A., R. Gupta, B. Singh, V. Singh, R. Mathur and S. Dhakate. 2014. Mechanical and electrical properties of high performance MWCNT/polycarbonate composites prepared by industrial viable twin screw extruder with back flow channel. RSC Adv. 4: 64649–64658.

[31] Gedler, G., M. Antunes, J. Velasco and R. Ozisik. 2016. Enhanced electromagnetic interference shielding effectiveness of polycarbonate/ graphene nanocomposites foamed via 1-step supercritical carbon dioxide process. Mater. Des. 90: 906–914.

[32] Shen, B., W. Zhai, M. Tao, D. Lu and W. Zheng. 2013. Enhanced interfacial interaction between polycarbonate and thermally reduced graphene induced by melt blending. Compos. Sci. Technol. 86: 109–116.

[33] Zeng, X., J. Yang and W. Yuan. 2012. Preparation of a poly (methyl methacrylate)-reduced graphene oxide composite with enhanced properties by a solution blending method. Eur. Polym. J. 48(10): 1674–1682.

[34] Malinskii, Y.M., V. Prokopenko, N. Ivanova and V. Kargin. 1970. Investigation of self-healing of cracks in polymers. Polym. Mech. 6(2): 240–244.

[35] Wool. R.P. 1980. Crack healing in semicrystalline polymers, block copolymers and filled elastomers. pp. 341–362. *In*: Adhesion and Adsorption of Polymers. Springer. Berlin.

[36] Wool, R. and K. O'connor. 1981. A theory crack healing in polymers. J. Appl. Phys. 52(10): 5953–5963.

[37] White, S.R., N. Sottos, P. Geubelle, J. Moore, M.R. Kessler, S. Sriram, et al. 2001. Autonomic healing of polymer composites. Nature. 409(6822): 794–797.

[38] Kessler, M. 2007. Self-healing: A new paradigm in materials design. Proceedings of the institution of mechanical engineers. Part G. J. Aerosp. Eng. 221(4): 479–495.

[39] Lanzara, G., Y. Yoon, H. Liu, S. Peng and W. Lee. 2009. Carbon nanotube reservoirs for self-healing materials. Nanotechnology 20(33): 335704.

[40] Sinham, R.S., D. Pelot, Z. Zhou, A. Rahman, X.-F. Wu and A.L. Yarin. 2012. Encapsulation of self-healing materials by coelectrospinning, emulsion electrospinning, solution blowing and intercalation. J. Mater. Chem. 22(18): 9138–9146.

[41] Mario, L. and N. Ahmed. 2011. Organic thermoelectrics: Green energy from a blue polymer. Nat. Mater. 10(6): 409–410.

[42] Blaiszik, B., S. Kramer, S. Olugebefola, J.S. Moore, N.R. Sottos and S.R. White. 2010. Self-healing polymers and composites. Annu. Rev. Mater. Res. 40: 179–211.

[43] Yuan, Y., T. Yin, M. Rong and M. Zhang. 2008. Self-healing in polymers and polymer composites. Concepts, realisation and outlook: A review. Polym. Lett. 2(4): 238–250.

[44] Herbst, F., D. Döhler, P. Michael and W.H. Binder. 2013. Self-healing polymers via supramolecular forces. Macromol. Rapid. Commun. 34(3): 203–220.

[45] Aissa, B., E. Haddad, W. Jamroz, S. Hassani, R. Farahani, P. Merle, et al. 2012. Micromechanical characterisation of single-walled carbon nanotube reinforced ethylidene norbornene nanocomposites for self-healing applications, Smart. Mater. Struct. 21(10): 105028.

[46] Bailey, B.M., Y. Leterrier, S. Garcia, S. Van Der Zwaag and V. Michaud. 2015. Electrically conductive self-healing polymer composite coatings. Prog. Org. Coat. 85: 189–198.

[47] Cui, W., J. Ji, Y.-F. Cai, H. Li and R. Ran. 2015. Robust, Anti-fatigue, and self-healing graphene oxide/hydrophobically associated composite hydrogels and their use as recyclable adsorbents for dye wastewater treatment. J. Mater. Chem. A 3(33): 17445–17458.

[48] Jyoti, J., S. Basu, B. Singh and S. Dhakate. 2015. Superior mechanical and electrical properties of multi-wall carbon nanotube reinforced acrylonitrile butadiene styrene high performance composites. Compos. B. Eng. 83: 58–65.

[49] Sharma, S., V. Gupta, R. Tandon and V. Sachdev. 2016. Synergic effect of graphene and MWCNT fillers on electromagnetic shielding properties of graphene–MWCNT/ABS nanocomposites. RSC Adv. 6(22): 18257–18265.

[50] Rowe, D.M. 1995. In Handbook of Electrics. CRC Press. Boca Raton. FL.; b) Zhao D., Tan G. 2014. A review of thermoelectric cooling: materials, modeling and applications. Appl. Therm. Eng. 66(1–2): 15–24.

[51] Riffat, S.B. and X. Ma. 2003. Thermoelectrics: a review of present and potential applications. Appl. Therm. Eng. 23(8): 913–935.

[52] DiSalvo, F.J. 1999. Thermoelectric cooling and power generation. Science 285(5428): 703–706.

[53] Chen, G., M.S. Dresselhaus, G. Dresselhaus, J.-P. Fleurial and T. Caillat. 2003. Recent developments in thermoelectric materials. Int. Mater. Rev. 48(1): 45–66.

[54] Snyder, G.J. and E.S. Toberer. 2008. Complex thermoelectric materials. Nat. Mater. 7(2): 105–114.

[55] Tee, B.C., C. Wang, R. Allen and Z. Bao 2012. An electrically and mechanically self-healing composite with pressure-and flexion-sensitive properties for electronic skin applications. Nat. Nanotechnol. 7(12): 825–832.

[56] Han, Y., T. Wang, X. Gao, T. Li and Q. Zhang. 2016. Preparation of thermally reduced graphene oxide and the influence of its reduction temperature on the thermal, mechanical, flame retardant performances of PS nanocomposites. Compos. A Appl. Sci. Manuf. 84: 336–343.

[57] Garg, P., B.P. Singh, G. Kumar, T. Gupta, I. Pandey, R. Seth, et al. 2010. Effect of dispersion conditions on the mechanical properties of multi-walled carbon nanotubes-based epoxy resin composites. J. Polym. Res. 18(6): 1397–1407.

[58] Singh, B.P, K. Saini, V. Choudhary, S. Teotia, S. Pande, P. Saini, et al. 2014. Effect of length of carbon nanotubes on electromagnetic interference shielding and mechanical properties of their reinforced epoxy composites. J. Nanopart. Res. 16(1): 1–11.

[59] Tang, L.-C., Y.-J. Wan, D. Yan, Y.-B. Pei, L. Zhao, Y.-B. Li, et al. 2013. The effect of graphene dispersion on the mechanical properties of graphene/epoxy composites. Carbon 60: 16–27.

[60] Mathur, R.B., B.P. Singh, T. Dhami, Y. Kalra, N. Lal, R. Rao, et al. 2010. Influence of carbon nanotube dispersion on the mechanical properties of phenolic resin composites. Polym. Compos. 31(2): 321–327.

[61] Ansari, M.O. and F. Mohammad. 2011. Thermal stability, electrical conductivity and ammonia sensing studies on p-toluenesulfonic acid doped polyaniline: titanium dioxide (pTSA/Pani: TiO_2) nanocomposites. Sensor and Actuators B 157: 122–129.

[62] Sun T., Y. Fan, P. Fan, F. Geng, P. Chen and F. Zhao. 2019. Use of graphene coated with ZnO nanocomposites for microextraction in packed syringe of carbamate pesticides from juice samples. J. Separ. Sci. 42: 2131–2139.

[63] Kondawar, S.B., P.T. Patil and S.P. Agrawal. 2014. Chemical vapour sensing properties of electrospun nanofibers of polyaniline/ ZnO nanocomposites. Adv. Mater. Lett. 5: 389–395.

[64] Rahman, M.M., M.A. Hussein, K.A. Alamry, F.M. Al-Shehry and A.M. Asiri. 2018. Polyaniline/graphene/carbon nanotubes nanocomposites for sensing environmentally hazardous 4-aminophenol. Nano-Struct. Nano-Objects. 15: 63–74.

[65] Zhang, J., X. Liu, S. Wu, H. Xu and B. Cao, Sens. 2013. One-pot fabrication of uniform polypyrrole/Au nanocomposites and investigation for gas sensing. Sens. Actuators B. 186: 695–700.

[66] Mane, A.T., S.T. Navale, S. Sen, D.K. Aswal, S.K. Gupta and V.B. Patil. 2015. Nitrogen dioxide (NO_2) sensing performance of p-Polypyrrole/n-Tungsten oxide hybrid nanocomposites at room temperature. Org. Electron. 16: 195–204.

[67] Rawool, C.R., N.S. Punde, A.S. Rajpurohit, S.P. Karna and A.K. Srivastava. 2018. High energy density supercapacitive material based on a ternary hybrid nanocomposite of cobalt hexacyanoferrate/carbon nanofibers/ polypyrrole. Electrochim. Acta 268: 411–423.

[68] Feng, L. K. Wang, X. Zhang, X. Sun, C. Li, X. Ge, et al. 2018. Flexible solid-state supercapacitors with enhanced performance from hierarchically graphene nanocomposite electrodes and ionic liquid incorporated gel polymer electrolyte. Adv. Funct. Mater. 28: 1704463–1704475.

[69] Idrees, M., S. Batool, J. Kong, Q. Zhuang, H. Liu, Q. Shao, et al. 2019. Polyborosilazane derived ceramics - Nitrogen sulfur dual doped graphene nanocomposite anode for enhanced lithium ion batteries. Electrochim. Acta 296: 925–937.

[70] Song, Z., Y.L. Ma and C.E. Li. 2019. The residual tetracycline in pharmaceutical wastewater was effectively removed by using MnO_2/ graphene nanocomposite. Sci. Total Environ. 651: 580–590.

[71] Ananthanarayanan, A., X. Wang, P. Routh, B. Sana, S. Lim, D.H. Kim, et al. 2014. Facile synthesis of graphene quantum dots from 3D graphene and their application for Fe^{3+} sensing. Adv. Funct. Mater. 24: 3021–3026.

[72] Cheng, Y., R. Wang, J. Sun and L. Gao. 2015. A Stretchable and highly sensitive graphene-based fiber for sensing tensile strain, bending, and torsion. Adv. Mater. 27: 7365–7371.

[73] Liu, Q., Y. Tang, X. Yang, M. Wei and M. Zhang. 2019. An ultra-low detection limit glucose sensor based on reduced graphene oxide-concave tetrahedral Pd NCs@CuO composite. J. Electrochem. Soc. 166(6): B381–387.

[74] Neri, G., S.G. Leonardi, M. Latino, N. Donato, S. Baek, D. E. Conte, et al. 2013. Sensing behavior of SnO_2/reduced graphene oxide nanocomposites toward NO_2. Sens. Actuators B. 179: 61–68.

[75] Zhang, H., J. Feng, T. Fei, S. Liu and T. Zhang. 2014. SnO_2 nanoparticles-reduced graphene oxide nanocomposites for NO_2 sensing at low operating temperature. Sens. Actuators B 190: 472–478.

[76] An, X., X. Yu, C. Y. Jimmy and G. Zhang. 2013. CdS nanorods/reduced graphene oxide nanocomposites for photocatalysis and electrochemical sensing. J. Mater. Chem. A 1: 5158–5164.

[77] Xu, F., M. Deng, G. Li, S. Chen and L. Wang. 2013. Electrochemical behavior of cuprous oxide–reduced graphene oxide nanocomposites and their application in nonenzymatic hydrogen peroxide sensing. Electrochim. Acta 88: 59–65.

[78] Reddy, Y.V.M., S. Bathinapatla, T. Łuczak, M. Osińska and H. Maseed. 2018. An ultra-sensitive electrochemical sensor for the detection of acetaminophen in the presence of etilefrine using bimetallic Pd–Ag/reduced graphene oxide nanocomposites. New J. Chem. 42(4): 3137–3146.

[79] Govindasamy, M., S.F. Wang, B. Subramanian, R.J. Ramalingam, H. Al-lohedan and A. Sathiyan. 2019. A novel electrochemical sensor for determination of DNA damage biomarker (8-hydroxy-2'-deoxyguanosine) in urine using sonochemically derived graphene oxide sheets covered zinc oxide flower modified electrode. Ultrason. Sonochem. 58: 104622–104623.

[80] Yu, L., F. Ma, X. Ding, H. Wang and P. Li. 2018. Silica/graphene oxide nanocomposites: Potential adsorbents for solid phase extraction of trace aflatoxins in cereal crops coupled with high performance liquid chromatography. Food Chem. 245: 1018–1024.

[81] Liu, H., J. Gao, W. Huang, K. Dai, G. Zheng, C. Liu, et al. 2016. Electrically conductive strain sensing polyurethane nanocomposites with synergistic carbon nanotubes and graphene bifillers. Nanoscale 8: 12977–12990.

[82] Le, X.V., T.L.A. Luu, H.L. Nguyen and C.T. Nguyen. 2019. Synergistic enhancement of ammonia gas-sensing properties at low temperature by compositing carbon nanotubes with tungsten oxide nanobricks. Vacuum 168: 108861–108861.

[83] Jalilian, N., H. Ebrahimzadeh and A.A. Asgharinezhad. 2018. Determination of acidic, basic and amphoteric drugs in biological fluids and wastewater after their simultaneous dispersive micro-solid phase extraction using multiwalled carbon nanotubes/magnetite nanoparticles@ poly(2-aminopyrimidine) composite. Microchem. J. 143: 337–349.

[84] Abbasghorbani. M. 2018. Fe_3O_4 loaded single wall carbon nanotubes and 1-methyl-3-octylimidazlium chloride as two amplifiers for fabrication

of highly sensitive voltammetric sensor for epirubicin anticancer drug analysis. J. Mol. Liq. 266: 176-180.

[85] Norouzi, P., B. Larijani, T. Alizadeh, E. Pourbasheer, M. Aghazadeh and M.R. Ganjali. 2019. Application of advanced electrochemical methods with nanomaterial-based electrodes as powerful tools for trace analysis of drugs and toxic compounds. Curr. Anal. Chem. 15(2): 143–151.

[86] Upadhyay S.S. and A.K. Srivastava. 2019. Hydroxypropyl β-cyclodextrin cross-linked multiwalled carbon nanotube-based chiral nanocomposite electrochemical sensors for the discrimination of multichiral drug atorvastatin isomers. New. J. Chem. 43: 11178–11188.

[87] Niezabitowska, E., J. Smith, M.R. Prestly, R. Akhtar, F.W. von Aulock, Y. Lavallée, et al. 2018. Hydroxypropyl β-cyclodextrin cross-linked multiwalled carbon nanotube-based chiral nanocomposite electrochemical sensors for the discrimination of multichiral drug atorvastatin isomers. RSC Adv. 8: 16444–16454.

[88] Wang, Y., J. Luo, J. Liu, X. Li, Z. Kong, H. Jin, et al. 2018. Electrochemical integrated paper-based immunosensor modified with multi-walled carbon nanotubes nanocomposites for point-of-care testing of 17β-estradiol. Biosens. Bioelectron. 107: 47–53.

[89] Miraki, M., H. Karimi-Maleh, M.A. Taher, S. Cheraghi, F. Karimi, S. Agarwal, et al. 2019. Voltammetric amplified platform based on ionic liquid/NiO nanocomposite for determination of benserazide and levodopa. J. Mol. Liq. 278: 672–676.

[90] Mazrouaa, A.M., N.A. Mansour, M.Y. Abed, M.A. Youssif, M.A. Shenashen and Md R. Awual. 2019. Nano-composite multi-wall carbon nanotubes using poly(p-phenylene terephthalamide) for enhanced electric conductivity. J. Environ. Chem. Eng. 7(2): 103002–103003.

Chapter 3

Nanostructured Silicon for Solar Energy Conversion Applications

Ragavendran Venkatesan[1], Jeyanthinath Mayandi[1]*,
Terje G. Finstad[2]*, J.M. Pearce[3] and
Vishnukanthan Venkatachalapathy[2,3]

[1]Department of Materials Science, School of Chemistry,
Madurai Kamaraj University, Madurai – 625 021, India.

[2]Department of Physics/Centre for Materials Science and Nanotechnology,
University of Oslo, P.O. Box 1048 Blindern, N-0316 Oslo, Norway.

[3]Department of Electrical and Computer Engineering,
Western University, London, ON, Canada.

[4]Department of Materials Science, National Research Nuclear University
"MEPhI", 31 Kashirskoe sh, Moscow, Russian Federation.

3.1 INTRODUCTION

Climate change and global warming caused by the burning of fossil fuels, resulting in carbon dioxide (CO_2) emissions, have led to an urgent need for abundant and cost-efficient renewable energy sources.

*For Correspondence: E-mail: jeyanthinath.chem@mkuniversity.org
Email: terje.finstad@fys.uio.no

The available renewable energy sources are wind energy, hydropower, geothermal, and solar energy. Among various renewable energies, solar energy is the most abundant and universally available. Sunlight strikes the Earth's surface at a rate of ~120,000 TW [1]. Sunlight can be captured and converted into electricity by solar photovoltaic (PV) cells. It can also be utilised to produce chemical fuels, such as methanol and hydrogen via CO_2 reduction [2] and water splitting [3] in photoelectrochemical cells. Solar energy, however, is diffuse (~100 mW cm^{-2}) and diurnal, only available during daytime [4], requiring large areas of solar collectors to harvest civilisational significant amounts of energy. PV energy generation is the fastest-growing energy source [5]. Until now, the high cost and low efficiency hinder the wide installation of these solar energy conversion systems. For instance, even though the solar cell industry has experienced a significant expansion in the past decade, the total installed capacity is only ~1.1 TW in 2022 [6]. The levelised cost of electricity from PV is already lower than conventional generation [7], and the value of solar is much higher than conventional electricity [8] because of the environmental benefits that make it a truly sustainable energy source [9]. Even though solar is the lowest-cost source of electric power [10] according to the EIA [11], the cost of producing electricity from solar must decrease further to accelerate the investment needed to eliminate all use of fossil fuels for energy. At present, photovoltaics is the most elegant method to produce electricity without moving parts, emissions, or noise, and all this by converting abundant sunlight without practical limitations. The relevance of solar energy specifically PV can be justified mainly by the factors like scalability, sustainability, environmental impact, and the security of the source [12]. The scalability means the abundant availability of solar radiation to be utilised for PV. Solar cells are zero-emission electricity generators, which proves their environmental friendliness. Therefore, it is crucial to develop even more cost-effective solar energy conversion systems with high light absorption/conversion efficiency. In order to decrease cost and increase power output simultaneously, novel processes must be introduced in the conventional production of solar cells. These processes must result in an increased power conversion efficiency of the solar cells while making the fabrication process itself less expensive. At the same time, only significantly low-cost and abundant materials may be used in cell production in order to be able to scale-up the production to the order of magnitude of the global power consumption, which is in the TW range [13].

Progress made in nanotechnology can be used as a driving tool to produce less-expensive and more efficient solar cells. Nanostructures have attracted much interest over the past few years as they offer very attractive features, which make them good candidates for solar cell applications [14]. In particular, porous silicon and silicon nanowires (SiNWs) have remarkable optical properties, such as antireflection and light-trapping effects as illustrated in Figure 3.1, allowing for efficient light harvesting capabilities [15]. A low-temperature and high-pressure oxidation process is done to passivate the formed porous silicon.

These excellent optical properties are partly due to the sub-wavelength structure offered by the porous silicon and SiNWs, acting like an antireflective layer. The enhanced absorption is also explained by the electromagnetic wave interaction with the SiNWs. Many papers have been published recently on NW-based solar cells. Some attempts have already been carried out using CdS, CdSe, CdTe, ZnO, or TiO_2 NWs [16–20]. Nevertheless, the toxicity of CdS, CdSe, and CdTe as well as the high energy gap ($E_g > 3$ eV) of ZnO and TiO_2 nanostructures make these materials unsuitable for efficient devices. In this chapter, we focus on silicon nanostructures as they are environmental-friendly, abundant, well-understood, and show excellent light trapping and antireflection properties and Si already dominates the PV industry.

Figure 3.1 shows that the light is effectively trapped within the Si nanostructures. One common strategy to reduce the cost of Si-based solar cells is to use thinner wafers so less material is necessary. The thickness reduction, however, is detrimental to the optical absorption. Indeed, solar cells based on c-Si are generally relatively thick because of the indirect band gap of Si, which requires large thickness to absorb most of the incoming light.

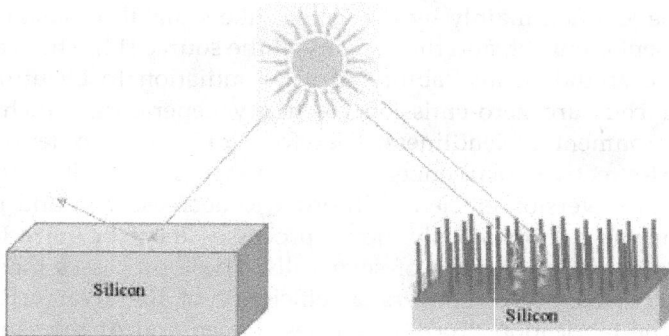

Figure 3.1 Illustration of the light-trapping effect of Si nanostructures.

Porous silicon and SiNWs can be used as a driving force to transfer the bulk c-Si-based solar cells to the thin films technology and cut the material and process costs of the PV modules.

Indeed, the perfect light-trapping properties of SiNW arrays enable high absorption in just 1 μm thick long NW arrays, rendering the transfer to the thin film technology highly desirable. By using NWs, we are able to produce high-quality materials for the thin film technology, and therefore high efficiency is expected. The ultimate goal will be to produce high-efficiency solar cells based on SiNWs.

One of the most important aspects to improve overall cell efficiency is by reducing the front surface reflectance of crystalline material. Compared with a 100% reflected mirror, polished silicon wafers have surface reflectivity from 70% to 30%, depending on the wavelength of incident light [21]. In shorter wavelength ranges, silicon wafers appear to have a higher average reflectivity and a higher average absorption coefficient than compared to longer wavelength ranges. In this case, reduced reflectivity in wavelength regions with high absorption will significantly enhance PV cell efficiency.

Reducing front surface reflectivity is a commonly used method in modern solar cell fabrication. Antireflection (AR) coating and surface textured structures are the most often employed technologies to reduce surface reflectivity. AR coating is achieved by depositing a thin layer of low-reflectivity material, on the front surface of the solar cell to reduce reflectivity at specific wavelengths [22]. The AR layer can only reduce reflectivity at specific wavelengths, however, which is not sufficient when solar cells are excited under the whole solar light spectra. A flat surface has higher reflectivity than a textured surface since more light can be instantly reflected from the surface. The surface textured structure methodology uses a chemical solution to etch the front surface of the solar cell to reduce reflectivity. Due to precise and controlled chemical etching, textured structures in the order of nano/micro scale can be fabricated [23]. Black silicon (b-Si) is one such type of surface textured structure. Compared with normal micron-sized surface texture structures, the black silicon structures are smaller to be in a nanometer scale and have a larger distribution angle to incident light.

3.2   REDUCED SURFACE REFLECTIVITY

Black silicon (b-Si) structures feature needle-like surface structures, where these needle-like shaped structures are made of crystalline silicon (c-Si) material. The b-Si solar cell is a c-Si solar cell that has the b-Si structure on the front surface. The reflectivity of untreated c-Si solar cells reduces the number of photons injected into the solar cell, which decreases cell efficiency at the same time. The average reflectivity of an untreated c-Si solar cell is around 40%. By using the b-Si structures on the front surface, the front surface reflectivity of the c-Si solar cell can be controlled below 10%, which means that an additional 20% of photons from incident light are injected into the solar cell substrate compared to the conventional c-Si solar cells [24].

The b-Si structures on the c-Si solar cell front surface can reduce the reflectivity according to diffraction effects and the "moth-eye effect" occurring under incident light. The reflectivity of c-Si solar cells is due to the abrupt transition of light from the surrounding medium to the solar cell. Increased refractivity difference between the solar cell and the surrounding media will increase reflectivity on the solar cell-surrounding media interface. The interface reflectivity can be reduced when the disparity between the two mediums is decreased. The b-Si structures on the solar cell are constructed at the nanometer scale so that the geometry of b-Si structures can reduce the reflectivity at different wavelengths by making a continuous transition between two materials [25]. When incident light is in a shorter wavelength region, the diffraction effects occur at the b-Si structures since the dimensions of the needle-like structures are comparable to the wavelength of incident light. As the incident light has been diffracted through the b-Si structures, the incident light can be diffracted to different angles, which would increase the path length of light in the substrate. When the incident light is at a longer wavelength region, the moth-eye effect occurs at the b-Si structures since the dimensions of the needle-like structures are close to the wavelength of incident light. The surface on the b-Si structures behaves like an effective medium, and the effective index of refraction changes continuously from that of the surrounding medium to that of the needle-like structures. Hence, the reflectivity on the b-Si structures is lower than the reflectivity on conventional c-Si solar cells [26]. The best b-Si can produce record PV efficiencies [27] and use dry etching [28], which can be cost-effective in mass production compared to conventional optical engineering strategies [29]. There are several approaches, however, to get to high-efficiency black silicon solar cells that we will review here.

3.3   SILICON NANOSTRUCTURES

Silicon nanostructures (SiNS) are classified into different types, but this chapter will focus on two structures used in solar cell applications:

(i)   Porous silicon
(ii) Silicon nanowires

3.3.1   Porous Silicon (PS)

The root of porous silicon (PS) can be traced back to the pioneering work of Arthur Uhlir Jr. and Ingeborg Uhlir in 1956 [30]. PS was discovered accidentally when they performed electropolishing of silicon wafers in hydrofluoric acid-containing electrolytes. This early work, however, attracted little attention because the PS films were mistakenly suggested to be silicon sub-oxide. It was not until the early 1970s, the porous features of Si were observed for the first time on the Si surface. Watanabe and Sakai demonstrated the first application of P-Si in electronics in 1971 [31]. They oxidised PS to achieve dielectric isolation of silicon devices. In 1986, Takai and Itoh first introduced silicon-on-insulator (SOI) technology to the integrated circuit industry [32]. In 1990, Leigh Canham discovered the tuneable photoluminescence of PS at room temperature due to quantum-confinement effects [33], which surged a tremendous amount of research effort to study PS and was believed to hold great promise in silicon-based integrated optoelectronics. In the late 1990s, interesting photonic structures (e.g., Bragg stacks, rugate filters, and microcavities) were conveniently achieved using PS, and the extraordinary biocompatibility of PS was demonstrated. Then, the focus of the PS researches slowly extended to other directions towards applications for antireflective coatings [34], interference filters, waveguides, solar cells [35], biomedical implants, drug delivery [36], chemical and biological sensors [37], and many other applications.

3.3.2 Silicon Nanowire (SiNW)

Nanostructures such as nanotubes and nanowires offer unique access to low-dimensional physics. Nanowires could be important, as they can be considered the building blocks of a technology, which harnesses the quantum size effects for useful device applications. Such one-dimensional (1-D) nanomaterials can play a key role in nanotechnology, as well as provide model systems to demonstrate quantum size effects. Silicon-based nanowires including silicon, silicate, and silicide nanowires are particularly attractive due to the central role of silicon in the semiconductor industry. Potential applications include the field of photonics, interconnects, sensors, and nano-electromechanical systems; perhaps most importantly, they can also be applied to photovoltaics.

For SiNWs, the carrier type and concentration can be controlled by doping, as in bulk silicon. Furthermore, silicon turns into a direct bandgap semiconductor at nanometer size, due to quantum confinement, opening an entirely new field of optoelectronic applications for such nano-sized silicon-based devices. For instance, Si-nanowire-based devices have shown stable electroluminescent at room temperature. In addition, Si-nanowires have also shown efficiency enhancement in test photovoltaic applications, arising from enhanced optical absorption in the nanostructured form of silicon. Even though Si-nanowire research has led to a clear improvement in device performance in terms of efficiency, there is still much scope in understanding the growth and behaviour of this material before viable device production becomes possible. Theoretical studies suggest that the band-gap character (direct or indirect) strongly depends on the crystallographic direction of the nanowire axis. Also, the magnitude of the direct or indirect band-gap due to quantum confinement depends on the nanowire growth direction and the nanowire diameter [38].

At the nano-scale, quantum-confinement effects are important, which affect the properties of materials. So, it is important to see the quantum-confinement effects and exciton effects at the nano-scale, which affects the optical properties of the nanowires. The orientation of the SiNWs grown depends on the diameter of the nanowires. Smaller SiNWs whose diameter range from 3 nm to 10 nm, are grown primarily along the (110) direction. Also, medium SiNWs whose diameter ranges from 10 nm to 20 nm, grow primarily along the (112) direction, and large SiNWs, whose diameter ranges from 20 nm to 30 nm, grow primarily along the (111) direction [39]. Several possible structures for SiNWs have been reported. These include fullerene-like cage structures (clathrate), tetrahedral structures, and polycrystalline nanowires [40–42] as shown in Figure 3.2.

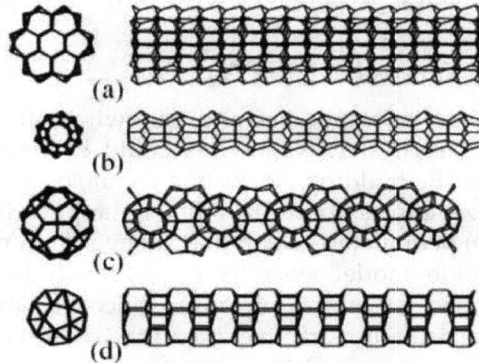

Figure 3.2 Schematic presentations of small-diameter silicon nanowires (a) tetrahedral (top), (b) Si34-clathrate, (c) Si46-clathrate, and (d) polycrystalline types of nanowires [42].

3.3.3 Physical Properties of Silicon Nanostructures

1-D silicon nanostructures have one free direction in which the electrical carriers can flow. Their physical properties are significantly different from those in bulk silicon, as explained below.

3.3.3.1 Electronic and Optical Properties

Silicon nanostructure's quantum confinement is manifested as a change in the band gap from the bulk material [43]. It is usually accepted that the band gap opens as a result of quantum confinement, which pushes the PL in the visible for crystallite sizes below 5 nm [44]. It has even been proposed that quantum confinement explains the entire PL from P-Si. However, many groups have reported that when the crystallite size decreases to a few nanometers, the PL in the air does not increase much beyond 2.1 eV, even when the crystallite size drops well below 3 nm [45]. This observation does not coincide with theory, which predicts a much larger opening of the band gap, in excess of 3 eV for sizes below 2 nm. For 1-D nanostructures, the band gap is directly related to the diameter of the nanowires [46].

$$\text{Band gap } \alpha \frac{1}{d^n}; 1 \leq n \leq 2$$

where d is the diameter of the 1-D nanostructure.

Silicon nanostructures have exhibited band-gap variations from 1.1 to 3.95 eV with diameters ranging from 7 nm to 1.3 nm, respectively [47]. The electronic properties such as the band gap, valley splitting, and effective mass also scale as a function of the diameter [48]. These affect the transport properties of the nanostructures like nanowires [49]. Hydrogen and oxygen-terminated SiNWs have been studied to gain an understanding of their optical and electronic properties. Confinement in SiNWs renders them optically active (direct band gap) and is independent of their specific orientation [50]. Quantum confinement in nanowires could play an important role in photovoltaic devices. SiNWs could also be used as light-emitting diodes and lasers at small diameters.

3.4 THERMAL AND MECHANICAL PROPERTIES

Mesoporous silicon has tuneable thermal properties that can be radically different from those of bulk silicon. The combination of very low thermal conductivity and heat capacity of high PS has led to a range of potential applications, such as thermal isolation of microdevices, chip-based ultrasound emission, thermoelectrics, and photothermal therapy.

SiNWs could be used in nano-scale thermoelectric power generator applications [51]. Therefore, it is important to study their thermal conductivity. Thermal conductivity in thin SiNWs having diameters between 1.4 and 8.3 nm has been studied. As the nanowire diameter decreases, the surface-to-volume ratio increases, which increases the surface scattering effects. This decreases the thermal conductivity of the nanowires. While at a very small diameter (<1.5 nm), quantum-confinement effects occur and these increase the thermal conductivity. The values of Young's modulus for SiNWs were estimated to be 186 and

207 GPa, respectively, for single- and double-clamped SiNWs and the values are close to the bulk value of 169 GPa for Si (111) [52].

3.4.1   Light Trapping

One of the common techniques used to decrease the front-face reflectance is by texturing the surface and thus enhancing the level of absorption of incident light by increasing the number of reflections, optical path length, and total internal reflection. In Si-wafer-based solar cells, texturing can also improve light absorption within the bulk of the cell by absorbing photons closer to the collection junction and by making use of weakly absorbed near band-gap photons. Depending on the texturing methodology and the crystallographic orientation of silicon surface planes, there are different types of texture morphologies produced on silicon surfaces. The most commonly being the upright pyramid structures demonstrated in Figure 3.3.

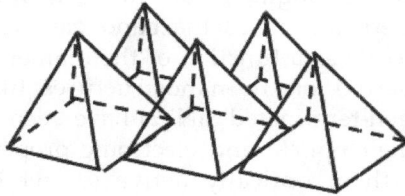

Figure 3.3   Pyramids as surface textures.

The surface textures decrease the overall reflectance by three distinct mechanisms. First, a portion of the light that is incident on the Si wafer is reflected from and onto the textured surface resulting in multiple reflections or multiple light bounces (Figure 3.4; case 1).

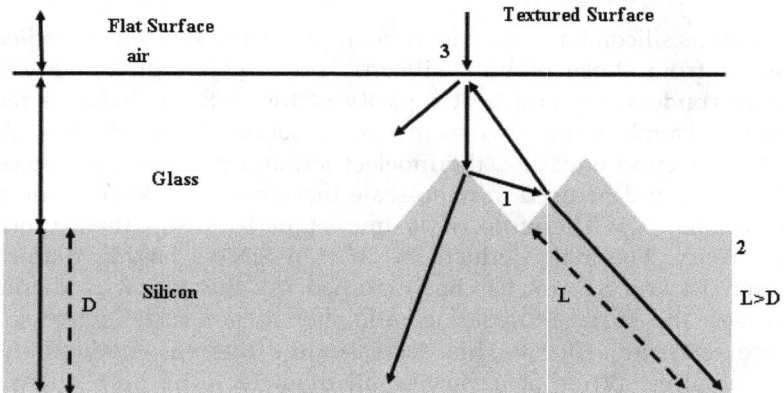

Figure 3.4   Normally incident light behaviour on a textured surface (on the right) comparison with a flat surface (on the left).

It increases, the amount of transmitted solar flux and thus the probability of photon absorption since initially reflected photons will have at least one other chance to enter the solar cell [53]. The longer path length of the refracted light carrying photons is the second reason for increased photon absorption and absorption closer to the collection junction. Normally incident light propagates within the bulk of the silicon at an angle in textured solar cells. This results in longer path lengths that light will travel before reaching the rear surface, as opposed to the path lengths of normally incident light refracted from a flat surface (see Figure 3.4, case 2) [54]. This is especially important if the diffusion lengths are comparable to or less than the cell thickness [55].

Finally, increased total internal reflection probability also increases the overall transmitted flux since it does not permit the light from escaping the silicon material (Figure 3.4; case 3). This phenomenon is also known as light-trapping. Long wavelength photons that the semiconductor material was unable to absorb are reflected from the back surface to the front.

Because of the poor optical properties of silicon material (absorption coefficient), the thickness of solar cells has to be relatively large in order to absorb 99% of the solar spectrum, which leads to an increase in weight, production cost, and recombination probability in the bulk. Since light absorption is enhanced in textured cells, the thickness of such cells can remain relatively small [56]. Some other techniques, such as antireflection coatings (ARC) can be used to reduce front-face reflection. ARC is a thin layer of an optimal thickness of dielectric material. This layer terminates the reflected beams by interference. In fact, texturing can be used in combination with an ARC. Such configuration can produce even lower reflectance than each of these individual layers separately.

3.5 NANOSTRUCTURED SILICON FABRICATION METHODS

Nanostructuring of silicon (Si) has received increasing substantial interest over the past few decades due to its ability to possess unique material properties that are not possible in bulk form [57, 58]. The fabrication of the nanostructures is often separated into two categories; bottom-up and top-down approaches.

3.5.1 Bottom-Up Silicon Nanostructure Formation

3.5.1.1 *Vapour-Liquid-Solid (VLS) Method*

Since the early 1990s, bottom-up techniques have been the more favorable single crystalline Si nanostructuring method. In particular, the vapour-liquid-solid (VLS) growth method has received substantial attention for the synthesis of Si nanowires (SiNWs). The VLS growth was first described by Wagner and Ellis [59], and the growth occurs via phase changes that are mediated through a catalyst particle. This means that the material to be grown, in this case, silicon is delivered in the gas phase, either molecularly or in the form of a gas compound into the growth chamber. Figure 3.5 shows a schematic sketch illustrating the widely adopted VLS growth. In the case of Si, molecular Si can be obtained either by evaporating Si in effusion cells, resulting in physical vapour deposition, or Si gas precursors resulting in chemical vapour deposition. Silicon gas precursors include mono-silane (SiH_4), trichlorosilane ($SiHCl_3$), or higher-order silanes. The catalyst metal particle acts as a collector of silicon. In the case of employing gas precursors, the catalytic nature leads to the lowering of the dissociation energies of the gas. Consequently, a high concentration of silicon atoms is found at the catalyst particles' surface, leading to diffusion into the cluster. For gold catalyst clusters, the gold-silicon system can become liquid at temperatures above the eutectic point (363°C), provided that a sufficient fraction of Si is present (e.g. 18.6 at. % at the eutectic temperature). In case of a constant silicon flow, silicon atoms will continue to diffuse into the Au–Si melt, although this is thermodynamically unstable. To reduce energies, the super-saturated or excess silicon atoms condense into the solid-phase. Nucleation of Si occurs at the footprint of the Au–Si catalyst layer by layer. Consequently, a silicon monolith or nanowire is formed.

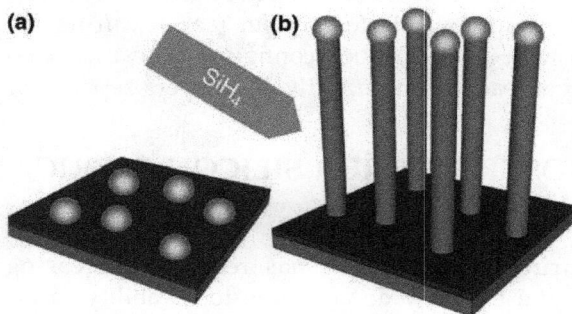

Figure 3.5   Schematic view of VLS growth of silicon nanowires (a) gold particles formed on the growth substrate and (b) VLS growth using silane as a silicon precursor.

The advantage of the VLS method is that the growth can be done on different types of substrates. It has been shown that thermodynamics dictates the crystal growth direction of the nanowire, even if growth occurs on amorphous substrates, i.e., without pre-defined lattice orientation. In addition, a correlation between crystal direction and nanowire diameter is clearly visible [60, 61]. Schmidt attributed this to the strong contribution of outer silicon atoms to the overall formation energies. When a silicon substrate is used, epitaxial growth can be achieved as well. Moreover, the grown wires can be subsequently transferred to any substrate even if direct growth on that substrate is not possible. In that case, however, the original vertical nanowire arrangement cannot be maintained. Although the VLS method is a well-developed synthesis method, it suffers from a few limitations and drawbacks.

3.5.1.2   Chemical Vapour Deposition (CVD)

SiNWs are grown in a chemical vapour deposition (CVD) chamber through the absorption of a Si-based gas precursor at a liquid catalytic alloy, typical gold (Au), at a temperature higher than the eutectic point (300–400°C) [62, 63]. The liquid catalyst serves as the preferred decomposition site of the Si atoms from the gas precursor until the catalyst becomes supersaturated with Si, at which point Si crystallises beneath the catalyst, forming SiNWs.

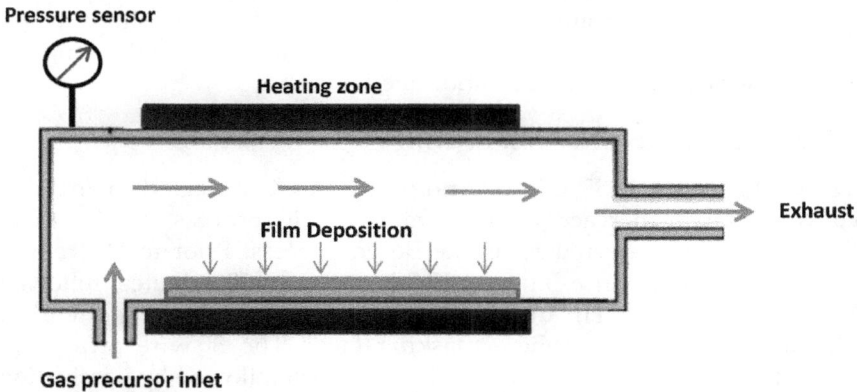

Figure 3.6   Schematic of a typical CVD film deposition unit.

Figure 3.6 shows the schematic of a typical CVD deposition unit. The CVD setup and growth process is rather expensive and is typically limited to small growth areas due to the chamber size. The SiNW density is relatively low due to the challenge of preventing the close-packed catalyst particles from merging at the growth temperature [64]. Achieving a high degree of vertical alignment is often a challenge. The SiNWs can be doped through a variety of methods, including with the metal catalyst or the addition of a dopant gas; however, accurately predicting the resulting dopant level is very challenging [60]. During the high-temperature growth process, the metal catalyst diffuses into the SiNWs forming crystallographic defects and trap sites that reduce the minority carrier lifetimes [65].

As an alternative to CVD growth, a wide variety of other growth methods can be applied using catalysts [66]. Among them are the annealing in a reactive atmosphere like hydrogen to form the nanowires directly from a silicon substrate [67], laser ablation [68], and molecular beam epitaxy [69].

3.5.2   Top-Down Silicon Nanostructure Formation

A top-down approach starts with a bulk Si wafer and involves removing material rather than adding material. These offer the potential to alleviate some of the limitations and drawbacks associated with the bottom-up synthesis of Si nanostructures. For instance, since the resulting SiNWs are etched from a bulk wafer, the doping concentration is based on the original wafer, and the high-temperature growth conditions are eliminated, removing the crystallographic defect sites. Furthermore, precise scalable patterning methods can be used to form highly aligned, dense Si nanostructures over a large area, making the top-down technique a very promising method. The top-down methods begin with the basics of forming porous Si and expanding to highly aligned SiNWs fabrication and modification methods.

3.5.2.1   *Deep Reactive Ion Etching*

Vertically aligned Si micro/nano wires are fabricated using the conventional deep reactive ion etching (DRIE) process, as shown in Figure 3.7, often referred to as the Bosch process. Prior to DRIE, a Si wafer was cleaned in a 3:1 (v/v) $H_2SO_4:H_2O_2$ for 10 minutes, followed by a 10-minute 2% HF dip [70]. The Si wires are patterned with a photoresist or a silica sphere masking layer. The Si wire arrays are formed by DRIE with cycles of etching with SF6 followed by passivation with C4F8 in radio-frequency generated plasma. The etch time will vary depending on the exposed Si area and the desired Si wire length. The Si wires are fabricated approximately 10–20 μm in length with diameters ranging from a few microns to approximately 300 nm depending on the patterning method. The Si wire diameter is limited to above 300 nm due to the on/off etch/passivation steps forming scalloped sidewalls along the length of the wire which is too severe in Si wires below 300 nm in diameter.

Si micro wires are patterned using conventional photolithography. The DRIE masking layer consists of a 1 μm thick photoresist layer that is exposed and developed, leaving behind an array of masked circular features.

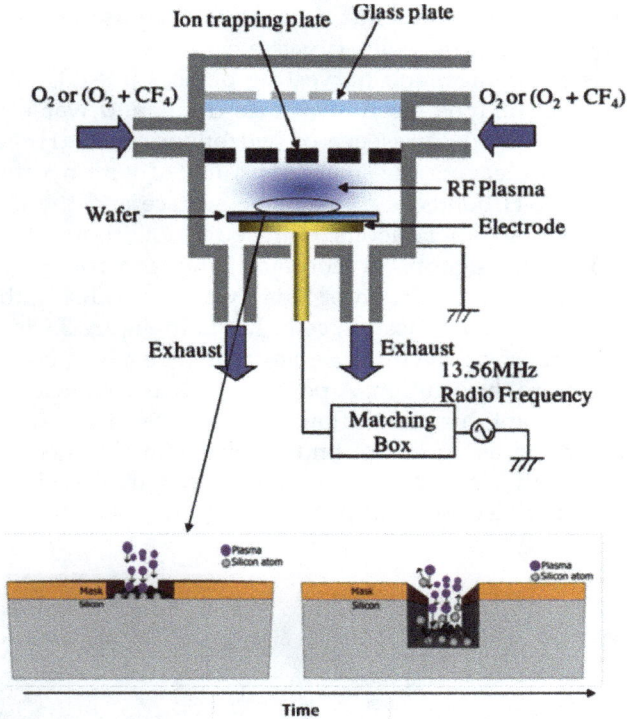

Figure 3.7 Schematic of a typical reactive ion etching unit [70].

3.6 FABRICATION METHODS

3.6.1 Electrochemical Etching

3.6.1.1 *PS Formation, Etching Chemistry, and Theory*

Porous Si was discovered accidentally in the mid-1950s while attempting to develop an electrochemical method to polish the surface of silicon (Si) and germanium (Ge) [71]. Instead of polishing the Si to form a reflective surface, a brittle dark film was found that contained porous holes propagating primarily in the <100> direction through the Si wafer. Although this was recorded in a technical report at Bell Labs, it did not receive substantial interest until the early 1990s when it was determined that porous Si can emit visible light [33, 72, 73]. A plethora of work focused on creating porous Si-based optoelectronics was reported during the mid-1990s until it slowly calmed down due to disappointing electroluminescence efficiency, and poor chemical and structural stability [74]. Since then, the fundamental understanding of porous Si has been improved and is currently being used in a wide range of applications from drug delivery systems, optoelectronics, sacrificial materials, energy storage, and harvesting devices [74–79].

Silicon is thermodynamically unstable in air and water, forming a native oxide layer. In the presence of hydrofluoric acid (HF), the oxide is rapidly removed leaving the surface terminated with hydrogen atoms (Si-H) [80]. The Si-H bonds passivated the surface and thereby prevent further oxidation in an aqueous HF medium without the presence of strong oxidising agents. A common and controlled method to promote the continued oxidation of Si is through anodising the Si in an HF solution. A typical anodisation cell, shown in Figure 3.8 is composed of a Teflon chamber with a circular opening where a clamped piece of Si wafer is exposed to the solution. A power supply is connected to the cell, supplying a constant current to flow through the Si wafer with metal back contact acting as an anode and to platinum (Pt) electrode acting as a cathode. Both electrodes being submerged in the HF electrolyte solution allow the electrochemical reaction to occur at the interface of the Si and electrolyte solution.

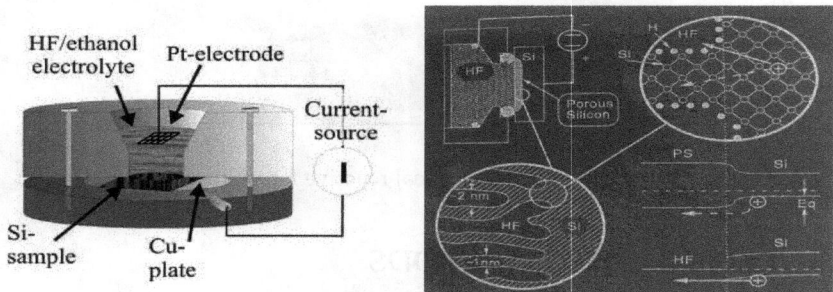

Figure 3.8   A sketch of the basic etch cell used and the schematics of PS formed [77].

Although porous Si forms in HF solutions diluted with deionised water, the hydrophobic properties of the hydrogen-terminated surface hinder the ability of the solution to infiltrate the pores and therefore induce lateral and in-depth inhomogeneity [77]. Thus, alcoholic solutions are commonly used to increase the wettability on the porous Si surface, allowing the solution to infiltrate the pores [74].

The dissolution of Si is obtained by supplying either an anodic current or potential; however, supplying a constant current allows for superior control and reproducibility [81]. The power source supplies holes to the Si valence band, oxidising the Si surface. The holes are driven to the surface by diffusion from the electric field generated within the wafer. At the Si surface, holes break apart the Si-H bonds causing competition between the formation of Si-H, Si-O, and Si-F bonds [74].

The Si-F bond has the highest electronegativity making it feasible over the formation of the Si-O bond, being the second most likely favorable one. The high electronegativity makes the Si-F bond highly polarised, so as soon a fluoride ion attaches to a Si atom, it becomes the preferred atom for the subsequent holes to attach to, leading to the rapid dissolution of the Si atom into the water as soluble SiF_6^{2-} as shown in Figure 3.9 [74].

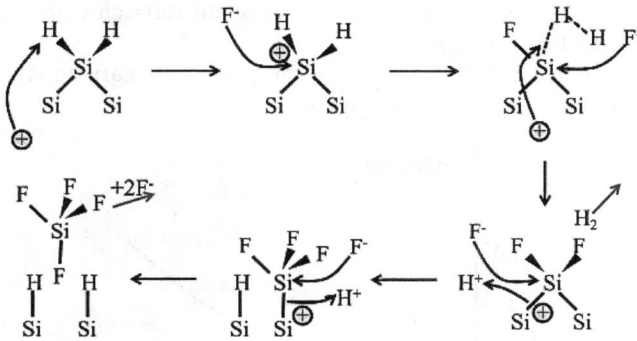

Figure 3.9 Dissolution mechanism of Si in the presence of hydrofluoric acid [74]. Overall Reaction at Silicon Surface: $Si + 6F^- + 2H^+ + 4h^+ \rightarrow SiF_6^{2-} + H_2$.

The preferred crystal plane direction for the removal of Si in HF electrolyte solution is related to the energy associated with cleaving the Si back-bonds [82]. The effected number density of back-bonds increases relative to the crystal plane in the order of (100) < (110) < (111); therefore, the pores typically propagate along the <100> direction [83]. The pore morphology, size, and overall porosity greatly depend on the wafer, solution, and applied conditions during the porous Si formation process. Varying the wafer type and doping level will affect the pore formation mechanism, which in turn affects the size and directionality of the pores [84]. For instance, micropores (pore width ≤ 2 nm) are often formed in low and moderately-doped p-type Si with the most uniform porosity. Mesopores (2 nm < pore width ≤ 50 nm) are often formed with highly doped p- and n-type wafers with pores perpendicular to the wafer surface due to the enhanced electric field driving the holes toward the base of the pore. Macropores (pore width > 50 nm) are often formed in low-doped n-type wafers with pores forming perpendicular to the surface due to the space charge region depleting the holes around the pore [74, 81]. The pore size and morphology can also be tuned by varying the solution or conditions. In general, increasing the current density, increasing the process temperature, or decreasing the HF concentration has a similar effect, leading to an increase in the pore size and porosity [82–85].

3.6.2 Electropolishing

As mentioned earlier, Uhlir discovered porous Si formed under certain conditions while trying to develop an electrochemical method to polish Si and Ge [1]. If the porous Si conditions are modified in a particular way, electropolishing occurs, which is very beneficial for removing thin porous layers from bulk Si wafers [86]. The key condition required for electropolishing is the rate at which the current injects holes into the Si surface has to exceed the rate at that fluorine ions can be transported to the pore tip [87–91].

The JV curves for n- as well as for p-type Si can be divided into two main regions as shown in Figure 3.10.

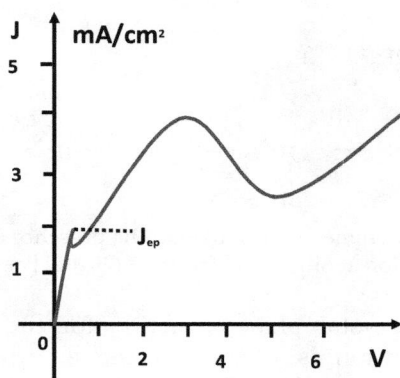

Figure 3.10 J-V characteristics of electro polishing and PS formation.

- Pore formation region.
- Electropolishing region.

Pore formation occurs in the initial rising part of the curve for $0 < V < V_{ep}$, with V_{ep} the potential of the small sharp peak. This peak, also called the electropolishing peak, has intensity J_{ep}, which depends mostly on the solution composition and little effect from the substrate. For $V > V_{ep}$ electro polishing occurs since the surface becomes covered by an oxide layer, whose composition and dielectric properties depend on the applied potential. By increasing the HF concentration, the first peak shifts to higher current values (higher electropolishing current), while increasing the substrate doping concentration shifts the first peak towards lower voltages.

3.6.3 Metal-Assisted Chemical Etching (MACE)

One of the earliest and most widely cited works on MACE was reported by Li and Bohn in 2000 [92], which characterised the chemical etching of silicon assisted by metal catalyst (gold-Au, platinum-Pt, or gold/palladium-Au/Pd alloys) to fabricate pores and wires. Later, several other reports started establishing the legitimacy of this process which resulted in the early foundation of this technique, which has since moved to less-expensive catalysts like silver. A notable work by Chartier et al. [93] characterised the silver-Ag catalyst system and discussed the relationship between metal-assisted chemical etching (MACE) and the electrochemistry of silicon. This report helped the researchers working in this field to approach MACE in a very systematic manner. These early reports pioneered the establishment of MACE and resulted in increased acceptance within the broader scientific community [94, 95].

3.6.4 Possible Mechanism for MACE of Silicon

MACE is a localised electrochemical reaction with the metal region acting as a microscopic cathode and the metal-semiconductor interface acting as the anode as shown in Figure 3.11. The oxidant used in the etching solution gets reduced locally on the surface of the catalyst, where the role of the metal is to reduce the activation energy required for the

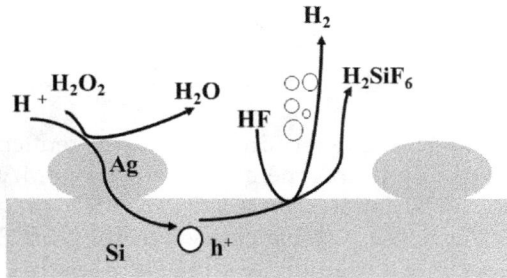

Figure 3.11 Schematic diagram showing the chemical process of MACE. The reduction of oxidant takes place only on top of the metal catalyst (cathode) and the dissolution of silicon takes place near the metal-silicon interface (anode). Hydrogen gas bubbles evolve during the reaction.

reduction of the oxidant (gain of electrons). Any noble metals like Ag, Au, Pt, etc., can act as a catalyst for the reduction of an oxidant, with the rate of reduction reaction directly related to the electronegativity of the metal used. The reduction of the oxidant (e.g. H_2O_2) in the presence of protonated hydrogen (H^+) coming from the acidic solution results in the injection of holes into the semiconductor region surrounding the metal layer. It should be noted that non-noble metals like chromium-Cr, aluminium-Al, and titanium-Ti [96] could not act as a catalyst for MACE as these metals do not reduce the activation energy for the reduction of oxidants.

At Cathode:

$$H_2O_2 + 2H^+ \rightarrow 2H_2O + 2h^+ \text{ (Local reduction)}$$
$$2H^+ \rightarrow H_2 + 2h^+ \text{ (Hydrogen gas formation)}$$

At Anode:

$$Si + 6HF + 4h^+ \rightarrow H_2SiF_6 + 4H^+ \text{ (Dissolution of silicon)}$$

Net reaction:

$$Si + 6HF + H_2O_2 \rightarrow H_2SiF_6 + 2H_2O + H_2$$

In addition, there is a second cathodic reaction resulting in the formation of hydrogen which is observed as gas bubbles during the reaction. On the other hand, the injection of holes or removal of electrons from the neutral silicon atom changes the silicon from a Si^0 to a Si^{4+} state. The oxidised silicon is dissolved by HF into (hexafluorosilicic acid) H_2SiF_6, which is a soluble product. Unlike wet chemical etching, the metal layer sinks to the bottom and travels along with the semiconductor which helps to achieve higher aspect ratios and novel surface morphologies. The lateral dimension of the structure formed is limited by the dimension of the metal mesh made using lithography or the diameter of the metal particle used and the vertical dimension is decided by the etch time.

3.7   ROLE OF CATALYST METALS

The catalyst used in MACE determines the rate of reduction of oxidant, H_2O_2, in the case of silicon. In general, the rate of the MACE process for the noble metal catalyst follows the trend Pd > Pt > Au > Ag. Also, we should note that Ag is not a stable catalyst in the MACE solution as it gets easily oxidised into a solution forming Ag^+ ions that diffuse a short distance before getting re-deposited after reduction at the HF/Si interface through the oxidation of silicon, which does not happen with Pt or Au catalysts. As the oxidant is reduced at the catalyst, holes are injected into the valence band. This charge transfer process is heavily affected by the surface bend banding of the Si at the metal catalyst contact interface. The calculations of the Schottky barrier height (SBH) and potential of valence band maximum (PVBM) made by Huang et al. [96] provided the first quantitative explanation of the localised galvanic etching taking place in MACE.

3.8   TYPES OF DEPOSITION METHOD

Ag, Au, Pt, and Pd are the most frequently used noble metals in MACE. Some of the commonly used methods for metal deposition on the Si substrate are thermal evaporation, sputtering, electron beam (e-beam) evaporation, electroless deposition, electrode deposition, focused ion beam (FIB)-assisted deposition, and spin-coating of particles, physical deposition in vacuum (e.g., thermal evaporation, sputtering, and e-beam evaporation) is favourable to obtain patterned structures of Si by MACE because the morphology of the resulting noble metal film can more easily be controlled in these methods. For electroless deposition, there are several different plating solutions containing noble metal ions that can be used to deposit these noble metals onto a Si substrate.

3.9 THE SHAPE OF THE METAL AND DISTANCE BETWEEN METALS

The shape of the metal catalyst generally defines the morphologies of the resulting etched structures as the Si under a metal catalyst is etched much faster than Si without metal coverage. With well-separated noble metal particles, we usually get well-defined pores, but the etched structures might evolve from pores into wall-like or wire-like structures as the distance between noble metal particles decreases. Discontinuous patches will result in wall-like or wire-like structures with a broad distribution of cross-sectional shapes and spacing. If the metal film contains orderly distributed pores with uniform diameters and cross-sectional shapes, the Si substrate will be etched into an array of Si nanowires with identical cross-sectional shapes and spacing [97]. Table 3.1 describes the method of Ag deposition, electrolyte composition, and the substrate used for MACE. MACE methods have also already been shown to be compatible with both single-and multicrystalline diamond-wire sawn Si solar cells, which can further reduce PV costs [98].

Let us consider an instance for the electrochemical etching process and MACE to know the details of optical and morphological changes in the silicon. In both cases, oxidation was done in a controlled manner to study the behaviour.

3.9.1 ECE

When the electrochemical etching was carried out by applying a constant current density (6 mA/cm^2) for 10 minutes, 20 minutes, and 30 minutes using aqueous HF (48 wt.%) and ethanol (1:2) as the electrolyte solution. After electrochemical etching, samples were rinsed with deionised water and ethanol sequence to avoid capillarity forces. For passivation, the prepared samples were oxidised using an in-house low-temperature

Table 3.1 Method of Ag deposition with electrolyte composition and types of Si substrates

S. No.	Metal Catalyst	Bath Composition [Oxidant]:[HF]	Substrate Types	References
1	Ag thin film by PVD	2%–49% HF: [H_2O_2] bubbling	c-Si (p) (100), Si (n) (100), and mc-Si (p)	[97]
2	Ag dots by ionic precursor reduction	[H_2O_2]:[HF]=1:4.16	c-Si (p) (2.7 Ω cm) (100)	[99]
3	Ag, dots by ionic precursor reduction	[H_2O_2]:[HF] = 0.4:4.8	Si (p) (100), (111) Si (n)(100), (111)	[100–102]
4	Ag, Pt dots by ionic precursor reduction	[H_2O_2]:[HF] =5.3:1.8	Si (p) (100)	[103]
5	Ag dots and antidots by PVD and electroless in Si/Ge	[H_2O_2]:[HF]=2.5:10 for (110) or=1:2 for (100)	Si (p) (100), (110) Also seen in Si/Ge superlattice	[104]
6	Ag nanoparticles by ionic precursor reduction	4.8M HF + 0.15M H_2O_2	Si (p) (100)	[105]
7	Ag nanoparticles by ionic precursor reduction	[HF]:[H_2O_2]:[H_2O]=25:10:4	Si (100)	[106]
8	Ag nanoparticles by ionic precursor reduction	4M HF + 0.10M $AgNO_3$	c-Si (p) (100) mc-Si (p)	[107]
9	Ag nanoparticles by ionic precursor reduction	[HF]:[H_2O_2]	Si (100), (110), (111)	[108]
10	Ag nanoparticles by ionic precursor reduction	[HF]:[H_2O_2]:[H_2O]=4:1.3:2.7	Si (p) (100)	[109]
11	Ag nanoparticles by ionic precursor reduction	[HF]:[H_2O_2]:[H_2O]=10:5:35	Si (p) (100)	[110]
12	Ag nanoparticles by ionic precursor reduction	[HF]:[H_2O_2]=4:15.6	Si (p) (100)	[111]
13	Ag dots by ionic precursor reduction	7.3M HF with Ar+O_2 bubbling	Si (p) (100)	[112]

and high-pressure setup. During the oxidation process, the pressure is maintained at 20–25 kg/cm^2 (30 Psi) for 2 hours. The as-prepared and oxidised PS samples were investigated to find the spectral changes and structure of PS after oxidation.

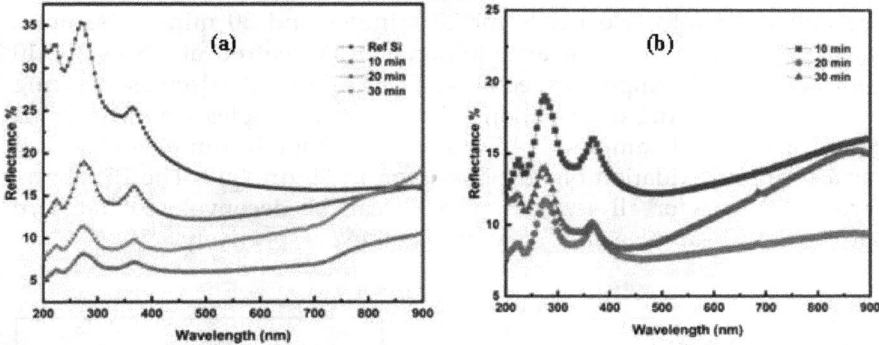

Figure 3.12 Diffused reflectance spectra of (a) as prepared and (b) oxidised PS made from UMG wafer with different etching times; 10 minutes, 20 minutes, and 30 minutes along with reference silicon.

The reflection measurements for as-prepared samples and oxidised PS samples were recorded in the wavelength range of (200–900 nm) illustrated in Figure 3.12(a) and (b) respectively. It can be seen that the reflectance of PS layers and oxidised PS samples decreased when compared with the Ref-Si sample. The maximum reduction in light reflection is around 200 nm to 500 nm and a slight increase in the reflection occurred from 500 to 900 nm, perhaps because of the random distribution of the pores and the increasing roughness on the PS surface. The PS layer reduces light reflection and leads to changes in the optical, vibrational, and electronic transitions [113]. The surface roughness of the Si samples is directly related to the percentage of its reflectivity. The variation in the optical properties of the PS layer appears to be a good result for the possibility of the use of these PS layers as an ARC for Si solar cells [114].

Figure 3.13 shows the photoluminescence spectra (PL) from the porous layers formed by varying the etching time. The presented spectra show the shift of the maximum emission towards shorter wavelengths as the porosity increases. The observed "blue shift" is in agreement with the quantum size effect model proposed by Canham [33]. The spectra exhibit some additional maxima and shoulders for the 10 minutes etched sample in Figure 3.13. A broad PL band, a structural character of the spectrum, and the continuous shift of the maximum emission by changing the etching conditions suggest the existence of various optical transition processes, possibly because of a dispersion of the

Si crystallites size. The PL of porous Si prepared by the electrochemical dissolution suffers from the problems of instability and degradation. The PL peak centred around 2.1 eV was observed for 20 minutes and 30 minutes etched samples, while a broad PL peak extending up to 1.7 eV was observed for 10 minutes etched samples. The PL can be deconvoluted into two peaks for 20 minutes and 30 minutes samples at 2.2 eV and 2.0 eV and an additional peak centred at 1.8 ev for 10 minutes etched sample, respectively. The PL intensity decreases during oxidation. Studying degradation effects in our samples we have found that for oxidised samples "a blue shift" of about 40 nm was observed as a result of oxidation of the Si skeleton in Figure 3.13. The PL occurs around 2.2 eV for all samples. The PL can be deconvoluted into two peaks for all samples at 2.25 eV and 2.05 eV, respectively.

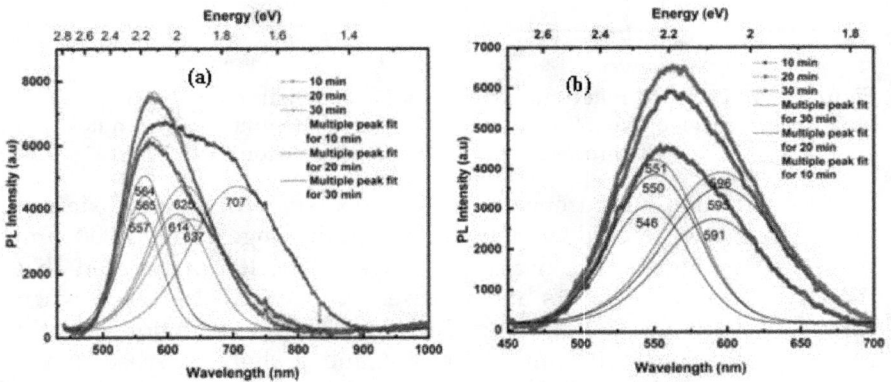

Figure 3.13 PL Spectra of (a) as prepared and (b) oxidised PS made from UMG wafer with different etching times; 10 minutes, 20 minutes, and 30 minutes along with their respective multiple peak fit.

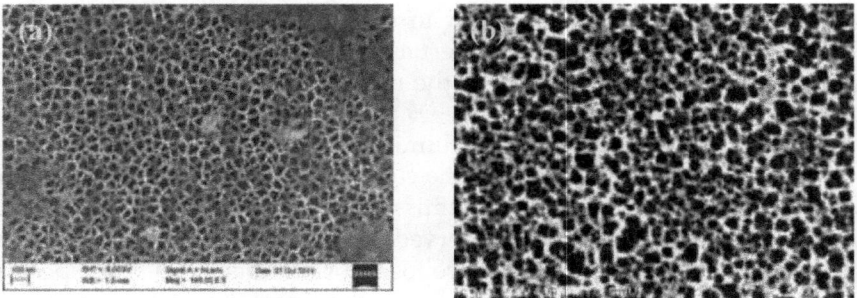

Figure 3.14 SEM image of (a) as prepared and (b) oxidised PS made from UMG wafer with 10 minutes etching time.

It was produced by non-uniform etching during the electropolishing of silicon with an electrolyte containing hydrofluoric acid. In the most basic sense, PS is a network of air holes within an interconnected silicon matrix as shown in Figure 3.14. The size of these air holes, called pores, can vary from a few nanometers to a few microns depending on the conditions of formation and the characteristics of the silicon. Some of the works done in ECE are listed in Table 3.2.

3.9.2   MACE

Then the silver nanoparticles (AgNPs) were coated on the freshly cleaned Si wafers by immersing in 3.6 ml of HF with 28 mg of silver nitrate ($AgNO_3$) aqueous solution (20 ml) for one minute. The excess Ag^+ ions present on the surface were washed with distilled water, then samples were immersed in an etching solution comprising 3.6 ml of HF in 20 ml of H_2O and 0.6 ml of H_2O_2 for different etching time, i.e., 15 minutes, 30 minutes, 45 minutes, 60 minutes, and 75 minutes. The residual AgNPs on the sample surface and in the pores of the Si were removed by immersing the samples in diluted nitric acid (HNO_3) for 60 minutes. Finally, the prepared samples were washed with distilled water and dried in nitrogen (N_2). The dry oxidation is done by annealing the samples in an oxygen atmosphere (100 cc/min) at 700°C for 30 minutes.

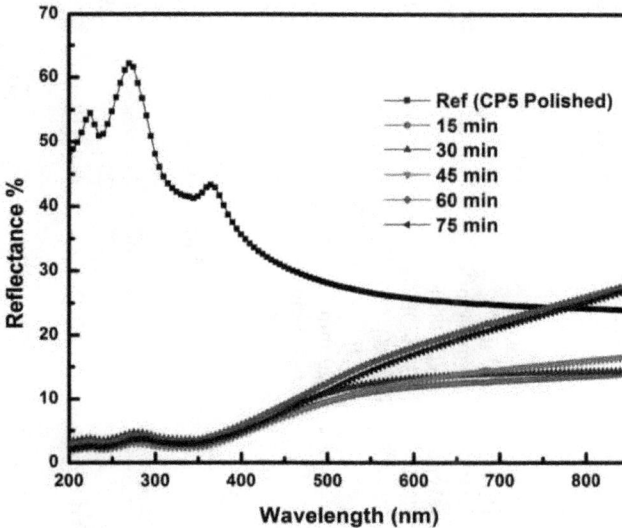

Figure 3.15   Reflectance spectra of the PS samples of different etching times (a) 15 minutes, (b) 30 minutes, (c) 45 minutes, (d) 60 minutes, and (e) 75 minutes including as-cut UMG Si wafer (Reference) without CP5 polishing.

Figure 3.15 shows the reflectance spectra for PS fabricated using different etching times. The measured reflectance of the PS sample

Table 3.2 Important literature review on the PS by anodisation

Author and Published Year	Silicon Wafer Details	Electrolyte Solution and Etching Details	Porous Silicon Details	ARC Materials and Coating Methods	I-V Measurements V_{oc} (mV)	J_{sc} (mA/m²)	FF%	References
K.A. Solman 2012	n-type [100] [111]; p-type [100] [111]	HF (48%): C_2H_5OH (98%) [1:4] 20 min 6 mA/cm²	Porosity 91% 14% 45% 27%	—	595 500 545 510	33 20.50 29.05 27.15	76.65 70.24 78.02 70.00	[115]
K.A. Solman 2011	n-type [100] 306–406 µm	HF (48%): C_2H_5OH (98%) [1:4] 20 min 6 mA/cm²	—	ZnO-135 nm Rf sputtering	627.3 595	35.50 33.90	81.50 76.40	[116]
Aisha Gakce 2016	n-type [100] 500 µm	HF (18%): C_2H_5OH (22%) 1 to 30 mA/cm²	Pore depth 250 nm 52% 390 nm 56%	—	286 303	8.42 8.97	44 50	[117]
C.S. Solanki 2004	—	—	20–30% 20 µm	ZnO as epitaxial layer	606	32.9	60	[86]
Juc Hyun Kim 2007	p-type [100] 1-10 Ω-cm	HF:$(CH_2)_2SO$: H_2O [2:5:15]	Aspect ratio 35 61 80.1 101	AZO-1.8 µm ALD method	530 520 517 502	19.5 25.3 26.2 18.3	49.9 54 48.7 49.2	[118]

Author and Published Year	Silicon Wafer Details	Electrolyte Solution and Etching Details	Porous Silicon Details	ARC Materials and Coating Methods	I-V Measurements			References
					V_{oc} (mV)	Jsc (mA/m^2)	FF%	
W.J. Aziz 2010	p-type [111] 0.75 Ω-cm 283 μm	HF: C$_2$H$_5$OH 60 mA/cm^2 30 minutes RCA	Pore size 8 nm	ZnO/TiO$_2$ ZnO-Rf Sputtering TiO$_2$-DC sputtering	440 370 340	12.4 6.04 5.1	82 79 77	[119]
Jae-Hong Kwon 2007	mc-si p-type 330 μm 0.5–2 Ω-cm RCA I RCA II	(49%)HF: C$_2$H$_5$OH [3:1] 10–270 mA/cm^2 1–35 sec	—	SiN$_X$ PECVD	604 585	30.5 28.9	72.1 67.6	[120]
Asmiet Ramizy 2010	n-type 283 μm [100]	HF: C$_2$H$_5$OH [1:4] 60 mA/cm^2 15 min	—	—	490 450	12.37 15.5	84 88	[121]
C.L. Clement 2003	mc-si p-type 1 Ω-cm	HF electrolyte 1 mA/cm^2 80 sec 140 sec	75% 73%	—	585 587	29.6 29.0	77 77	[122]
Asmiet Ramizy 2010	n-type [111] 283 μm 0.17 μm	HF: C$_2$H$_5$OH [1:4] 60 mA/cm^2 15 min	—	—	310 430 490 450	6.72 8.83 12.37 15.5	83 78 84 88	[123]
O. Tobail 2009	p-type [100] 0.01–0.016 Ω-cm	HF:H$_2$O: [1:1:1]	380 μm 47.7 μm 38 μm	—	611.3 601.2 615	38.4 29.2 28.9	70.9 69.2 76.6	[124]

(Contd.)

Table 3.2 Important literature review on the PS by anodisation (*Cont.*)

Author and Published Year	Silicon Wafer Details	Electrolyte Solution and Etching Details	Porous Silicon Details	ARC Materials and Coating Methods	I-V Measurements			References
					V_{oc} (mV)	Jsc (mA/m²)	FF%	
Hyukyong Kwon 2011	p-type [100] 0.5–3 Ω-cm	HF electrolyte	—	ZnO as epitaxial layer KOH texture	560 560 570	0.13 0.15 0.16	55.27 64.33 65.77	[125]
B.C. Chakrovarty 2007	p-type 300 μm 5–10 μm [100] [111] [110]	C₂H₅OH: HF [1:1] 20 mA/cm² 1–16 min	70 nm	—	578.4 580.3	20.9 26.3	74.1 73.7	[126]
P.N. Vinod 2009	p-type [100] 1 Ω.cm	HF: C₂H₅OH:H₂O [1:1:5] 25 mA/cm² 5 minutes	—	—	580.1 582.3 589.2	24.2 25.5 27.4	70.6 72.2 74.1	[127]
C.L. Clement 1997	p-type [100] [111]	HF electrolyte 50 mA/cm² 3.5 s	—	SiO₂+TiO₂	595 583	7.90 5.83	72.5	[128]

varies from 15% in longer wavelengths (800 nm) and gradually decreases to below 5% at shorter wavelengths (200 nm) for 15-minute, 30-minute, and 45-minute etched samples. The reflectance increases further up to 25% for 60-minute and 75-minute samples near longer wavelengths (800 nm). On the other hand, the reference Si surface has reflectance from 35% to 25% in the UV and around 20% to 15% visible region. The average surface reflectance of the Si substrate in the measured range decreased dramatically from 35% for the reference to less than 5% in the lower wave length region for all samples throughout the UV wavelength range. In the present chapter, all the etched PS samples exhibit reflectance <5% in 200–400 nm wavelength.

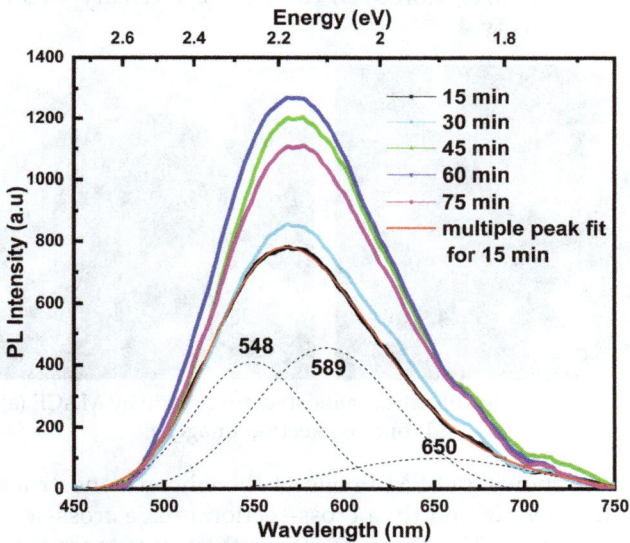

Figure 3.16 Photoluminescence spectra for the porous Si samples of different etching times (a) 15 minutes, (b) 30 minutes, (c) 45 minutes, (d) 60 minutes, and (e) 75 minutes with simulated peak for 15 minutes sample.

The photoluminescence (PL) spectra of the mesoporous silicon with different etching times are shown in Figure 3.16. The PL spectra can be deconvoluted into three peaks centred around 550, 590, and 660 nm. Efficient red luminescence, often referred to as S-band was first reported from free-standing Si quantum wires and attributed to 2-dimensional (2-D) quantum confinement of carriers widening the Si band gap. Later, a weak luminescence band in the blue spectral range (F-band) was observed in aged porous Si and was attributed to Γ–Γ transitions in Si nanocrystals. Similar observations were made in Si oxidised at high temperatures, photoanodised Si, etc. Generally, a PL signature in a semiconductor consists of near-band-edge emission and defects-related luminescence. Near-band-edge emission results from recombination between free electrons (holes) with bound holes (electrons), known as free-to-bound (FB) transitions. The defects related to PL occur from recombination between (i) electrons bound in donors and holes bound in acceptors, known as DAP, (ii) electrons from the conduction band to holes bound in acceptors, and (iii) electrons from donors to valance band. From PL studies, they identified their dominant processes, (i) blue emission from oxygen-related recombination, (ii) green emission from Γ–Γ transition, and (iii) NIR emission from Γ–X excitonic transition. However, the hot PL (Γ–Γ transition) was found to red-shift with a decrease in SiNC size, white the excitonic PL (Γ–X transition) blue shifts with a decrease in size. Moreover, the hot PL intensity was dependant on excitation intensity.

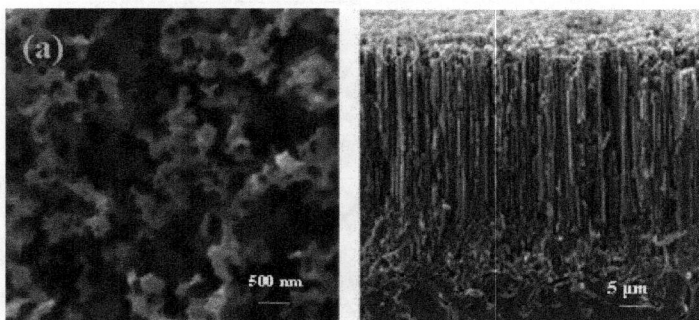

Figure 3.17   SEM image of silicon nanostructure made by MACE (a) top view and (b) cross-section image.

Figure 3.17 shows an SEM image of the silicon nanostructure made by MACE (a) top view and (b) a cross-section image cross-section of the mesoporous Si layers. The 11 μm to 30 μm thick mesoporous Si layer can be easily distinguished. It can be observed that in all the samples the MACE process has created vertical nanostructures. However, porous/defect structures were also formed on these vertical nanowires. The AgNPs did not sink homogenously on the silicon surface. Another major obstacle in the etching experiment is the formation of gas bubbles, which interfere with the fabrication of homogenous PS layers. From the results obtained, we observe that during the MACE process, the Ag catalyst morphology has a significant influence on the formation of PS. The formation of porous Si layers was not uniform because of the non-homogeneous formation of AgNPs on the Si surface. The extended etched regions were formed, where the AgNPs' sizes were larger continued the etching process. However, it should be noted that the irregular distribution of the pores could enhance photo conversion few works in MACE are listed in Table 3.3.

Table 3.3 Specific literature review on the PS by MACE

Junction Type and Year	Device Structure	Device Output Parameters	References
Substrate p-n homojunction (2005)	Conventional phosphorous diffusion on SiNWs fabricated by wet etching on p-type crystalline (100) and polycrystalline substrate.	**Crystalline** η = 9.31% V_{oc} = 0.5485 V I_{sc} = 26.06 mA/cm² FF = 65.12% **Polycrystalline** η = 4.73% V_{oc} = 0.4756 V I_{sc} = 20.9 mA/cm² FF= 47.4%	[129]
Substrate p-n homojunction (2008)	Standard phosphorous diffusion on SiNWs fabricated by wet etching on p-type c-Si crystalline (111) wafers. SiNWs length was about 3–5 μm	η = 11.37% V_{oc} = 0.58 V I_{sc} = 27.1 mA/cm² FF= 72.22% V_{oc} = 0.58 V I_{sc} = 27.1 mA/cm² FF= 72.22%	[130]
Substrate p-n homojunction (2011)	Conventional phosphorous diffusion on SiNWs fabricated by wet etching on p-type c- Si (100) wafers.	η = 13.7% V_{oc} = 0.544 V I_{sc} = 37 mA/cm² FF = 68%	[131]
Substrate p-n homojunction (2013)	SiNWs were fabricated by MACE on texturised p-type c-Si with pyramids. After completion of the cell, SiNWs were passivated by SiO₂, SiO₂- SiNx, or SiNx only. The area of the cell is 154.83 cm².	η = 17.11% V_{oc} = 0.5485 V I_{sc} = 35.8 mA/cm² FF = 77.2%	[132]

(Contd.)

Table 3.3 Specific literature review on the PS by MACE (*Cont.*)

Junction Type and Year	Device Structure	Device Output Parameters	References
Substrate p-n homojunction (2012)	Black silicon was fabricated via wet etching on a p-type (100) c-Si wafer. The junction was formed via a conventional phosphorous diffusion process. The completed cell was capped by SiO_2 obtained by thermal oxidation to passivate the nanostructure. The area of the cell is 0.8081 cm^2.	$\eta = 18.2\%$ $V_{oc} = 0.628$ V $I_{sc} = 36.5$ mA/cm^2 $FF = 79.6\%$	[133]
IBC substrate p-n homojunction (2015)	Black silicon was fabricated by using cryogenic DRIE. The black silicon was passivated by a thin Al_2O_3 layer. The design of the structure was an interdigitated back-contact junction (IBC). The area of the cell is 9 cm^2.	$\eta = 22.1\%$ $V_{oc} = 0.665$ V $I_{sc} = 42.2$ mA/cm^2 $FF = 78.7\%$	[134]
Axial p-n homojunction (2009)	MACE on 2.7 μm multicrystalline p^+nn^+ doped silicon layers deposited on glass. The multicrystalline layer was obtained through the laser crystallisation of a (p) a-Si:H film deposited by electron beam evaporation (EBE). Doping was realised by a standard phosphorus diffusion process.	$\eta = 1.4$–4.4% $Voc = 0.45$ V $FF = 30\%$	[135]

3.10 CONCLUSION

The efficient conversion of solar radiation to electrical energy is a challenging problem with many facets and potential solutions. The people working in this area are approaching the problem from two angles. On one side, working to reduce the amount of silicon material required in crystalline silicon photovoltaics to reduce their overall cost while maintaining high efficiencies. To do this, developing surface structures designed to trap incident photons in thin silicon films to overcome the weak absorption of silicon that becomes problematic as the silicon thickness is reduced. Silicon nanostructures can be an alternative to the (ARC) silicon nitride coating on solar cells. Si nanostructures in various forms can be controllably synthesised easily using electrochemical etching and MACE methods as well. Si nanostructures (nanowires and quantum dots) exhibit unique and interesting structural, optical, electronic, and chemical properties, which are being exploited for myriad exciting applications. For instance, energy devices based on Si-nanowire arrays or network structures can achieve efficiencies as high as 12% for solar energy conversion. Additionally, Si nanodots and hybrid nanostructures can serve as efficient photo-catalysts for the redox reactions of organics. This chapter has highlighted the recent works in developing silicon nanostructures for green, high-efficiency, and low-cost solar energy harvesting and catalysis applications.

REFERENCES

[1] Letcher, T.M. 2022. Global warming, greenhouse gases, renewable energy, and storing energy. pp. 3–12. *In*: Letcher, T.M. (ed.). Storing Energy: with Special Reference to Renewable Energy Sources, 2nd Ed. Elsevier.

[2] Kumar, B., M. Llorente, J. Froehlich, T. Dang, A. Sathrum and C.P. Kubiak. 2012. Photochemical and photoelectrochemical reduction of CO_2. Annu. Rev. Phys. Chem. 63: 541–569.

[3] Mayer, T.M., L. Yongjing, Y. Guangbi and D. Wang. 2013. Forming heterojunctions at the nanoscale for improved photoelectrochemical water splitting by semiconductor materials: case studies on hematite. Acc. Chem. Res. 46(7): 1558–1566.

[4] Moore, F.G. and W.B. Gary. 2011. Energy conversion in photosynthesis: a paradigm for solar fuel production. Annu. Rev. Condens. Matter. Phys. 2(1): 303–327.

[5] https://eu.usatoday.com/story/news/world/2022/03/30clean-energy-wind-solar-2021/7219298001/published 3:05 pm ET 30 March 2022

[6] Mahmoud, N., Z. Issaabadi, M. Sajjadi, S.M. Sajadi and M. Atarod. 2019. Types of nanostructures. Interface Sci. Technol. 28: 29–80.

[7]   Branker, K., M.J.M. Pathak and J.M. Pearce. 2011. A review of solar photovoltaic levelized cost of electricity. Renewable Sustainable Energy Rev. 15(9): 4470–4482.

[8]   Hayibo, K.S. and J.M. Pearce. 2021. A review of the value of solar methodology with a case study of the US VOS. Renewable Sustainable Energy Rev. 137: 110599.

[9]   Pearce, J. and A. Lau. 2002. Net energy analysis for sustainable energy production from silicon based solar cells. In International Solar Energy Conference 16893: 181–186.

[10] https://www.unsustainablemagazine.com/solar-is-now-the-cheapest -energy/Published on March 11, 2022

[11] https://www.iea.org/topics/world-energy-outlook/Published October 2022

[12] Pearce, J.M. 2002. Photovoltaics—a path to sustainable futures. Futures 34(7): 663–674.

[13] Andrea, F. and A. Freundlich. 2008. Material considerations for terawatt level deployment of photovoltaics. Renewable Energy 33(2): 180–185.

[14] Narendra, S., R. Arya, A. Chaubal, K. Deshmukh, P. Ghosh, A. Kottantharayil, et al. 2022. Recent developments in solar manufacturing in India. Solar Compass. 1: 100009.

[15] Zhongliang, G., G. Lin, Y. Zheng, N. Sang, Y. Li, L. Chen and M. Li. 2020. Excellent light-capture capability of trilobal SiNW for ultra-high J_{SC} in single-nanowire solar cells. Photonics Res. 8(6): 995–1001.

[16] Tom, B., G. Zoppi, L. Bowen, T.P. Shalvey, S. Mariotti, K. Durose, et al. 2018. Incorporation of CdSe layers into CdTe thin film solar cells. Sol. Energy Mater. Sol. Cells. 180: 196–204.

[17] Chuang, L., G. Hu, X. Hao, C. Li, B. Tan, Y. Wang, et al. 2021. Performance improvement of CdS/CdTe solar cells by incorporation of CdSe layers. J. Mater. Sci.: Mater. Electron. 32: 19083–19094.

[18] Isshiki, M. and J. Wang. 2017. II–IV Semiconductors for Optoelectronics: CdS, CdSe, CdTe. pp. 853–863. *In*: S. Kasap and P Capper (eds). Springer Handbook of Electronic and Photonic Materials. Springer Handbooks. Springer, Cham.

[19] Smida, A., Z. Zaaboub, N. Bel Haj Mohamed, M. Hassen, F. Laatar, H. Maaref, et al. 2018. Photoluminescence behavior in the synthesized CdSe thin films deposited on ITO substrates. J. Lumin. 194: 686–691.

[20] Samuel, A. and K. Dellinger. 2022. ZnO and TiO_2 nanostructures for surface-enhanced Raman scattering-based biosensing: a review. Sens. Bio-Sens. Res. 100499.

[21] Ammar Mahmoud Al-Husseini and Bashar Lahlouh. 2017. Silicon pyramid structure as a reflectivity reduction mechanism. J. Appl. Sci. 17(8): 374–383.

[22] Ali Samet, S., N. Ekren and Ş. Sağlam. 2020. A review of anti-reflection and self-cleaning coatings on photovoltaic panels. Sol. Energy 199: 63–73.

[23] Zhou, Z., Z. Zhang, J. Jing, R. Gao, Z. Liao, W. Zhang, et al. 2023. Black silicon for near-infrared and ultraviolet photodetection: a review. APL Mater. 11(2): 021107.

[24] Jicheng, Z., Y. Tan, W. Liu, X. Cai, H. Huang and Y. Cao. 2020. Effect of front surface light trapping structures on the PERC solar cell. SN Appl. Sci. 2: 1–10.

[25] Kuan, W.A.C., Z. Tang, H. Lu and F. Huang. 2018. Anti-reflective structures for photovoltaics: Numerical and experimental design. Energy Rep. 4: 266–273.

[26] Lucia, V.M. and Paola D. Veneri. 2020. Silicon solar cells: materials, technologies, architectures. Solar Cells and Light Management 35–57.

[27] Chai, J.Y.-H., B.T. Wong and S. Juodkazis. 2020. Black-silicon-assisted photovoltaic cells for better conversion efficiencies: a review on recent research and development efforts. Mater. Today Energy 18: 100539.

[28] Hele, S., P. Repo, G.V. Gastrow, P. Ortega, E. Calle, M. Garín, et al. 2015. Black silicon solar cells with interdigitated back-contacts achieve 22.1% efficiency. Nat. Nanotechnol. 10(7): 624–628.

[29] Chiara, M., H.S. Laine, T.P. Pasanen, H. Savin and J.M. Pearce. 2018. Economic advantages of dry-etched black silicon in passivated emitter rear cell (PERC) photovoltaic manufacturing. Energies 11(9): 2337.

[30] Uhlir, J.A. 1956. Electrolytic shaping of germanium and silicon. The BSTJ 35(2): 333–347.

[31] Watanabe, Y., Y. Arita, T. Yokoyama and Y. Igarashi. 1975. Formation and properties of porous silicon and its application. J. Electrochem. Soc. 122(10): 1351.

[32] Ito, T., A. Yamama, A. Hiraki and M. Satou. 1990. Silicidation of porous silicon and its application for the fabrication of a buried metal layer. Appl. Surf. Sci. 41: 301–305.

[33] Canham, L.T. 1990. Silicon quantum wire array fabrication by electrochemical and chemical dissolution of wafers. Appl. Phys. Lett. 57(10): 1046–1048.

[34] Violeta, P., V. Rădițoiu, A. Raditoiu, F.M. Raduly, R. Manea, A. Frone, et al. 2021. Bilayer coatings based on silica materials and iron (III) phthalocyanine–Sensitized TiO_2 photocatalyst. Mater. Res. Bull. 138: 111222.

[35] Mouafki, A.M., F. Bouaicha, A. Hedibi and A. Gueddim. 2022. Porous Silicon Antireflective Coatings for Silicon Solar Cells. Eng. Technol. Appl Sci. Res. 12(2): 8354–8358.

[36] Wei, L., Z. Liu, F. Fontana, Y. Ding, D. Liu, J.T. Hirvonen, et al. 2018. Tailoring porous silicon for biomedical applications: from drug delivery to cancer immunotherapy. Adv. Mater. 30(24): 1703740.

[37] Farid, A.H. 2014. Porous silicon chemical sensors and biosensors: a review. Sens. Actuators, B: Chemical. 202: 897–912.

[38] Noor, S.M. 2014. Understanding quantum confinement in nanowires: basics, applications and possible laws. J. Phys.: Condens. Matter. 26(42): 423202.

[39] Ma, D.D.D., C.S. Lee, F.C.K. Au, S.Y. Tong and S.T. Lee. 2003. Small-diameter silicon nanowire surfaces. Science. 299(5614): 1874–1877.

[40] Renkun, C., A.I. Hochbaum, P. Murphy, J. Moore, P. Yang and A. Majumdar. 2008. Thermal conductance of thin silicon nanowires. Phys. Rev. Lett. 101(10): 105501.

[41] Alan, C., A. Fasoli, P. Beecher, P. Servati, S. Pisana, Y. Fu, et al. 2007. Thermal and chemical vapor deposition of Si nanowires: Shape control, dispersion, and electrical properties. J. Appl. Phys. 102(3): 034302.

[42] Hofmann, S., C. Ducati, R.J. Neill, S. Piscanec, A.C. Ferrari, J. Geng, et al. 2003. Gold catalyzed growth of silicon nanowires by plasma enhanced chemical vapor deposition. J. Appl. Phys. 94(9): 6005–6012.

[43] Madan, S., G. Monika and D. Kamal. 2018. Size and shape effects on the band gap of semiconductor compound nanomaterials. J. Taibah Univ. Sci. 12(4): 470–475.

[44] Gilles, L., O. Guillois, D. Porterat, C. Reynaud, F. Huisken, B. Kohn, et al. 2000. Photoluminescence properties of silicon nanocrystals as a function of their size. Phys. Rev. B 62(23): 15942.

[45] Ruoxue, Y., D. Gargas and P. Yang. 2009. Nanowire photonics. Nat. Photonics 3(10): 569–576.

[46] Jawdat, A.B. 2022. Size effect of band gap in semiconductor nanocrystals and nanostructures from density functional theory within HSE06. Mater. Sci. Semicond. Process 137: 106214.

[47] Zheng, Yun, Cristian Rivas, Roger Lake, Khairul Alam, T.B. Boykin and Gerhard Klimeck. 2005. Electronic properties of silicon nanowires. IEEE Trans. Electron Devices 52(6): 1097–1103.

[48] Sacconi, F., M.P. Persson, M. Povolotskyi, L. Latessa, A. Pecchia, A. Gagliardi, et al. 2007. Electronic and transport properties of silicon nanowires. J. Comput. Electron. 6: 329–333.

[49] Jiansheng, J., W. Zhang, K. Peng, G. Yuan, C.S. Lee and S.-T. Lee. 2008. Surface-dominated transport properties of silicon nanowires. Adv. Funct. Mater. 18(20): 3251–3257.

[50] Mehedhi, H., M.F. Huq and Z.H. Mahmood. 2013. A review on electronic and optical properties of silicon nanowire and its different growth techniques. Springer Plus 2(1): 1–9.

[51] Tianzhuo, Z., R. Yamato, S. Hashimoto, M. Tomita, S. Oba, Y. Himeda, et al. 2018. Miniaturized planar Si-nanowire micro-thermoelectric generator using exuded thermal field for power generation. Sci. Technol. Adv. Mater. 19(1): 443–453.

[52] Stefan, W., E. Bertagnolli, S. Kawase, Y. Isono and A. Lugstein. 2014. Electrostatic actuated strain engineering in monolithically integrated VLS grown silicon nanowires. Nanotechnology 25(45): 455705.

[53] Schneider, B.W., N.N. Lal, S.B. Finch and T.P. White. 2014. Pyramidal surface textures for light trapping and antireflection in perovskite-on-silicon tandem solar cells. Optics Express 22(106): A1422–A1430.

[54] Basore, P.A. 1990. Numerical modelling of textured silicon solar cells using PC-1D. IEEE Transactions on Electron Devices 37(2): 337–343.

[55] Biao, S., B. Liu, J. Luo, Y. Li, C. Zheng, X. Yao, et al. 2017. Enhanced light absorption of thin perovskite solar cells using textured substrates. Sol. Energy Mater. Sol. Cells 168: 214–220.

[56] Yubo, C., Q. Wang and J. Xie. 2011. Online social interactions: a natural experiment on word of mouth versus observational learning. J. Market. Reser. 48(2): 238–254.

[57] Ibrahim, K., K. Saeed and I. Khan. 2019. Nanoparticles: properties, applications and toxicities. Arabian J. Chem. 12(7): 908–931.

[58] Oktaviani, O. 2021. Nanoparticles: properties, applications and toxicities. Jurnal Latihan 1(2): 11–20.

[59] Wagner, A.R.S. and S.W.C. Ellis. 1964. Vapor-liquid-solid mechanism of single crystal growth. Appl. Phys. Lett. 4(5): 89–90.

[60] Volker, S., S. Senz and U. Gösele. 2005. Diameter-dependent growth direction of epitaxial silicon nanowires. Nano Lett. 5(5): 931–935.

[61] Marolop, S., K. Usami, T. Kodera, K. Uchida and S. Oda. 2011. Growth of narrow and straight germanium nanowires by vapor–liquid–solid chemical vapor deposition. Jpn. J. Appl. Phys. 50(10R): 105002.

[62] Cui, Y.L., J. Lauhon, M.S. Gudiksen, J. Wang and C.M. Lieber. 2001. Diameter-controlled synthesis of single-crystal silicon nanowires. Appl. Phys. Lett. 78(15): 2214–2216.

[63] Lu, W. and C.M. Lieber. 2006. Semiconductor nanowires. J. Phys. D: Appl. Phys. 39(21): R387.

[64] Hsu, C.M., S.T. Connor, M.X. Tang and Y. Cui. 2008. Wafer-scale silicon nanopillars and nanocones by Langmuir–Blodgett assembly and etching. Appl. Phys. Lett. 93(13): 133109.

[65] Huang, Z., N. Geyer, P. Werner, J. De Boor and U. Gösele. 2011. Metal-assisted chemical etching of silicon: a review: in memory of Prof. Ulrich Gösele. Adv. Mater. 23(2): 285–308.

[66] Peng, K.Q., X. Wang, L. Li, Y. Hu and S.T. Lee. 2013. Silicon nanowires for advanced energy conversion and storage. Nano Today 8(1): 75–97.

[67] Sandulova, A.V., P.S. Bogoyavlenskii and M.I. Dronyuk. 1964. The obtaining of, and certain properties of filiform and acicular single crystals of germanium and silicon, and their solid solutions. Sov. Phys. Sol. State 5(9): 1883.

[68] Zhang, Y.F., Y.H. Tang, N. Wang, D.P. Yu, C.S. Lee, I. Bello, et al. 1998. Silicon nanowires prepared by laser ablation at high temperature. Appl. Phys. Lett. 72(15): 1835–1837.

[69] Werner, P., N.D. Zakharov, G. Gerth, L. Schubert and U. Gösele. 2022. On the formation of Si nanowires by molecular beam epitaxy. Int. J. Mater. Res. 97(7): 1008–1015.

[70] Sivakov, V., G. Andra, A. Gawlik, A. Berger, J. Plentz, F. Falk, et al. 2009. Silicon nanowire-based solar cells on glass: synthesis, optical properties, and cell parameters. Nano Lett. 9(4): 1549–1554.

[71] Uhlir, J.A. 1956. Electrolytic shaping of germanium and silicon. The Bell System Technical Journal 35(2): 333–347.

[72] Cullis, A.G. and L. T. Canham, Visible light emission due to quantum size effects in highly porous crystalline silicon. Nature 353(6342): 335–338.

[73] Lehmann, V. and U. Gösele. 1991. Porous silicon formation: a quantum wire effect. Appl. Phys. Lett. 58(8): 856–858.

[74] Sailor, M.J. 2012. Porous Silicon in Practice: Preparation, Characterization and Applications. John Wiley and Sons.

[75] Lee, J.H., G.A. Galli and J.C. Grossman. 2008. Nanoporous Si as an efficient thermoelectric material. Nano Lett. 8(11): 3750–3754.

[76] Anglin, E.J., L. Cheng, W.R. Freeman and M.J. Sailor. 2008. Porous silicon in drug delivery devices and materials. Adv. Drug Delivery Rev. 60(11): 1266–1277.

[77] Bisi, O., S. Ossicini and L. Pavesi. 2000. Porous silicon: a quantum sponge structure for silicon-based optoelectronics. Surf. Sci. Rep. 38(1–3): 1–126.

[78] Foll, H., M. Christophersen, J. Carstensen and G. Hasse. 2002. Formation and application of porous silicon. Mater. Sci. Eng.: R: Rep. 39(4): 93–141.

[79] Kim, H., B. Han, J. Choo and J. Cho. 2008. Three-dimensional porous silicon particles for use in high-performance lithium secondary batteries. Angew. Chem. 120(52): 10305–10308.

[80] Trucks, G.W., K. Raghavachari, G.S. Higashi and Y.J. Chabal. 1990. Mechanism of HF etching of silicon surfaces: a theoretical understanding of hydrogen passivation. Phys. Rev. Lett. 65(4): 504.

[81] Cullis, A.G., L..T. Canham and P.D.J. Calcott. 1997. The structural and luminescence properties of porous silicon. J. Appl. Phys. 82(3): 909–965.

[82] Smith, R.L. and S.D. Collins. 1992. Porous silicon formation mechanisms. J. Appl. Phys. 71(8): R1–R22.

[83] Kim, J., H. Han, Y.H. Kim, S.H. Choi, J.C. Kim and W. Lee. 2011. Au/Ag bilayered metal mesh as a Si etching catalyst for controlled fabrication of Si nanowires. ACS Nano 5(4): 3222–3229.

[84] Zhang, X.G. 2003. Morphology and formation mechanisms of porous silicon. J. Electrochem. Soc. 151(1): C69.

[85] Zhang, X.G., S.D. Collins and R.L. Smith. 1989. Porous silicon formation and electropolishing of silicon by anodic polarization in HF solution. J. Electrochem. Soc. 136(5): 1561.

[86] Solanki, C.S., R.R. Bilyalov, J. Poortmans, J.P. Celis, J. Nijs and R. Mertens. 2004. Self-standing porous silicon films by one-step anodizing. J. Electrochem. Soc. 151(5): C307.

[87] Bocking, T., K.A. Kilian, P.J. Reece, K. Gaus, M. Gal and J.J. Gooding. 2012. Biofunctionalization of free-standing porous silicon films for self-assembly of photonic devices. Soft. Matter. 8(2): 360–366.

[88] Ghulinyan, M., C.J. Oton, G. Bonetti, Z. Gaburro and L. Pavesi. 2003. Free-standing porous silicon single and multiple optical cavities. J. Appl. Phys. 93(12): 9724–9729.

[89] Von Behren, J., L. Tsybeskov and P.M. Fauchet. 1995. Preparation and characterization of ultrathin porous silicon films. Appl. Phys. Lett. 66(13): 1662–1664.

[90] Xiao, Y., X. Li, H.D. Um, X. Gao, Z. Guo and J.H. Lee. 2012. Controlled exfoliation of a heavily n-doped porous silicon double layer electrochemically etched for layer-transfer photovoltaics. Electrochim. Acta. 74: 93–97.

[91] Archer, R.J. 1960. Stain films on silicon. J. Phys. Chem. Solids. 14: 104–110.

[92] Li, X. and P.W. Bohn. 2000. Metal-assisted chemical etching in HF/H_2O_2 produces porous silicon. Appl. Phys. Lett. 77(16): 2572–2574.

[93] Chartier, C., S. Bastide and L.C. Clement. 2008. Metal-assisted chemical etching of silicon in HF/H_2O_2. Electrochim. Acta 53(17): 5509–5516.

[94] Ragavendran, V., A. Muthu Kumar, V. Vishnukanthan, J.M. Pearce and J. Mayandi. 2018. Effects of silver catalyst concentration in metal assisted chemical etching of silicon. Mater. Lett. 221: 206–210.

[95] Ragavendran, V., J. Mayandi, J.M. Pearce and V. Vishnukanthan. 2019. Influence of metal assisted chemical etching time period on mesoporous structure in as-cut upgraded metallurgical grade silicon for solar cell application. J. Mater. Sci.: Mater. Electron. 30: 8676–8685.

[96] Huang, Z.P., N. Geyer, L.F. Liu, M.Y. Li and P. Zhong. 2010. Metal-assisted electrochemical etching of silicon. Nanotechnology 21(46): 465301.

[97] Toor, F., J.B. Miller, L.M. Davidson, L. Nichols, W. Duan, M.P. Jura, et al. 2016. Nanostructured silicon via metal assisted catalysed etch (MACE): chemistry fundamentals and pattern engineering. Nanotechnology 27(41): 412003.

[98] Kexun, C., J. Zha, F. Hu, X. Ye, S. Zou, V. Vahanissi, et al. 2019. MACE nano-texture process applicable for both single-and multi-crystalline diamond-wire sawn Si solar cells. Sol. Energy Mater. Sol. Cells 191: 1–8.

[99] Oh, J., H.C. Yuan and H.M. Branz. 2012. An 18.2%-efficient black-silicon solar cell achieved through control of carrier recombination in nanostructures. Nat. Nanotechnol. 7(11): 743–748.

[100] Jung, J.Y., Z. Guo, S.W. Jee, H.D. Um, K.T. Park and J.H. Lee. 2010. A strong antireflective solar cell prepared by tapering silicon nanowires. Optics Express 18(103): A286–A292.

[101] Peng, K.Q., X. Wang, L. Li, Y. Hu and S.T. Lee. 2013. Silicon nanowires for advanced energy conversion and storage. Nano Today 8(1): 75–97.

[102] Chartier, C., S. Bastide and C.L. Clement. 2008. Metal-assisted chemical etching of silicon in HF/H_2O_2. Electrochim. Acta 53(17): 5509–5516.

[103] Tsujino, K. and M. Matsumura. 2005. Helical nanoholes bored in silicon by wet chemical etching using platinum nanoparticles as catalyst. Electrochem. Solid-State Lett. 8(12): C193.

[104] Korotcenkov, G. (ed.). 2016. Porous silicon: from formation to application: formation and properties. CRC Press.

[105] Zhang, M.L., K.Q. Peng, X. Fan, J.S. Jie, R.Q. Zhang, S.T. Lee, et al. 2008. Preparation of large-area uniform silicon nanowires arrays through metal-assisted chemical etching. J. Phys. Chem. C 112(12): 4444–4450.

[106] Abdullah, C.A.C., D.F.A. Razak, M.B.M. Yunus, M. Zaki and M. Yusoff, 2019. Structural and optical properties of N-Type and P-Type porous

silicon produced at different etching time. Int. J. Electroactive Mater. 7: 28–37.

[107] Yaoping, L., T. Lai, H. Li, Y. Wang, Z. Mei, H. Liang, et al. 2012. Nanostructure formation and passivation of large-area black silicon for solar cell applications. Small 8(9): 1392–1397.

[108] Han, H., Z. Huang and W. Lee. 2014. Metal-assisted chemical etching of silicon and nanotechnology applications. Nano Today 9(3): 271–304.

[109] Hildreth, O.J. and D.R. Schmidt. 2014. Vapor phase metal-assisted chemical etching of silicon. Adv. Funct. Mater. 24(24): 3827–3833.

[110] Huang, Z., X. Zhang, M. Reiche, L. Liu, W. Lee, T. Shimizu, et al. 2008. Extended arrays of vertically aligned sub-10 nm diameter [100] Si nanowires by metal-assisted chemical etching. Nano Lett. 8(9): 3046–3051.

[111] Hildreth, O.J., W. Lin and C.P. Wong. 2009. Effect of catalyst shape and etchant composition on etching direction in metal-assisted chemical etching of silicon to fabricate 3D nanostructures. ACS Nano 3(12): 4033–4042.

[112] Yae, S., Y. Morii, M. Enomoto, N. Fukumuro and H. Matsuda. 2013. MACE of silicon using oxygen as an oxidizing agent: influence of HF concentration on etching rate and pore morphology. ECS Transactions 50(37): 31.

[113] Naderi, N. and M.R. Hashim. 2012. A combination of electroless and electrochemical etching methods for enhancing the uniformity of porous silicon substrate for light detection application. Appl. Surf. Sci. 258(17): 6436–6440.

[114] Ragavendran, V., P. Jayabal and J. Mayandi. 2014. Investigation on low temperature high pressure oxidation of porous silicon J. NanoScience and NanoTechnology 2: 711.

[115] Salman, K.A., Z. Hassan and K. Omar. 2012. Effect of silicon porosity on solar cell efficiency. Int. J. Electrochem. Sci. 7(9): 376–386.

[116] Salman, K.A., K. Omar and Z. Hassan. 2011. The effect of etching time of porous silicon on solar cell performance. Superlattices Microstruct. 50(6): 647–658.

[117] Gokce, A.G. and E. Akturk. 2015. A first-principles study of n-type and p-type doping of germanium carbide sheet. Appl. Surf. Sci. 332: 147–151.

[118] Kim, J. 2007. Formation of a porous silicon anti-reflection layer for a silicon solar cell. J. Korean Phys. Soc. 50(4): 1168–1171.

[119] Aziz, W.J., A. Ramizy, K. Ibrahim, Z. Hassan and K. Omar. 2011. The effect of anti-reflection coating of porous silicon on solar cells efficiency. Optik 122(16): 1462–1465.

[120] Kwon, J.H., S.H. Lee and B.K. Ju. 2009. Thin film silicon substrate formation using electrochemical anodic etching method. Surf. Eng. 25(8): 603–605.

[121] Ramizy, A., W.J. Aziz, Z. Hassan, K. Omar and K. Ibrahim. 2010. The effect of porosity on the properties of silicon solar cell. Microelectron. Int. 27(2): 117–120.

[122] Clement, C.L., S. Lust, M. Mamor, J. Rappich and T. Dittrich. 2005. Investigation of p-type macroporous silicon formation. Physica. Status Solidi (a) 202(8): 1390–1395.

[123] Aziz, W.J., A. Ramizy, K. Ibrahim, Z. Hassan and K. Omar. 2011. The effect of anti-reflection coating of porous silicon on solar cells efficiency. Optik 122(16): 1462–1465.

[124] Tobail, O., Z. Yan, M. Reuter and J.H. Werner. 2008. Lateral homogeneity of porous silicon for large area transfers solar cells. Thin Solid Films 516(20): 6959–6962.

[125] Kwon, H., J. Lee, M. Kim and S. Lee. 2011. Investigation of antireflective porous silicon coating for solar cells. International Scholarly Research Notices. 2011: Article ID 716409.

[126] Chakraborty, B.R., D. Kabiraj, K. Diva, J.C. Pivin and D.K. Avasthi, 2006. Mixing behaviour of buried transition metal layer in silicon due to swift heavy ion irradiation. Nucl. Instrum. Methods Phys. Res., Sect. B. 244(1): 209–212.

[127] Vinod, P.N. 2007. Application of power loss calculation to estimate the specific contact resistance of the screen-printed silver ohmic contacts of the large area silicon solar cells. J. Mater. Sci.: Mater. Electron. 18: 805–810.

[128] Clement, C.L., A. Lagoubi, R. Tenne and M.N. Spallart. 1992. Photoelectrochemical etching of silicon. Electrochim. Acta. 37(5): 877–888.

[129] Coakley, K.M. and M.D. McGehee. 2003. Photovoltaic cells made from conjugated polymers infiltrated into mesoporous titania. Appl. Phys. Lett. 83(16): 3380–3382.

[130] Fang, H., X. Li, S. Song, Y. Xu and J. Zhu. 2008. Fabrication of slantingly-aligned silicon nanowire arrays for solar cell applications. Nanotechnology 19(25): 255703.

[131] Kumar, D., S.K. Srivastava, P.K. Singh, M. Husain and V. Kumar. 2011. Fabrication of silicon nanowire arrays based solar cell with improved performance. Sol. Energy Mater. Sol. Cells 95(1): 215–218.

[132] Lin, X.X., X. Hua, Z.G. Huang and W.Z. Shen. 2013. Realization of high-performance silicon nanowire based solar cells with large size. Nanotechnology 24(23): 235402.

[133] Oh, J., H.C. Yuan and H.M. Branz. 2012. An 18.2%-efficient black-silicon solar cell achieved through control of carrier recombination in nanostructures. Nat. Nanotechnol. 7(11): 743–748.

[134] Savin, H., P. Repo, G.V. Gastrow, P. Ortega, E. Calle, M. Garin, et al. 2015. Black silicon solar cells with interdigitated back-contacts achieve 22.1% efficiency. Nat. Nanotechnol. 10(7): 624–628.

[135] Sivakov, V., G. Andra, A. Gawlik, A. Berger, J. Plentz, F. Falk, et al. 2009. Silicon nanowire-based solar cells on glass: synthesis, optical properties, and cell parameters. Nano Lett. 9(4): 1549–1554.`

Selenium-Based Metal Chalcogenides Thin Films on Flexible Metal Foils for PEC Water-Splitting Applications

Bheem Singh[1,2], Sudhanshu Gautam[1,2],
Vishnu Aggarwal[1,2], Rahul Kumar[1,2],
Vidya Nand Singh[1,2] and Sunil Singh Kushvaha[1,2]*

[1]CSIR-National Physical Laboratory, Dr. K.S. Krishnan Marg,
New Delhi, India 110012.

[2]Academy of Scientific and Innovative Research (AcSIR),
Ghaziabad, India 201002.

4.1 INTRODUCTION

The search for renewable and environment-friendly (non-polluting) energy resources has become indispensable for a sustainable society to overcome global energy and environmental depletion over the last few decades. Hydrogen gas (H_2) is widely considered a sophisticated future fuel. It is probable to substitute traditional fossil fuels with sustainable H_2 fuel [1–3]. Nonrenewable energy sources, such as coal, oil, gasoline, etc.,

*For Correspondence: Sunil Singh Kushvaha (kushvahas@nplindia.org)

are supplied to the world for a decade to fulfill the demands of energy needs. However, to produce H_2, these processes use a massive number of natural resources. Meanwhile, with the increments in world population and improvement in living standards, fossil fuels are also depleting rapidly. Moreover, nonrenewable energy sources harm human health and our earth's climate due to the emission of harmful gases such as CO_2 [4, 5]. To confront this global problem, we need a technology that can replace fossil fuel or nonrenewable energy sources to save our planet from ecological deprivation.

Among all available green fuels, hydrogen (H_2) has the highest energy yield per unit mass of 121 KJ/g compared to fossil fuels, liquefied petroleum gases (LPG gases), and diesel, as these all have specific energy densities of less than 40 KJ/g. As an energy carrier, H_2 is highly adorable for conveyance, storage, and power replacement in places where other energy sources are difficult or expensive to access [6, 7]. Even though sustainable energy sources such as solar, wind, tidal, geothermal, and hydrothermal have been widely used to replace fossil fuels by supplying sustainable energy for a long time. However, the most developed renewable energy sources have limitations due to their enormous electricity production cost and low efficiency, which remains a challenge [8].

In this way, the abundant water supply and sunlight offer us an affordable alternate source to produce hydrogen from fossil fuel and biomass products. The production of usable energy from abundant solar energy, and water can be an ideal solution to our future energy needs [9]. H_2 production by solar-driven water-splitting process emerged as an efficient way to solve the scarcity of fossil fuels and overcome the environmental degradation situation by zero carbon emission. H_2 production by water splitting is an environment-friendly technique without the emission of harmful gases as it is based on water and sunlight, renewable energy sources [10, 11].

Water splitting is the process whereby the water splits into hydrogen (H_2) and oxygen (O_2) molecules. There are various processes of water splitting, including photocatalytic (PC), electrochemical, thermochemical, radiolysis, photobiological, and photoelectrochemical (PEC), etc. [12]. Exclusive to PEC, the rest of the techniques are limited due to cost and efficiency, which are responsible for their limited use. For example, in the photobiological process, the solar-to-hydrogen conversion performance is not more than 1%, which is not desired for large-scale implementation [13]. In the solar thermochemical process, to split water, we need to heat the water to 2,000 °C. Therefore, designing a robust reactor is also challenging, and poor H_2 efficiency is also the leading cause of their limited use [14]. The by-product of radiolysis is nuclear waste, which is also not desired. The electrochemical process is limited by the cost of electricity and by its production method [15]. The more efficient, most straightforward, and inexpensive way to split water is by photocatalytic (PC) and photoelectrochemical (PEC) processes. However, there are some disadvantages of the PC process over PEC, such as the separation of generated H_2 and O_2 is required immediately, which will consume additional energy, causing low efficiency of the water-splitting process. Besides this, a gas separator is also required in the PC system to separate H_2 and O_2 gases as the water oxidation and reduction process occur at the same surface of the photocatalyst [16, 17].

Hydrogen production by PEC water splitting has been considered a promising approach for solar energy conversion to chemical energy (H_2). By PEC water splitting, HER and OER can be readily achieved at a potential below 1.23 V and above 0 V, compared to many photovoltaic cells that require appropriate voltage for electrocatalytic water splitting [18]. PEC can willingly achieve high photocatalytic activity on the sensible selection of photoanode and photocathode. The pure form of H_2 and O_2 via PEC water splitting can be attained, which is environmentally friendly. There is no need for high-temperature treatment and several steps like solar thermochemical and hydrothermal processes in PEC water-splitting process [2, 19]. In 1972, Honda and Fujishima first demonstrated water splitting using a TiO_2 single crystal as a photoanode and Pt as a cathode in an aqueous electrolyte solution [18]. Since then, researchers worldwide have been working toward cumulative solar-to-hydrogen conversion (STH) efficiency of photocatalysts through several methods like doping, ion implantation, dye sensitization, and bilayered system, etc. [20, 21].

The natural photosynthesis process allows us to think about the direct conversion of sunlight into chemical fuel (H_2), where solar radiation converts H_2O and CO_2 into oxygen and carbohydrates. Splitting water into H_2 and O_2 molecules via PEC is one of the most suitable approaches under solar radiation, also called the artificial photosynthesis process. The water-splitting process is readily be achieved with a net Gibbs free energy of 238 KJ/mol or 1.23 eV, as shown below [16, 17, 19]:

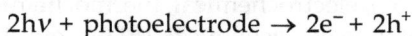

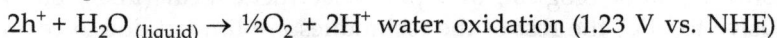

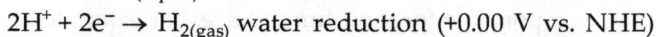

$2h\nu + \text{photoelectrode} \rightarrow 2e^- + 2h^+$

$2h^+ + H_2O_{\text{(liquid)}} \rightarrow \frac{1}{2}O_2 + 2H^+$ water oxidation (1.23 V vs. NHE)

$2H^+ + 2e^- \rightarrow H_{2(\text{gas})}$ water reduction (+0.00 V vs. NHE)

$2H_2O \rightarrow 2H_2 + O_2$ (Overall water splitting)

The PEC water-splitting process is based on the conversion of solar energy into electrical energy within a cell equipped with two electrodes: a photoelectrode to absorb light and another metal electrode (counter electrode) to collect the carriers; the whole system is immersed in an aqueous electrolyte solution and the generated electricity used for water decomposition [22]. The schematic of the PEC water-splitting working process is shown in Figure 4.1. The main components of a PEC process are its working electrode and the counter electrode. The counter electrode is usually inert Pt, while the working electrode is a semiconductor photoelectrode, wherein the third electrode is termed a reference electrode (Ag/AgCl). When light is exposed to a semiconductor photoelectrode, it can generate excitons (electron-hole pair), hole reacts with water molecules and oxidizes them in O_2 molecules and H^+ ions. These H^+ ions transport through the electrolyte to the Pt electrode, electrons will come through the external circuit to the Pt electrode, and then H^+ and e^- reduce to $H_{2(gas)}$. Thus, for n-type semiconducting materials, oxygen evolution occurs at the photoanode and hydrogen evolution at the counter electrode (Pt cathode). An n-type semiconductor produces an anodic current.

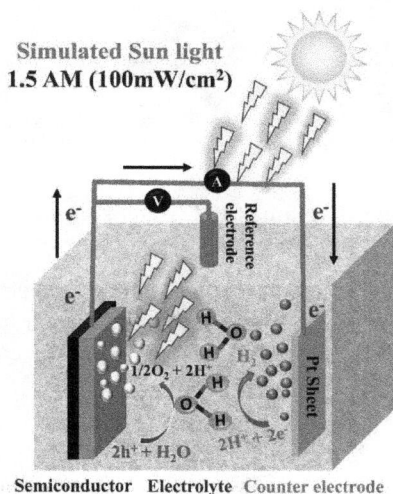

Figure 4.1 Schematic diagram of the PEC water spitting process in which three electrodes, namely the photoelectrode, the counter electrode, and the reference electrode.

Similarly, a p-type semiconductor photoelectrode can generate excitons and efficiently reduce water (hydrogen evolution reaction; HER) and is referred to as a photocathode, and OER takes place at the cathode and a cathodic current flow by a p-type semiconductor. Moreover, n-type and p-type materials are used as the photoanode and photocathode in the bi-photoelectrode PEC system. The advantage of such a system is that the photovoltages are developed on both electrodes, resulting in the formation of an overall voltage sufficient for water decomposition without applying a bias [23, 24]. The comparison of these processes has been shown in Figure 4.2. The conventional three-electrode PEC cells are commonly used to measure PEC performance. A three-electrode cell offers subsidiary information about the semiconductor photoelectrodes, such as the relationship between an applied bias on the photoelectrodes and the capability of photoelectrodes to convert solar energy into valuable fuels such as H_2 and O_2 at this applied bias. This conversion efficiency is referred to as applied bias photon-to-current conversion efficiency (ABPE) [25–27].

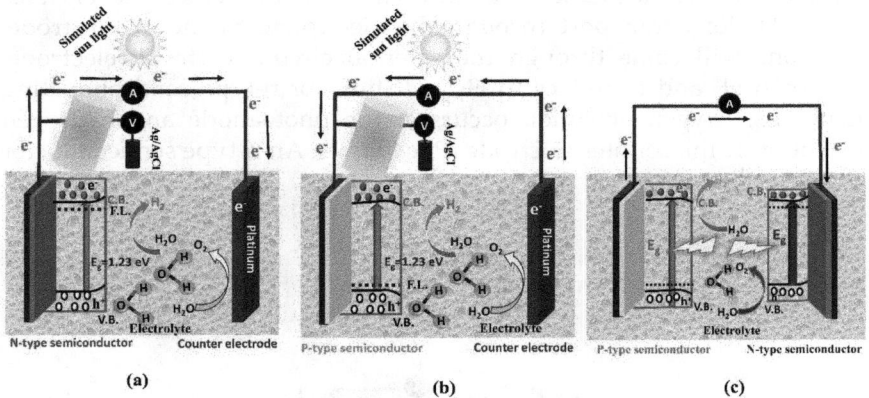

Figure 4.2 Schematic diagram of PEC water spitting process for (a) n-type semiconductor, (b) p-type semiconductor, (c) bi-photoelectrode PEC system.

Since the first demonstration of water photoelectrolysis by Fujishima and Honda in the early 1970s, various kinds of material have been explored that fulfill the essential requirement to convert solar energy into helpful H_2 fuel, as discussed following:

1. The Band gap of the semiconductors should be more than or equivalent to 1.23 eV or $\Delta G^0 = 237.1$ kJ·mol^{-1} to split water in H_2 and O_2 gas. The overpotential loss at the electrode and ionic conductivity loss in the electrolyte also have to be taken into consideration.

2. Semiconductors should have the capability to create photo excitons (electron-hole pairs) from the absorbed solar light (effective charge transport).

3. For the spontaneous water-splitting process, semiconducting materials should have proper conduction band and valence band edge positions. For the water redox process, the C.B. position of the semiconductor should be above the water redox potential 0 V NHE [E (H_2O/H_2)] and for water oxidation, the V.B. position should be below the water oxidation potential 1.23 V NHE [E (H_2O/H_2)]. The conduction and valance band edge position of various semiconducting materials has been presented in Figure 4.3 [28–33].

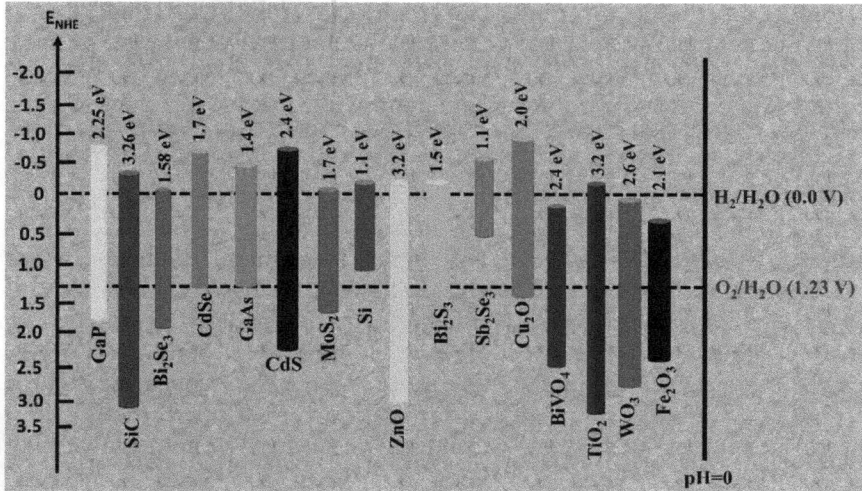

Figure 4.3 Band gap and band edge positions (conduction and valence band) of various semiconducting materials in the aqueous electrolyte at pH=0 for PEC water applications.

4. To achieve efficient charge separation and collection, semiconductors should be more crystalline for easy charge carrier transport and a large surface area to produce more HER and OER activities.

5. Stability is of the utmost importance to the solar water-splitting application. Selected materials should be stable in an aqueous electrolyte solution of a specific pH range and possess corrosion-resistance properties (long-term stability).

6. Semiconducting material also should be cost-effective and earth-abundant elements to enable large-scale implementation; for instance, α-Fe_2O_3, $ZnSiN_2$, etc.

The semiconductor materials that meet all these criteria can be considered for the commercial production of solar fuels. The efficiency and performance of semiconductors are mainly determined by their morphology, tidiness, and homogeneity. Various key factors affect the PEC performance, including electrolyte pH, electrolyte resistance, particle size and defects, shape and size of nanoclusters, band gap energy and intensity of light, etc. All these factors should be considered to achieve better PEC performance of photoelectrode materials.

The hunt for an efficient semiconductor photocatalyst is a major driving force in artificial photosynthesis. After the discovery of TiO_2 in the water-splitting process, several semiconducting materials have been explored for the same purpose. Various p-type materials, such as CuO (1.2–1.8 eV), Cu_2O (1.9–2.2 eV), $CuFeO_2$ (1.5–1.6 eV), NiO (3.6–4.0 eV),

CaFe$_2$O$_4$ (1.9 eV), etc., [2, 29, 30, 34] has been widely investigated due to their significant photocathode performance and excellent photovoltaic properties. At the same time, excellent photoconversion efficiency has been achieved for compound semiconductors, such as GaAs, GaInP, AlGaAs, etc. Furthermore, various n-type semiconducting materials, such as TiO$_2$ (3.1–3.2 eV), ZnO (3.2 eV), WO$_3$ (2.5–2.7 eV), α-Fe$_2$O$_3$ (2.1–2.3 eV), BiVO$_4$ (2.4–2.5 eV), Ta$_3$N$_5$ (2.1 eV), ZnS (3.6 eV), TaON (2.4 eV), CdS (2.4 eV), etc., have been used as a photoanode for the PEC water-splitting process [35–37]. Currently, transition metal dichalcogenides (TMDCs) materials such as MoS$_2$ (1.2–1.8 eV), MoSe$_2$ (0.9–1.5 eV), WSe$_2$ (1–1.7 eV), etc., have been explored for artificial photosynthesis due to their excellent electronic, optical, and photocatalytic activities [38–41]. However, each semiconductor has its pros and cons concerning earth abundance, toxicity, stability, and fabrication cost. Thus, there is still a need for an efficient, stable, and robust semiconductor photocatalyst that can help to provide energy carriers for future generations. Therefore, there is still demand for the search for high-performance photocatalysts to hunt artificial photosynthesis by fabricating various kinds of compositions, nanostructures, heterostructures, hybrid structures, etc. Another aspect is to couple the low band gap material, such as binary chalcogenides materials: Sb$_2$S$_3$, Bi$_2$S$_3$, and Sb$_2$S$_3$, Sb$_2$Se$_3$, Bi$_2$Se$_3$, Bi$_2$Te$_3$, etc., with large band gap materials (e.g., stacked dual-electrode PEC cell), which can absorb a wide range of solar spectrum and can enhance the photocatalyst performance in water-splitting process [31, 42].

Recently, V–VI group binary chalcogenides semiconducting materials (such as Bi$_2$Se$_3$, Bi$_2$Te$_3$, Sb$_2$Se$_3$, Sb$_2$Te$_3$, etc.) emerged as efficient photocatalysts for water-splitting applications due to their favorable conduction band edge position for facile H$_2$ production [42–44]. Although binary chalcogenides are a vast family of compound materials, many elements belong to chalcogen, i.e., S, Se, and Te. However, out of several compounds, Bi and Sb-based chalcogenides have excellent properties and tremendous application in energy devices. These materials have been extensively used in the field of thermoelectric devices, infrared photodetectors, spintronic devices, terahertz detection, optical recording, electrical transport devices, etc., due to their high electron mobility, good electrical conductivity, photoresponsivity, and electrochemical properties as shown in Figure 4.4 [45, 46]. These materials also come in a particular class of material, i.e., topological insulators (TIs). TIs materials are the materials that exhibit insulating properties in the interior but have conducting surface states. These materials have robust metallic surface states property against nonmagnetic impurities and disorders protected by time-reversible symmetry. In addition, elastic backscattering is forbidden and electron transport is spin momentum locked, where a super current (resistance less) flows through the surface edges of these materials. These unique properties of TIs give birth to various fundamental quantum aspects such as opto-spintronic devices, quantum Hall effect, quantum computers, etc. [47–51].

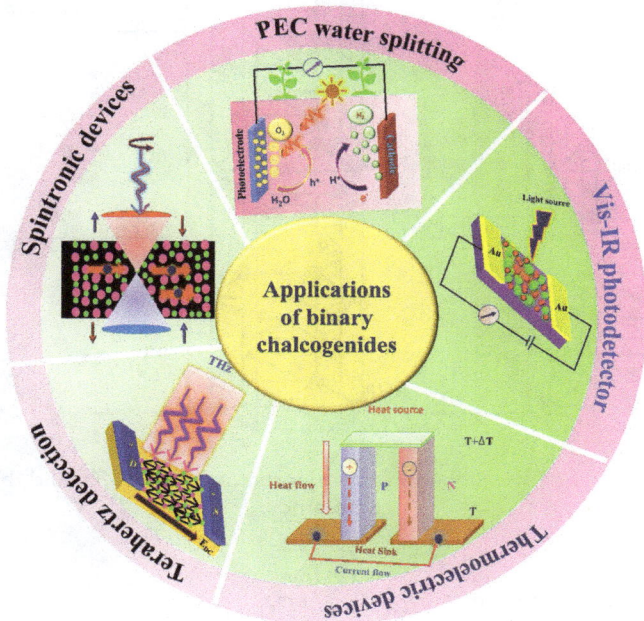

Figure 4.4 Schematic diagram of binary chalcogenides materials in various applications.

The crystal structure representation of Bi_2Se_3, Bi_2Te_3, Sb_2Se_3, and Sb_2Te_3 materials is shown in Figures 4.5 and 4.6. Figure 4.5(a) portrays the rhombohedral crystal structure of Bi_2Se_3, with space group D^5_{3d} (R3m) with lattice parameters: $a = b = 4.14$ Å and $c = 28.64$ Å. Usually, it is presented in quintuple layers with two equivalent Se and Bi atoms: Se (1)–Bi-Se (2)–Bi-Se (1) arrangement along the z-direction, where one quintuple layer is 0.955 nm thick linked with another quintuple layer with weak van der Waals *(vdW)* forces. One single layer of Bi_2Se_3 is composed of hexagonal and trigonal lattice planes. The most stable phase of Bi_2Se_3 is hexagonal at room temperature [52]. Figure 4.5(b) also shows the rhombohedral crystal structure for Bi_2Te_3 with quintuple layer arrangement with lattice parameters: $a = b = 4.38$ Å, $c = 30.5$ Å. It consists of one QLs layer thickness ~1 nm bonded with weak van der Waals forces. The crystal structure of antimony selenide (Sb_2Te_3) also has a rhombohedral crystal structure similar to that of Bi_2Te_3 and Bi_2Se_3 with lattice constant $a = 4.32$ Å and $c = 31.5$ Å belonging to the R3m (166) space group.

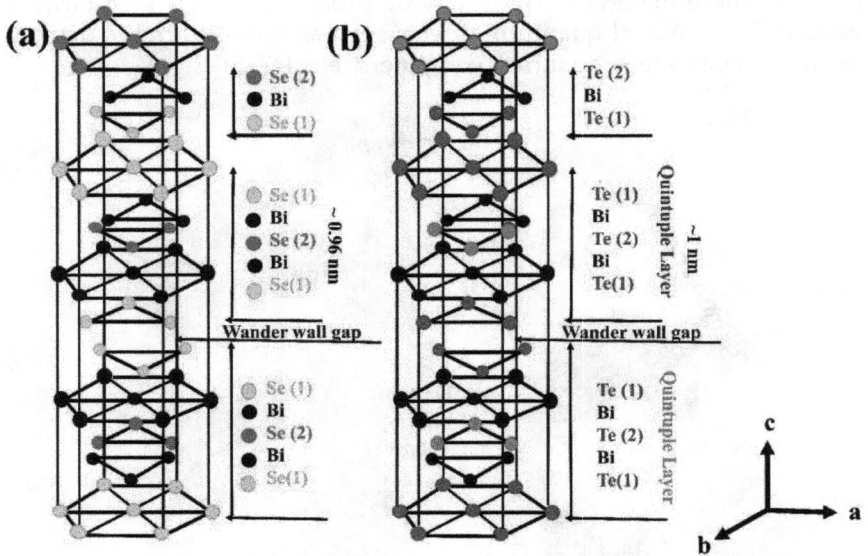

Figure 4.5 The Rhombohedral crystal structure of (a) Bi_2Se_3 and (b) Bi_2Te_3 compounds.

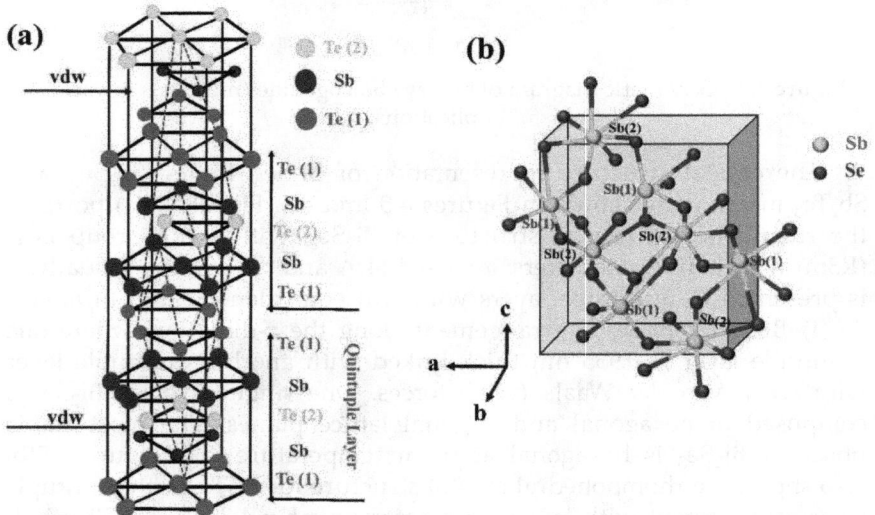

Figure 4.6 (a) Rhombohedral structure of Sb_2Te_3 compound (b) Orthorhombic crystal structure of Sb_2Se_3 compound.

The crystal structure of Sb_2Te_3 also contains five atomic layers structure in the primitive cell leading to the formation of a rhombohedral structure, where Sb and Te atoms are covalently stacked within the QLs along the c-axis as shown in Figure 4.6(a) [53, 54]. Figure 4.6(b) shows the orthorhombic P_{nma} crystal structure of Sb_2Se_3 made up of AB_6 and AB_{6+1} polyhedra with constant lattice values of a = 11.78 Å, b = 3.97 Å, c = 11.63 Å, where strong covalent Sb-Se bond keeps the cells together in the (001) direction at room temperature. The physical and electronic properties of these materials have been mentioned in Table 4.1 [51–55].

Table 4.1 Physical, electronic, and optical properties of V–VI group binary chalcogenides materials

Materials	Band gap (eV)	Mobility $(cm^2V^{-1}s^{-1})$	Melting point (K)	Thermal conductivity $(W\ m^{-1}\ K^{-1})$	Density ρ $(g\ cm^{-3})$
Bi_2Se_3	0.3	1407	979	1.70	7.51
Bi_2Te_3	0.1	481	858	1.37	7.85
Sb_2Te_3	0.21	675	893	2.2	6.5
Sb_2Se_3	1.1	15 (for electron), 42 (for hole)	884	—	5.81

Out of these materials, Bi_2Se_3 and Sb_2Se_3 semiconductors have been widely used as a photocatalyst for PEC water splitting due to their massive electronic and optical properties with suitable conduction band edge position for hydrogen evolution. Bi_2Se_3 is an n-type direct band gap semiconducting material with fascinating electronic and optical properties. This small and optimum band gap (0.3–2.0 eV) semiconductor has the potential to absorb visible and near-infra-red radiation, which can be beneficial to achieve high solar-to-chemical energy conversion efficiency [31, 43]. The high electron mobility (1,407 $cm^2V^{-1}\ s^{-1}$) of Bi_2Se_3 is beneficial for better device performance. The high thermal conductivity of Bi_2Se_3 (1.70 $W\ m^{-1}\ K^{-1}$) can more effectively transfer heat and readily take up heat from the environment. The high melting point (979 K) gives the advantage to the growth of Bi_2Se_3 thin film at elevated temperatures to use in relatively high-temperature applications [56–61]. Bi_2Se_3 also comes from the topological insulator family, where the strong metallic surface states of Bi_2Se_3 can play an essential role by strongly ceasing the elastic backscattering of electrons. As the surface electronic structure of the semiconductors has a significant role in PEC water-splitting performance, the fast and low dissipation electron transport can enhance the electrical conductivity of photoelectrode in the water-splitting process [42, 44]. Even though Bi_2Se_3 was not explored more as a photocatalyst in PEC water splitting due to the rapid recombination of charge carriers. However, some recent reports of Bi_2Se_3 semiconductor-based photocatalysts draw great attention by getting excellent solar-to-hydrogen conversion efficiency as summarized in Table 4.2 [31, 62–65].

Table 4.2 Literature survey on PEC water splitting of Bi_2Se_3 and Sb_2Se_3- based photoelectrodes

Photoelectrode	Photocurrent Density	Electrolyte	Year	References
Bi_2Se_3/TiO_2/FTO (NFs)	1.76 mA/cm^2 at 1.23 V_{RHE}	0.5 M Na_2SO_3	2021	[31]
Bi_2Se_3/TiO_2/FTO (QDs)	1 mA/cm^2 at 1.6 V_{RHE}	0.1 M Na_2SO_4	2022	[62]
Bi_2Se_3/TiO_2/ITO	6 μA/cm^2 0.5 $V_{Ag/AgCl}$	0.5, 1 M Na_2SO_4	2021	[63]
$MoSe_2$/Bi_2Se_3/FTO (Hybrids)	85 mA/cm^2 at −0.6 V_{RHE}	0.5 M H_2SO_4	2017	[64]
Bi_2Te_3/TiO_2/FTO (QDs)	0.86 mA/cm^2 at 1.6 V_{RHE}	0.1 M Na_2SO_4	2021	[62]
Sb_2Se_3/$CuSbS_2$/FTO	18.0 mA/cm^2 at 0 V_{RHE}	1 M H_2SO_4	2022	[71]
Sb_2Se_3 (Se-annealed)/ CdS/TiO_2/Pt on soda lime glass	−8.6 mA/cm^2 at 0 V_{RHE}	0.5 Na_2SO_4, 0.25 M Na_2HPO_4	2017	[70]
Au/Sb_2Se_3/MoS_x/FTO	16 mA/cm^2 at 0 V_{RHE}	1 M H_2SO_4	2018	[72]
Sb_2Se_3/MoS_2/FTO	10 mA/cm^2 at 0 V_{RHE}	0.5 M Na_2SO_4	2019	[73]
Bi_2Se_3/FTO thin film (4 nm)	10 mA /cm^2 at 385 mV overpotential	0.5 M Na_2SO_4	2021	[65]
Bi_2Te_3@CoNiMo/Ni foam	−60 mA cm^2 at −0.1 V_{RHE}	0.9 M KOH	2015	[78]
MoS_2/Bi_2Te_3/$SrTiO_3$	10 mA/cm^2 at -0.4 V_{RHE}	0.5 M H_2SO_4	2021	[79]

Sb_2Se_3 is also a member of the binary chalcogenide family that has been investigated as an efficient material for thermoelectric and photovoltaic areas; it is also used to fabricate PEC photocathodes, which profit from its low toxicity, appropriate band gap, and exceptional photo corrosion stability [66, 67]. Sb_2Se_3 shows p-type conductivity, which is suitable for constructing a suitable photocathode for hydrogen evolution reaction. Sb_2Se_3 is a low band gap semiconducting material that can be a promising candidate for a stacked bilayered system because of the following advantage:

1. Sb_2Se_3 has a proper conduction band edge position, i.e. −0.29 V NHE at pH = 0, which makes it suitable for hydrogen production.

2. It has an optical band gap (1.1–1.4 eV) and also contains a high optical absorption coefficient (~10^5 cm^{-1}), which permits to absorption of visible near-infrared sunlight within a thickness of ~500–600 nm.

3. The constituent elements Sb and Se are relatively earth-abundant and low-toxic, which also have the advantage of fabricating PEC energy devices at a large scale.

4. A low melting point (884 K) also permits us to prepare high-quality Sb_2Se_3 thin films at a relatively low temperature using thermal sublimation of a powder source.

5. It consists of good electron and hole mobility with high relative permittivity of 15. The high relative permittivity of Sb_2Se_3 can be beneficial to lower exciton binding energy that infers an instant separation of electrons and holes upon light irradiation.

6. Sb_2Se_3 consists of only a single phase, orthorhombic Sb_2Se_3, which allows us to control the phase and defects in Sb_2Se_3 much more manageable than other chalcogenides materials [68–73], although Sb_2Se_3 has a small band gap in the range of 1.0–1.3 eV. However, in many pieces of literature, the optical band gap of Sb_2Se_3 has been reported as 1.4 to 1.8 eV depending on the annealing temperature after film growth [74–77]. In various kinds of literature, Sb_2Se_3 and also Bi_2Te_3-based heterostructures have been showing excellent photocurrent density, as summarized in Table 4.2 [70–73, 78, 79]. However, these hard substrate-based photoelectrodes (e.g., Si, glass, quartz, etc.) have limitations in the use of flexible energy device applications due to their brittleness, and unbending nature, etc.

Nowadays, flexible electronics draw great attention due to their excellent transportability, lightweight, stretchability, bendability, and human-friendly interfaces to provide new avenues for next-generation electronics. Flexible electronics have been developed in flexible transistors, supercapacitors, displays, batteries, nanogenerators, light-emitting diodes (LEDs), solar cells, thermoelectric devices, etc. [80, 81]. To open new prospects in the advancement of solar cells or fuel applications, the search for innovative, flexible PEC devices is highly in demand. For a long time, FTO or ITO doped glass, silicon (Si) based rigid and heavyweight substrates have been widely used to construct photoelectrode, which has limitations in flexible PEC devices due to their hardness and unbinding nature. Selecting a flexible substrate is challenging to design flexible electrodes since it depends on several parameters such as excellent flexibility, robustness, pliability, thermal and electric conductivity, chemical and thermal stability at elevated temperatures, etc. [82]. The various organic and polymer-based substrates such as polyethylene terephthalate (PET), polyethylene naphthalate (PEN), and polyimide carbon cloth are not suitable for the growth of high-quality photocatalysts at high temperatures as well as high penetrability of oxygen and humidity also unfavorable to fabricate energy devices [83]. This way, flexible metal foils like Mo, Ti, Al, stainless steel, etc., have the potential advantage due to their excellent thermal and chemical stability at elevated temperatures. Flexible metal foils have been widely used in flexible solar cells due to their robustness, lightweight, outstanding flexibility, etc. Flexible metal foils are bearable at high-temperature sulfurization and selenization processes, whereas polymer substrates have their limitation at high temperatures [84]. There are some potential advantages of flexible metal foils. (1) They can be used to fabricate large-scale roll-to-roll devices; (2) the total weight of devices will be lowered, and easy for transportation; (3) the excellent flexibility of substrate can be beneficial for integrating irregular surfaces, such as an electric vehicle, ship, drone, etc., and this feature could help enhance the photocatalytic performance; (4) on the requirement of the specific application, metal foil-based thin-film cells can be easily molded/twisted into various design and shape; (5) there is no need for metallic back contact for photoelectrode if we select metal foils which also has a particular advantage over polymer substrates [85, 86]. The usefulness of flexible metal foil in various flexible energy device applications has been shown in Figure 4.7. Shiyani et al. prepared a flexible zinc oxide photoelectrode for PEC energy conservation have generated photocurrents of about 1.89 $\mu A/cm^2$ [87]. Quynh et al. prepared Fe_2O_3/ZnO heterostructure on mica for PEC water splitting and obtained a photocurrent density of 0.38 mA/cm^2 [88]. In another literature, 1.37 mA/cm^2 photocurrent density was achieved for flexible $PVDF/Cu/PVDF-NaNbO_3$ photoanode [89]. These results offer a fundamental understanding of flexible photoelectrode, which can be used to develop hybrid solar-based devices to generate solar fuels. However, there is a limited report for flexible metal foil-based photoelectrode for PEC water-splitting applications.

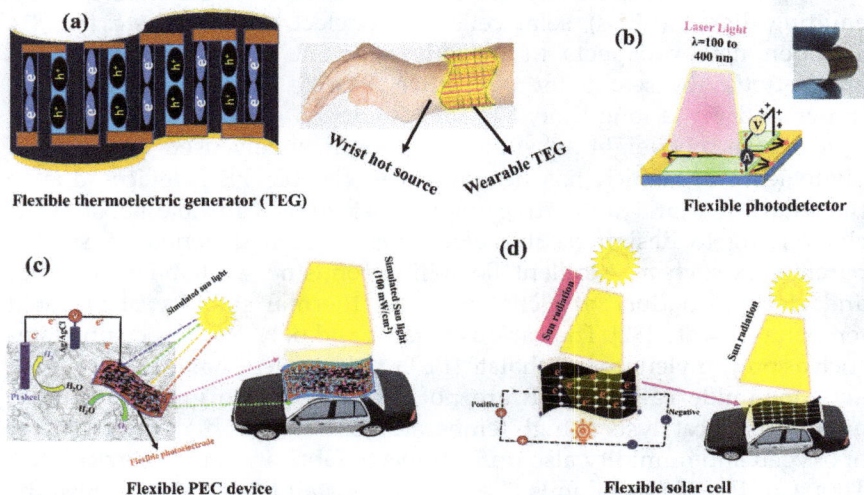

Figure 4.7 Role of flexible metal foils as futuristics: (a) flexible and wearable thermoelectric devices, (b) flexible photodetectors, (c) flexible PEC devices on the irregular surface such as electric vehicles, airplanes, ships, drones, etc., and (d) flexible solar cell.

Various growth technique has been used to grow Bi_2Se_3 thin films, such as magnetron sputtering, thermal evaporation, molecular beam epitaxy (MBE), pulse laser deposition (PLD), etc. For instance, Park et al. adopted the MBE technique to grow Bi_2Se_3 thin film, which showed high crystallinity, epitaxially oriented Bi_2Se_3 film on an h-BN substrate with atomically sharp interfaces [90]. Tabor et al. used this technique and obtained good crystallinity and optimal stoichiometric Bi_2Se_3 thin film on Al_2O_3 (110) substrate with epitaxial layer-by-layer growth in the c-axis oriented [91]. Among these techniques, the magnetron sputtering deposition technique offers large-area deposition thin films, low cost, good film-forming uniformity, and a relatively simple process. Using magnetron sputtering techniques, large-area Bi_2Se_3 thin films can be achieved on various substrates with good crystalline quality [45, 92–96]. Wei et al. group adopted the magnetron sputtering technique to grow Bi_2Se_3 thin films on Si (111) and found the highly c-axis-oriented Bi_2Se_3 thin films after post-annealing under a Se-rich environment [97]. Tang et al. adopted this technique to produce stoichiometric Sb_2Se_3 thin films with desired crystallinity and orientation for solar cell application [98]. In another report, Chen et al. showed that r.f. Magnetron sputtered Sb_2Se_3 thin film has a highly crystalline order with large crystal grains after *in situ* heat treatment [99]. These works of literature disclose the importance of magnetron sputtering systems for depositing high-quality and large-area Bi_2Se_3 thin films for various applications. We have adopted this technique to grow Bi_2Se_3 thin film on Ti foil. The post-selenization process in the Se-rich environment promotes obtaining a suitable stoichiometry Bi_2Se_3 thin film. The structural and crystalline properties and PEC performances of prepared Bi_2Se_3 thin film on metal foil-based substrates were investigated.

4.2 EXPERIMENTAL SECTION

The Bi_2Se_3 thin film was deposited on Ti foil using r.f. Magnetron sputtering technique. A magnetron sputtering system is ornamented with a high-temperature sustainable substrate heater (<1,000°C), high vacuum (base pressure < 2×10^{-7} Torr), confocal sputtering targets, and argon (Ar) gas mass flow controller. It consists of two isolated chambers connected by a manual gate valve; the main deposition chamber (growth chamber) is equipped with a turbo molecular pump to provide a high vacuum and a load lock chamber connected with a rotary pump. The load lock chamber has a transfer arm for transferring the samples inside the main chamber. First, Ti foil was cleaned with acetone and IPA and dried with nitrogen gas. A commercially high pure Bi_2Se_3 (purity 99.99%; ACI Alloys) material target was used as the sputtering source. The growth process was carried out in a pure Ar (99.9999%) environment with an Ar gas flow rate of 20 sccm and working pressure of ~2.3 × 10⁻³ torr. The radio frequency (rf) of 10 W and substrate temperature of 400°C were kept for the growth of the thin film. Moreover, the post-selenization process was carried out in Se-rich environment to obtain suitable stoichiometry Bi_2Se_3 thin film, as there is Se-deficiency in sputtered Bi_2Se_3 thin film due to the momentum mismatch between Bi and Se atoms.

Raman spectroscopy was used to disclose the formation of thin films and structural quality by using an excitation laser source with a wavelength of 514 nm in the backscattering mode. To study the crystalline structure of prepared thin films, an X-ray diffraction (XRD) pattern was performed with $CuK_{\alpha 1}$ x-ray source having a wavelength of 0.15406 nm. The surface morphology of Bi_2Se_3 thin film was tested by field emission scanning electron microscopy (FESEM) at an operating voltage of 15 kV. X-ray photoelectron spectroscopy (XPS) technique with an $AlK\alpha$ x-ray source having an energy of 1,486.7 eV has confirmed the chemical composition of Bi_2Se_3 compounds.

To investigate the photocatalyst performance of prepared samples, PEC measurement was carried out in sodium sulfate Na_2SO_4 (0.5 M) aqueous electrolyte solution under 100 mW/cm² (1.5AM) light intensity. A PEC system equipped with three standard electrodes, where Bi_2Se_3/ Ti (working electrode) is used as a photoanode to absorb light, where oxygen evolution takes place, second a platinum (Pt) sheet as a cathode (counter electrode) where hydrogen evolution takes place and Ag/AgCl as a reference electrode for potential measurement. The samples were excited with the light source AM 1.5 G simulated sunlight using a 300 W Xenon lamp. Transient photocurrent *(I-t)* curves were measured at intervals of 30 s according to the light ON–OFF cycling process at a given bias of 0.3 and 0.6 V vs Ag/AgCl. EIS measurements were performed in the frequency range of 0.01 Hz to 1 M Hz at an applied ac voltage of 5 mV for the sample. All the measurements were recorded on the CHI6054E electrochemical workstation.

4.3 RESULT AND DISCUSSION

Figure 4.8(a) depicts the schematic diagram for the relative motion of Bi and Se atoms for three A^1_{1g}, A^2_{1g}, and E^2_g observed phonon vibrational mode. The A^1_{1g} and A^2_{1g} vibrational modes correspond to out-of-the-plane, and E^2_g corresponds to the in-plane vibrational motion of Bi and Se atoms. Another low-frequency vibrational mode E^1_g at low wavenumber has not been observed generally due to high Rayleigh scattering. Figure 4.8(b) shows the Raman spectrum for Bi_2Se_3 thin film on polished Ti metal foil. It has shown three fingerprint Raman peaks of Bi_2Se_3 thin film, which were located at 71.3, 131.4, and 173.7 cm^{-1} peak positions corresponding to A^1_{1g} (out-of-plane), E^2_g (In-plane), A^2_{1g} (Out-of-plane) mode, respectively, which confirmed the formation of crystalline Bi_2Se_3 thin film on Ti foil [100–102]. Figure 4.9 shows the XRD pattern in which all signature peaks of Bi_2Se_3 correspond to rhombohedral crystal structure. Diffraction XRD peaks of Bi_2Se_3 thin film are found at 9.1°, 18.5°, 29.4°, 35.05°, 38.2°, 43.6°, 47.7°, 57.5°, 60.8°, and 68.8° peak positions which are indexed to (0003), (0006), (015), (018), (00012), (0111), (00015), (00018), and (00021) lattice planes, respectively [95, 103, 104]. The observed peaks were found oriented in various planes due to the polycrystalline nature of Ti metal foil. The remaining XRD peaks at position 40.3°, 53.2°, 63.1°, 70.7°, and 76.1° can be indexed respectively to (101), (102), (110), (103), and (112) lattice planes of Ti foil [95, 105, 106]. The surface morphology of the deposited thin film has been characterized by FESEM characterization, as shown in Figure 4.10. The FESEM image of pure Ti foil is presented in Figure 4.10(a), which clearly shows the large grains having cracks. Figure 10(b) revealed the layered hexagonal nanoflakes morphology of Bi_2Se_3 with the calculated grain size of 250–300 nm. It also has been seen that grains were found randomly oriented in different directions, which was also confirmed by oriented XRD peaks in Figure 4.9.

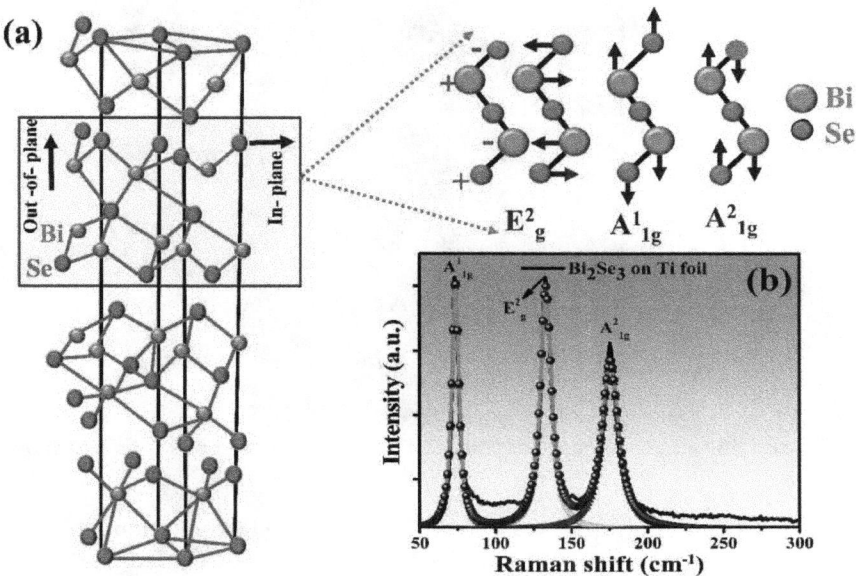

Figure 4.8 (a) Schematic diagram of vibration Raman mode of Bi_2Se_3 and (b) lorentzian fitted Raman spectrum of Bi_2Se_3 thin film on flexible Ti foil.

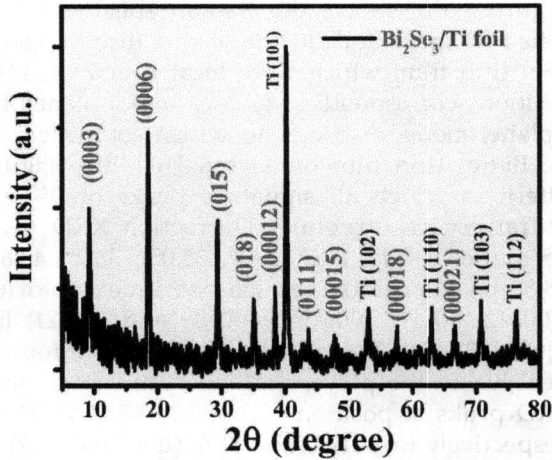

Figure 4.9   XRD pattern of Bi_2Se_3 thin films on Ti metal foil.

Figure 4.10   FESEM images of (a) bare Ti foil (b) sputtered Bi_2Se_3 thin film on Ti foil.

The XPS technique was executed to reveal the chemical and electronic composition of the Bi_2Se_3 thin film on Ti foil. The core level scan for Bi 4f spin states in the binding energy range of 153–168 eV is shown in Figure 4.11(a). It shows the two dominant peaks for Bi 4f states, located at binding energy 163.2 and 158.0 eV, corresponding to $4f_{5/2}$ and $4f_{7/2}$ valence states, respectively. These two spin-orbit coupled peaks are found to shift slightly to a high-energy region compared to elemental bulk Bi $4f_{5/2}$ and $4f_{7/2}$ peaks at 161.9 and 156.6 eV, respectively. The other four peaks correspond to the Bi-Se oxidation state situated at 165.1, and 159.3 eV peaks position correspond to Bi^{+5} states as well as 157.2 and 162.4 eV peaks related to Bi^{+3} oxidation states, which may occur due to air exposure after the deposition of the film. Figure 4.11(b) shows the deconvoluted Se 3d spectra in the binding energy range of 57–51 eV. The two highly core level peaks at binding energies 54.5 and 53.2 eV were assigned to Se $3d_{3/2}$ and Se $3d_{5/2}$ peak positions attributed to Se^{-2} states in Bi_2Se_3. One single peak at peak position 55.3 eV is attributed to the Se-Se bond. Bi and Se's binding energies shifted opposite, which infers the formation of the Bi_2Se_3 thin film [93, 107].

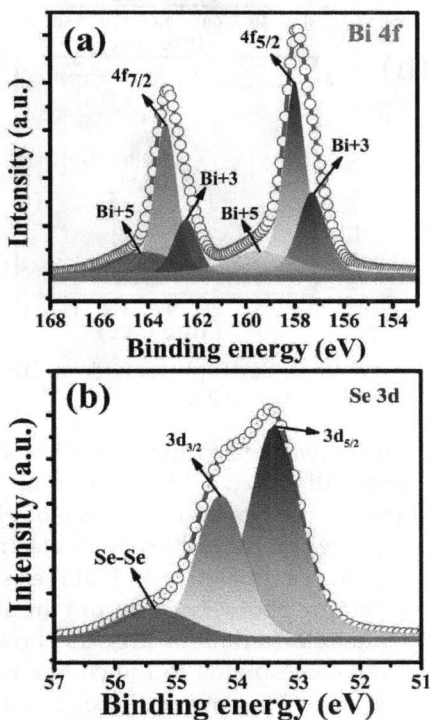

Figure 4.11   XPS core level spectra for Bi_2Se_3 thin film on Ti foil: (a) Bi 4f and (b) Se 3d core level.

After the successfully deposited of Bi_2Se_3 thin film on Ti foil, we investigated the performance of Bi_2Se_3/Ti foil-based photocatalyst in 0.5 M Na_2SO_4 electrolyte solution under 100 mW/cm^2 light intensity toward PEC water splitting. Figure 4.12 shows the *I-t* curve of Bi_2Se_3/Ti foil for each 30-second ON and OFF operation for Bi_2Se_3/Ti foil under 0.3 and 0.6 V vs Ag/AgCl bias voltages. The *I-t* curve shows the excellent stability of photoanode in an electrolyte solution and resultant photocurrent density of 2.2 and 4.3 $\mu A/cm^2$ for Bi_2Se_3/Ti foil at 0.3 and 0.6 V vs Ag/AgCl, respectively.

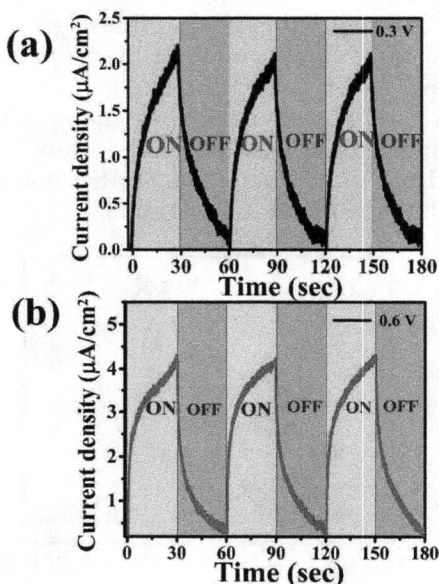

Figure 4.12 The *I-t* curve for Bi_2Se_3/Ti photoanode at: (a) 0.3 and (b) 0.6 V vs. Ag/AgCl bias v.

The EIS measurement was performed on Bi_2Se_3/Ti photoanode in a frequency range from 0.01 Hz to 1 M Hz at an applied ac voltage of 5 mV to understand the charge transport kinetics of photoanodes. EIS is the method of investigating the properties of an electrochemical system through the lens of impedance. In Figure 4.13, the resistance between the surface of the semiconducting photoelectrode and the electrolyte interface is followed by the semicircle (radius of arc), as shown in EIS spectra. Generally, the radius arc corresponds to the charge carrier's origin and transportation [108–110]. The EIS data of Bi_2Se_3/Ti foil is fitted with an equivalent circuit shown as the inset of Figure 4.13. The constant phase elements (CPE) describe the frequency-dependent impedance caused by surface roughness or nonuniformly properties of the electrode surface. R_s is the electrolyte resistance, and R_{ct} corresponds to charge transfer interfacial resistance across the semiconductor electrolyte interface [111, 112]. The EIS spectrum shows charge transfer resistance (R_{ct}) ~238 kΩ

for Bi_2Se_3/Ti foil. It indicates that Bi_2Se_3/Ti photoanode is facing high interfacial resistance. In Figure 4.14, the Bode plot reveals the electron lifetime for Bi_2Se_3/Ti foil. The lifetime (τ) value can be evaluated using the equation $\tau = 1/(2\pi f_{max})$, where f_{max} is the maximum frequency [113]. The τ values calculated for Bi_2Se_3/Ti foil is ~2.5 μs.

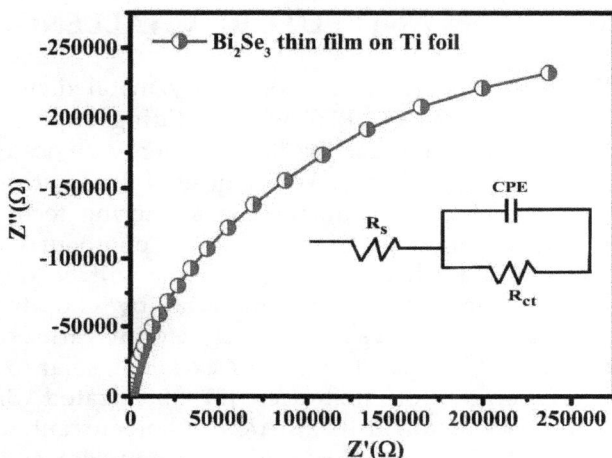

Figure 4.13   EIS Nyquist plot of Bi_2Se_3/Ti and photoanodes under applied ac voltage of 5 mV.

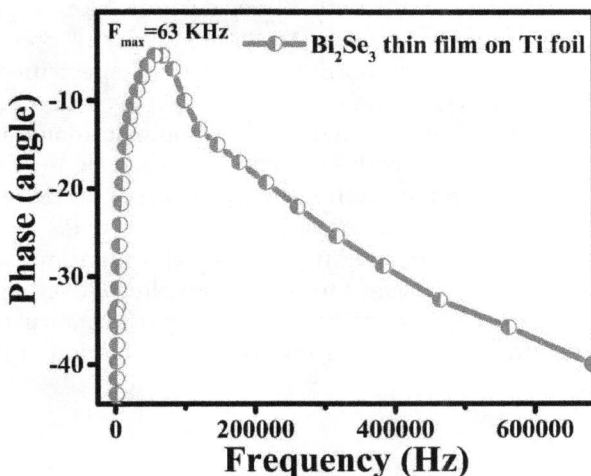

Figuer 4.14   The Bode plot of Bi_2Se_3/Ti photoanode.

Our study revealed that metal foil-based photoelectrode might benefit futuristic, flexible energy device applications, even though the obtained photocurrent density was not enough. To enhance the performance of Bi_2Se_3 photoelectrode, it needs to couple with various kinds of photocatalysts, which can enhance the photocurrent density by enabling the absorption of more sunlight and suppressing the charge carriers' recombination.

4.4 CONCLUSION AND FUTURE CHALLENGES

We have discussed the importance of binary metal dichalcogenides semiconducting materials for PEC water-splitting applications. The advantage of thin metal foils for the flexible energy device application has been discussed thoroughly. We deposited Bi_2Se_3 thin film on functionalized Ti foil using a magnetron sputtering technique. The PEC measurements show that the obtained photocurrent density of Bi_2Se_3/Ti foil was 4.2 $\mu A/cm^2$. Future work will be dedicated to increasing the efficiency of Bi_2Se_3 material by coupling it with various photocatalysts, such as TiO_2, Fe_2O_3, etc. In various literature, Bi_2Se_3 functionalized TiO_2/Ti foil shows excellent solar-to-hydrogen conversion efficiency; for instance, [31] fabricated Bi_2Se_3/TiO_2 heterostructure and obtained 1.7 mA/cm^2 photocurrent density. In another report, [62] designed Bi_2Se_3/TiO_2 quantum dot and obtained 0.8 mA/cm^2 photocurrent density. These reports infer us the conductivity of Bi_2Se_3/Ti photoanode can be increased by coupling it with TiO_2. We will fabricate heterostructure with Bi_2Se_3 such as MoS_2/Bi_2Se_3, $MoSe_2$/Bi_2Se_3, WSe_2/Bi_2Se_3, etc. to enhance the conductivity of photoelectrodes by absorbing the maximum portion of the solar spectrum, as TMDC materials are suitable photocatalyst for hydrogen evolution process. Our focus will also be on earth-abundant and nontoxic materials, such as Sb_2Se_3, Fe_2O_3, etc., that can be beneficial for large-scale implementation.

Photobiological-inspired methods are a new tool for designing novel photoelectrochemically active materials, focusing on PEC technology's advantages. Solar water splitting using earth-abundant materials is one of the most lucrative approaches to finding an alternative to fossil fuels. Lastly, the search for cost-effective and high-performance materials is the primary driving parameter for the commercialization of PEC energy technology.

ACKNOWLEDGMENTS

The authors would like to thank the Director of CSIR-NPL for providing constant encouragement and support. They would also like to acknowledge J.S. Tawale, S. Sharma of CSIR-NPL, Prof. Somnath C. Roy from IIT Madras, and Dr. R. Ganesan from BITS Pilani Hyderabad for their help with different sample characterization. This work is supported under the early career research award scheme (ECR/2017/001852) from the Science and Engineering Research Board (SERB-DST), Government of India.

REFERENCES

[1] Bak, T., J. Nowotny, M. Rekas and C.C. Sorrell. 2022 . Photoelectrochemical hydrogen generation from water using solar energy: materials-related aspects. Int. J. Hydrogen Energy 27: 991–1022.

[2] Bandar, Y.A., H. Ullah, S. Alfaifi, A.A. Tahir and T.K. Mallick. 2018. Photoelectrochemical solar water splitting: from basic principles to advanced devices. Veruscript Functional Nanomaterials 2(12):BDJOC3.

[3] Gratzelm, M. 2001. Photoelectrochemical cells. Nature 414: 338.

[4] Jacobsson, T.J. 2018. Photoelectrochemical water splitting: an idea heading towards obsolescence. Energy and Env. Sci. 11: 1977.

[5] Li X., Li. Zhao, J. Yu, X. Liu, X. Zhang, H. Liu, et al. 2020. Water splitting: from electrode to green energy system. Nano-Micro Lett. 12: 131.

[6] Vanka, S., G. Zeng, T.G. Deutsch, F.M. Toma and Z. Mi. 2022. Long-Term stability metrics of photoelectrochemical water splitting. Front. Energy Res. 10: 840140.

[7] Maeda, K. and K. Domen. 2010. Photocatalytic water splitting: recent progress and future challenges. J. Phys. Chem. Lett. 1: 2655–2661.

[8] Landman, A., H. Dotan, G.E. Shter, M. Wullenkord, A. Houaijia, A. Maljusch, et al. 2017. Photoelectrochemical water splitting in separate oxygen and hydrogen. Nat. Mater. 16: 646–651.

[9] Walter, M.G., E.L. Warren, J.R. McKone, S.W. Boettcher, Q. Mi, E.A. Santori, et al. 2010. . Solar water splitting cells. Chemi. Rev. 110: 6446–6473.

[10] Brown, L.F. 2001. A comparative study of fuels for on-board hydrogen production for fuel-cell-powered automobiles. Int. J. Hydrogen Energy 26: 381–397.

[11] Wu, H., H.L. Tan, C.Y. Toe, J. Scott, L.Z. Wang, R. Amal, et al. 2019. Photocatalytic and photoelectrochemical systems: similarities and differences. Adv. Mat. 32: 1904717.

[12] Akkerman, I., M. Janssen, J. Rocha and R.H. Wijffels. 2002. Photobiological hydrogen production: photochemical efficiency and bioreactor design. Int. J. Hydrogen Energy 27: 1195–208.

[13] Lede, J., F. Lapicque and J. Villermaux. 1983. Production of hydrogen by direct thermal decomposition of water. Int. J. Hydrogen Energy 8: 675–679.

[14] Cecal, A., A. Paraschivescu, K. Popa, D. Colisnic, G. Timco and L. Singerean. 2003. Radiolytic splitting of water molecules in the presence of some supramolecular compounds. J. Serbian Chemi. Soc. 68: 593–598.

[15] Yu, J.M., J. Lee, Y.S. Kim, J. Song, J. Oh, S.M. Lee, et al. 2020. High-performance and stable photoelectrochemical water splitting cell with organic-photoactive-layer-based photoanode. Nat. Comm. 11: 5509.

[16] Fujishima, A. and K. Honda. 1972. Electrochemical photolysis of water at a semiconductor electrode. Nature 238: 37–38.

[17] Marwat, M.A., M. Humayun, M.W. Afridi, H. Zhang, M.R.A. Karim, M. Ashtar, et al. 2021. Advanced catalysts for photoelectrochemical water splitting. ACS Appl. Energy Mat. 4: 12007–12031.

[18] Gai, Y., J. Li, S.S. Li, J.B. Xia and S.H. Wei. 2009. Design of narrow-gap tio_2: a passivated codoping approach for enhanced photoelectrochemical activity. Phys. Rev. Lett. 102: 036402.

[19] Shi, X., K. Zhang, K. Shina, M. Maa, J. Kwona, I.T. Choib, et al. 2015. Unassisted photoelectrochemical water splitting beyond 5.7% solar-to-hydrogen conversion efficiency by a wireless monolithic photoanode/dye-sensitized solar cell tandem device. Nano Energy 13: 182–191.

[20] Li, X., Z. Wang and L. Wang. 2021. Metal-organic framework-based materials for solar water splitting. Small Sci. 1:2000074.

[21] Pengtao, X., N.S. McCool and T.E. Mallouk. 2017. Water-splitting dye-sensitized solar cells. Nano Today 14: 1–17.

[22] Parihar, N.S. and S.K. Saraswat. 2020. Photoelectrochemical water splitting: an ideal technique for pure hydrogen production. J. Indian Chemi. Soc. 97: 1099–1103.

[23] Chen, S., D. Huang, P. Xu, W. Xue, L. Lei, M. Cheng, et al. 2020. Semiconductor-based photocatalysts for photocatalytic and photo-electrochemical water splitting: will we stop with photo corrosion? J. Mat. Chem. A 8: 2286–2322.

[24] Dotan, H., N. Mathews, T. Hisatomi, M. Gratzel and A. Rothschild. 2014. On the solar to hydrogen conversion efficiency of photoelectrodes for water splitting. J. Phys. Chem. Lett. 5: 3330–3334.

[25] Joy, J., J. Mathew and S.C. George. 2018. Nanomaterials for PEC water splitting: a review. Int. J. Hydrogen Energy 43: 4804–48017.

[26] Wilke, T., D. Schricker, J. Rolf and K. Kleinermanns. 2012. Solar water splitting by semiconductor nanocomposites and hydrogen storage with quinoid systems. Open J. Phys. Chem. 2:195–203.

[27] Tamirat, A.G., J. Rick, A.A. Dubale, W.N. Sub and B.J. Hwang. 2016. Using hematite for photoelectrochemical water splitting: a review of current progress and challenges. Nanoscale Horiz. 1: 243–267.

[28] Lu, Q., Y. Yu, Q. Ma, B. Chen and H. Zhang. 2015. 2D Transition-metal-dichalcogenide-nanosheet-based composites for photocatalytic and electrocatalytic hydrogen evolution reactions. Adv. Mat. 28: 1917–1933.

[29] Li, Y., J. Li, W. Yang and X. Wang. 2020. Implementation of ferroelectric materials in photocatalytic and photoelectrochemical water splitting. Nanoscale Horiz. 5: 1174.

[30] Martín, S.S., M.J. Rivero and I. Ortiz. 2020. Unravelling the mechanisms that drive the performance of photocatalytic hydrogen production. Catalysts 10: 901.

[31] Subramanyam, P., B. Meena, D. Suryakala and C. Subrahmanyam. 2021. TiO_2 photoanodes sensitized with Bi_2Se_3 nanoflowers for visible near-Infrared photoelectrochemical water splitting. ACS Appl. Nano Mat. 4: 739–745.

[32] Subramaniam, M.N., P.S. Goh, D. Kanakaraju, J.W. Lim, W.J. Lau and A.F. Ismail. 2022. Photocatalytic membranes: a new perspective for persistent organic pollutants removal. Env. Sci. and Poll. Res. 29: 12506–12530.

[33] Wang, J., T.V Ree., Y. Wu, P. Zhang and L. Gao. 2018. Metal oxide semiconductors for solar water splitting. Metal Oxides in Energy Techn. 205–249.

[34] Cho, S., J.W. Jang, K.H. Lee and J.S. Lee. 2014. Research update: strategies for efficient photoelectrochemical water splitting using metal oxide photoanodes. APL Materials 2: 010703.

[35] Szymanski, P. and M.A. El-Sayed. 2012. Some recent developments in photoelectrochemical water splitting using nanostructured TiO_2: a short review. Theor. Chem. Acc. 131: 1202.

[36] Concina, I. and Z.H. Ibupoto, A. Vomiero. 2017. Semiconducting metal oxide nanostructures for water splitting and photovoltaics. Adv. Energy Mat. 7: 1700706.

[37] Xia, X., L. Wang, N. Sui, V.L. Colvin and W.W. Yu. 2020. Recent progress in transition metal selenide electrocatalysts for water splitting. Nanoscale 12: 12249–12262.

[38] Hou, X., Y. Liu, H. Ju, B. Pan, J. Zhu, T. Ding, et al. 2016. Design and epitaxial growth of $MoSe_2$-NiSe vertical hetero nanostructures with electronic modulation for enhanced hydrogen evolution reaction. Chem. Mat. 28: 1838–1846.

[39] Haque, F., T. Daeneke, K.K. Zadeh and J.Z. Ou. 2018. Two-dimensional transition metal oxide and chalcogenide-based photocatalysts. Nano-Micro Lett. 10: 4–27.

[40] Wu, L., S. Shi, Q. Li, X. Zhang and X. Cui. 2019. TiO_2 nanoparticles modified with 2D $MoSe_2$ for enhanced photocatalytic activity on hydrogen evolution. Int. J. Hydrogen Energy 44: 720–728.

[41] Zhang, J.Z. 2011. Metal oxide nanomaterials for solar hydrogen generation from photoelectrochemical water splitting. MRS Bulletin 36: 48–55.

[42] Rajamathi, C.R., U. Gupta, K. Pal, N. Kumar, H. Yang, Y. Sun, et al. 2017. Photochemical water splitting by bismuth chalcogenide topological insulators. Chem. Phys. Chem. 18: 2322–2327.

[43] Jayachitra, S., P. Ravi, P. Murugan and M. Sathish. 2022. Super critically exfoliated Bi_2Se_3 nanosheets for enhanced photocatalytic hydrogen production by topological surface states over TiO_2. J. Colloid and Inter. Sci. 605: 871–880.

[44] Ranjbar, A., H. Mirhosseini and T.D. Kühne. 2021. On topological materials as photocatalysts for water splitting by visible light. J. Phys. Mater. 5: 1–12.

[45] Gautam, S., A.K. Verma, A. Balapure, B. Singh, R. Ganesan, M.S. Kumar. 2022. Structural, electronic, and thermoelectric properties of Bi_2Se_3 thin films deposited by RF magnetron sputtering. J. Elect. Mat. 51: 2500–2509.

[46] Yang, X.X., Wang and Z. Zhang. 2005. Synthesis and optical properties of single crystalline bismuth selenide nanorods via a convenient route. J. Crys. Growth. 276: 566–570.

[47] Yang, S.D., L. Yang, Y.X. Zheng, W.J. Zhou, M.Y. Gao, S.Y. Wang, et al. 2017. Structure-dependent optical properties of self-organized Bi_2Se_3 Nanostructures: From Nanocrystals to Nanoflakes, ACS Appl. Mat. Inter. 9: 29295–29301.

[48] Fu, L., C.L. Kane and E.J. Mele. 2007. Topological insulators in three dimensions. Phys. Rev. Lett. 98: 106803.

[49] Zhang, H., C.X. Liu, X.L. Qi, X. Dai, Z. Fang and S.C. Zhang. 2009. Topological insulators in Bi_2Se_3, Bi_2Te_3, and Sb_2Te_3 with a single dirac cone on the surface. Nat. Phy. 5: 438–442.

[50] Kane, C.L. and E.J. Mele. 2005. Z2 topological order and the quantum spin hall effect. Phys. Rev. Lett. 95: 146802.

[51] Tokura, Y., K. Yasuda and A. Tsukazaki. 2019. Magnetic topological insulators. Nat. Rev. Phy. 1: 126–143.

[52] Chege, S., P. Ning, J. Sifuna and G.O. Amolo. 2020. Origin of band inversion in topological insulators. AIP Adv. 10: 095018.

[53] Park, K.H., M. Mohammad, Z. Aksamija and U. Ravaioli. 2008. Phonon scattering due to van der Waals forces in the lattice thermal conductivity of Bi_2Te_3 thin films. J. Appl. Phy. 103: 024314.

[54] Singh, M.P., M. Mandal, K. Sethupathi, M.S. Ramachandra Rao and P.K. Nayak. 2021. Study of thermometry in two-dimensional Sb_2Te_3 from temperature-dependent Raman Spectroscopy. Nanoscale Res. Lett. 16: 22.

[55] Ko, T.Y., M. Shellaiah and K.W. Sun. 2016. Thermal and thermoelectric transport in highly resistive single Sb_2Se_3 nanowires and nanowire bundles. Sci. Rep. 6: 35086.

[56] Garcia, V.M., M.T.S. Nair, P.K. Nair and R.A. Zingaro. 1997. Chemical deposition of bismuth selenide thin films using N,N -dimethylselenourea. Semi. Sci. Tech. 12: 645–653.

[57] Adam, A.M., E. Lilov, E.M.M. Ibrahim, P. Petkov, L.V. Panina and M.A. Darwish. 2019. Correlation of structural and optical properties in as-prepared and annealed Bi_2Se_3 thin films. J. Mat. Proc. Tech. 264: 76–83.

[58] Wang, F.K., S.J. Yang and T.Y. Zhai. 2021. 2D Bi_2Se_3 materials for optoelectronics. iScience. 24: 103291.

[59] Lai, H.D., S.R. Jian, L.T.C. Tuyen, P.H. Le, C.W. Luo and J.Y. Juang. 2018. Nanoindentation of Bi_2Se_3 thin films. Micromachines 9: 518.

[60] Chen, X., H.D. Zhou, A. Kiswandhi, I. Miotkowski, Y.P. Chen, P.A. Sharma, et al. 2011. Thermal expansion coefficients of Bi_2Se_3 and Sb_2Te_3 crystals from 10 K to 270 K. Appl. Phy. Lett. 99: 261912.

[61] Q. Chen, J. Chen, X. Xu, Z. Wang, Y. Ding, L. Xiong, et al. 2021. Morphology optimization of Bi_2Se_3 thin films for enhanced thermoelectric performance. Cryst. Growth Des. 21: 6737–6743.

[62] G. Zhou, T. Zhao, O. Wang, X. Xia and J.H. Pan. 2021. Bi_2Se_3, Bi_2Te_3 quantum dots-sensitized rutile TiO_2 nanorod arrays for enhanced solar photo electrocatalysis in azo dye degradation. J. Phy. Energy 3: 014003.

[63] Guo, C., R. Hu, H. Qiao, C. Dua and X. Qi. 2021. TiO_2 nanoparticles anchoring on two-dimensional Bi_2Se_3 nanosheet as an enhanced visible light catalyst. J. Mat. Sci. Mat. Elect. 32: 19424–19433.

[64] Yang, J., C. Wang, H. Ju, Y. Sun, S. Xing, J. Zhu, et al. 2017. Integrated quasi plane hetero nanostructures of $MoSe_2$/ Bi_2Se_3 hexagonal nanosheets: synergetic electrocatalytic water splitting and enhanced supercapacitor performance. Adv. Func. Mat. 1703864: 1–10.

[65] Razzaque, Sh., M.D. Khan, M. Aamir, M. Sohail, S. Bhoyate, R.K. Gupta, et al. 2021. Selective synthesis of bismuth or bismuth selenide nanosheets from a metal-organic precursor: investigation of their catalytic performance for water splitting. Inorganic Chem. 60:1449–1461.

[66] Liu, X., J. Chen, M. Luo, M. Leng, Z. Xia, Y. Zhou, et al. 2014. Thermal evaporation and characterization of Sb_2Se_3 thin film for substrate Sb_2Se_3/ CdS solar cells. ACS Appl. Mat. Inter. 6: 10687–10695.

[67] Yang, W., J. Ahn, Y. Oh, J. Tan, H. Lee, J. Park, et al. 2018. Adjusting the Anisotropy of 1D Sb_2Se_3 Nanostructures for Highly Efficient Photoelectrochemical Water Splitting. Adv. Energy Mat. 8:1702888.

[68] Chen, S., T. Liu, Z. Zheng, M. Ishaq, G. Liang, P. Fan, et al. 2021. Recent progress and perspectives on Sb_2Se_3 -based photocathodes for solar hydrogen production via photoelectrochemical water splitting. J. Energy Chem. pp. 1–49.

[69] Yang, W., J.H. Kim, O.S. Hutter, L.J. Phillips, J. Tan, J. Park, et al. 2020. Benchmark performance of low-cost Sb_2Se_3 photocathodes for unassisted solar overall water splitting. Nat. Comm. 11: 861.

[70] Zhang, L., Y. Li, C. Li, Q. Chen, Z. Zhen, X. Jiang, et al. 2017. Scalable low-band-gap Sb_2Se_3 thin-film photocathodes for efficient visible–near infrared solar hydrogen evolution. ACS Nano 11: 12753–12763.

[71] Ran, F., P. Li, X. Yuan, J. Zhang, D. Zhang and S. Chen. 2022. Fabrication of a Sb_2Se_3/$CuSbS_2$ heterojunction photocathode for photoelectrochemical water splitting. J Phy. Chem. C. 126: 8581–8587.

[72] Prabhakar, R.R., W. Septina, S. Siol, T. Moehl, R. Wick-Joliat and S.D. Tilley. 2017. Photocorrosion-resistant Sb_2Se_3 photocathodes with earth-abundant MoSx hydrogen evolution catalyst. J. Mat. Chem. 5: 23139–23145.

[73] Guo, L., P.S. Shinde, Y. Ma, L. Li, S. Pan and F. Yan. 2019. Scalable core-shell MoS_2/Sb_2Se_3 nanorod array photocathode for enhanced PEC water splitting. Solar RRL. 4: 1900442.

[74] Jain, A.K., C. Gopalakrishnan and P. Malar. 2022. Study of pulsed laser deposited antimony selenide thin films. J. Mat. Sci.: Mat. Elect. 33: 10430–10438.

[75] Liu, C., Y. Yuan, L. Cheng, J. Sua, X. Zhang, X. Lia, et al. 2019. A study on optical properties of Sb_2Se_3 thin films and resistive switching behavior in Ag/Sb_2Se_3/W heterojunctions. Results in Phy. 13: 102228.

[76] Hamrouni, R., N.E.H. Segmane, D. Abdelkader, A. Amara, A. Drici, M. Bououdina, et al. 2018. Linear and nonlinear optical properties of Sb_2Se_3 thin films elaborated from nano-crystalline mechanically alloyed powder. App. Phy. A 124: 861.

[77] Singh, Y., M. Kumar, R. Yadav, A. Kumar, S. Rani, Shashi, et al. 2022. Enhanced photoconductivity performance of microrod-based Sb_2Se_3 device. Solar Energy Mat. Solar Cells 243: 111765.

[78] Yin, K., Z.D. Cui, X.R. Zheng, X.J. Yang, S.L. Zhu, Z.Y. Li, et al. 2015. Bi_2Te_3@CoNiMo composite as a high-performance bifunctional catalyst for the hydrogen and oxygen evolution reactions. J. Mat. Chem. A 3: 22770–22780.

[79] Li, G., J. Huang, Q. Yang, L. Zhang, Q. Mu, Y. Sun, et al. 2021. MoS_2 on topological insulator Bi_2Te_3 thin films: activation of the basal plane for hydrogen reduction. J. Energy Chem. 62: 516–522.

[80] Bahk, J.H., H. Fang, K. Yazawa and A. Shakouri. 2015. Flexible thermoelectric materials and device optimization for wearable energy harvesting. J. Mat. Chem. C 3: 10362.

[81] Corzo, D., G.T. Blázquez and D. Baran. 2020. Flexible electronics challenges and opportunities. Front. Elec. 1: 594003.

[82] Sircar, A. and H. Kumar. 2022. An introduction to flexible electronics: manufacturing techniques, types, and future. J. Phys. Conf. Ser. 1913: 012047.

[83] Lee, Y., B. Gupta, H.H. Tan, C. Jagadish, J. Oh and S. Karuturi. 2021. Thin silicon via crack-assisted layer exfoliation for photoelectrochemical water splitting. iScience 24: 102921.

[84] Cordill, M.J., P. Kreiml and C. Mitterer. 2022. Materials engineering for flexible metallic thin film applications. Materials 15: 926.

[85] Hassan, M., G. Abbas, N. Li, A. Afzal, Z. Haider, and . Ahmed, et al. 2021. Significance of flexible substrates for wearable and implantable devices: recent advances and perspectives. Adv. Mat. Tech. 7: 2100773.

[86] Khalil, M.I., R. Bernasconi, A. Lucotti, A.L. Donne, R.A. Mereu, S. Binetti, et al. 2021. CZTS thin-film solar cells on flexible molybdenum foil by the electrodeposition-annealing route. J. Appl. Electrochem. 51: 209–218.

[87] Shiyani, T., S.K. Mahapatra and A.K. Ray. 2021. Flexible zinc oxide photoelectrode for photoelectrochemical energy conversion. Journal of Materials Science: Mater. Electron. 32: 15386–15392.

[88] Quynh, L.T., C.N. Van, W.Y. Tzeng, C.W. Huang, Y.H. Lai and J.W. Chen. 2018. Flexible heteroepitaxy photoelectrode for photoelectrochemical water splitting. ACS App. Energy Mat. 1: 3900–3907.

[89] Singh, S. and N. Khare. 2017. Flexible $PVDF/Cu/PVDF-NaNbO_3$ Photoanode with ferroelectric properties: an efficient tuning of photoelectrochemical water splitting with electric field polarization and piezophototronic effect. Nano Energy 42: 173–180.

[90] Park, J.Y., G.H. Lee, J. Jo, A.K Cheng, H. Yoon, K. Watanabe, et al. 2016. Molecular beam epitaxial growth and electronic transport properties

of the high-quality topological insulator Bi_2Se_3 thin films on hexagonal boron nitride. 2d Mater. 3: 035029.

[91] Tabor, Ph., C. Keenan, S. Urazhdin and D. Lederman. 2011. Molecular beam epitaxy and characterization of thin Bi_2Se_3 films on Al_2O_3 (110). App. Phy. Lett. 99: 013111.

[92] Gautam, S., B. Sin,gh, V. Aggarwal, M. Senthil Kumar, V.N. Singh and S.P. Singh, et al. 2022. Thickness-dependent optical properties of sputtered Bi_2Se_3 films on mica. Mater. Today 64: 1725–1731.

[93] Gautam, S., V. Aggarwal, B. Singh, V.P.S. Awana, R. Ganesan and S.S. Kushvaha. 2022. Signature of weakantilocalization in sputtered topological insulator Bi_2Se_3 thin films with varying thickness. Sci. Rep. 12: 9770.

[94] Gautam, S.V. Aggarwal, B. Singh, R. Kumar, J.S. Tawale, B.S. Yadav, et al. 2022. Structural and optical properties of sputtered Bi_2Se_3 thin films on sapphire (0001), quartz, and GaN/sapphire (0001). J. Mat. Res. 1–12.

[95] Singh, B., S. Gautam, V. Aggarwal, J.S. Tawale and S.S. Kushvaha. 2022. Growth and characterization of crystalline Bi_2Se_3 thin films on flexible metal foils by magnetron sputtering system. Mater. Today 64: 1701–1706.

[96] Gautam, V., S. Gautam, G.K. Maurya, K. Kandpal, B. Singh, R. Ganesan, et al. 2022. Investigation of RF sputtered, n-Bi_2Se_3 heterojunction on p-Si for enhanced NIR optoelectronic applications. Solar Energy Mat. and Solar Cells. 248: 112028.

[97] Wei, Z.T., M. Zhang, Y. Yan, X. Kan, Z. Yu, Y.L. Chen, et al. 2015. Transport properties of Bi_2Se_3 thin films grown by magnetron sputtering. Funct. Mate. Lett. 8: 1550020.

[98] Tang, R., X. Chen, Y. Luo, Z. Chen, Y. Liu, Y. Li, et al. 2020. Controlled sputtering pressure on high-quality Sb_2Se_3 thin film for substrate configurated solar cells. Nanomater. 10: 574.

[99] Chen, Sh., Z. Zheng, M. Cathelinaud, H. Ma, X. Qiao, Z. Su, et al. 2019. Magnetron sputtered Sb_2Se_3-based thin films towards high-performance quasi-homojunction thin film solar cells. Sol. Energy Mater. Sol. Cells 203: 110154.

[100] Wang, S., Y. Li, A. Ng, Q. Hu, Q. Zhou, X. Li, et al. 2020. 2D Bi_2Se_3 van der Waals epitaxy on mica for optoelectronics applications. J. Nano 10: 1653.

[101] Yin, C., C. Gong, S. Tian, Y. Cui, X. Wang, Y. Wang, et al. 2021. Low-energy oxygen plasma injection of 2D Bi_2Se_3 realizes highly controllable resistive random-access memory. Adv. Func. Mat. 32: 2108455.

[102] Schöenherr, P., A.A. Baker, P. Kusch, S. Reich and T. Hesjedal. 2014. Engineering of Bi_2Se_3 nanowires by laser cutting. European Phys. J. Appl. Phy. 66: 10401.

[103] Zang, C., X. Qi, L. Ren, G. Hao, Y. Liu, J. Li, et al. 2014. Photoresponse properties of ultrathin Bi_2Se_3 nanosheets synthesized by hydrothermal intercalation and exfoliation route. Appl. Sur. Sci. 316: 341–347.

[104] Kim, S., S. Lee, J. Woo and G Lee. 2018. Growth of Bi_2Se_3 topological insulator thin film on Ge (1 1 1) substrate. Appl. Sur. Sci. B 432: 152–155.

[105] Ramesh, C., P. Tyagi, G. Gupta, M.S. Kumar and S.S. Kushvaha. 2019. Influence of surface nitridation and an AlN buffer layer on the growth of GaN nanostructures on a flexible Ti metal foil using laser molecular beam epitaxy. Jpn. J. Appl. Phy. 58: 079301.

[106] Ramesh, C., P. Tyagi, J. Kaswan, B.S. Yadav, A.K. Shukla, M. Senthil Kumar, et al. 2020. Effect of surface modification and laser repetition rate on growth, structural, electronic, and optical properties of GaN nanorods on flexible Ti metal foil. RSC Adv. 10: 2113–2122.

[107] Nascimento, V.B., V.E. Carvalho, R. Paniago, E.A. Soares, L.O. Ladeira, H.D. Pfannes. 1999. XPS and EELS study of the bismuth selenide. J. Electron. Spect. Related Phen. 104: 99–107.

[108] Xue, D., J. Luo, Z. Li, Y. Yin and J. Shen. 2020. Enhanced photoelectrochemical properties from mo-doped TiO_2 nanotube arrays film. Coating 10: 75.

[109] Ali, A., X. Li, J. Song, S. Yang, W. Zhang, Z. Zhang, et al. 2017. Nature-mimic ZnO nanoflowers architecture: chalcogenide quantum dots coupling with $ZnO/ZnTiO_3$ nano heterostructures for efficient photoelectrochemical water splitting. J. Phys. Chem. C 121: 21096–21104.

[110] Ali, A., F.A. Mangrio, X. Chen, Y. Dai, K. Chen, X. Xu, et al. 2019. Ultrathin MoS_2 nanosheets for high-performance photoelectrochemical applications via plasmonic coupling with Au nanocrystals. Nanoscale 11: 7813.

[111] Shahrezaei, M., S.M.H. Hejazi, Y. Rambabu, M. Vavrecka, A. Bakandritsos, S. Oezkan, et al. 2020. Multi-leg TiO_2 nanotube photoelectrodes modified by platinized cyanographene with enhanced photoelectrochemical performance. Catalysts 10: 717.

[112] Khatun, N., S. Dey, G.C. Behera and S.C. Roy. 2022. $Ti_3C_2T_x$ MXene functionalization induced enhancement of photoelectrochemical performance of TiO_2 nanotube arrays. Mat. Chem. Phy. 278: 125651.

[113] Hu, L., Y. Li, W. Chen, X. Liu, Sh. Liang, Z. Cheng, et al. 2022. Controlled synthesis and photoelectrochemical performance enhancement of Cu_2–xSe decorated porous Au/Bi_2Se_3 Z-scheme plasmonic photoelectrocatalyst. Catalysts 12: 359.

Chapter 5

Quantum-Cutting Phosphors for Thermal Sensor Applications

Abhijit Jadhav

Hyderabad Laboratories Private Limited, Hyderabad, India.
Email: ajadhav@hyderabadlaboratories.com

5.1 INTRODUCTION

Luminescence is a phenomenon of spontaneous light emission by a substance that is not due to heat or cold light. It involves the promotion of ground-state electrons to higher energy states/levels, eventually emitting light during the relaxation process. Luminescence induced by light energy is termed photoluminescence and is formally divided into two categories: fluorescence and phosphorescence. Phosphorescence has longer excited state lifetimes than fluorescence; it is usually used to determine the temperature in a thermographic phosphor system [1]. Phosphors are inorganic oxides that are usually white and show luminescence upon excitation. Down-conversion phosphors absorb high-energy UV photons and convert them to lower-energy visible photons. Upconversion phosphors absorb lower-energy IR or near IR (NIR) photons and convert them to higher-energy visible photons. At the same time, quantum-cutting (QC) phosphors absorb high-energy UV photons and convert them into lower-energy visible and NIR photons. That means that with QC phosphors, the obtained quantum efficiency (QE) is more than unity. These phosphors find various applications in various technologies, such as CRT tubing, plasma display, light bulbs, and X-ray conversion screens.

Temperature is the most essential thermodynamic parameter to describe any physical, chemical, and biological process. In various applications, knowledge of temperature plays an essential role in understanding exact process conditions and maintaining proper reaction conditions. The accurate temperature measurement during combustion helps to understand the heat transfer phenomenon. An infrared camera is used during thermometry to identify abnormally hot or cold areas on a component operating under normal conditions. The phosphors are also known to be thermographic if they show emission-changing characteristics with varying temperature conditions. The present chapter will overview QC phosphors used for thermal sensor applications.

5.2   QUANTUM-CUTTING PHENOMENON

Quantum cutting is a type of luminescence observed in different types of dopant ions host materials. The phosphors with singly, doubly, and sometimes triply doped ions show a quantum-cutting phenomenon, which is a conversion of photons from high to low energy and vice versa during the process of down-conversion and upconversion process. The process has been explained through the energy level diagram shown in Figure 5.1.

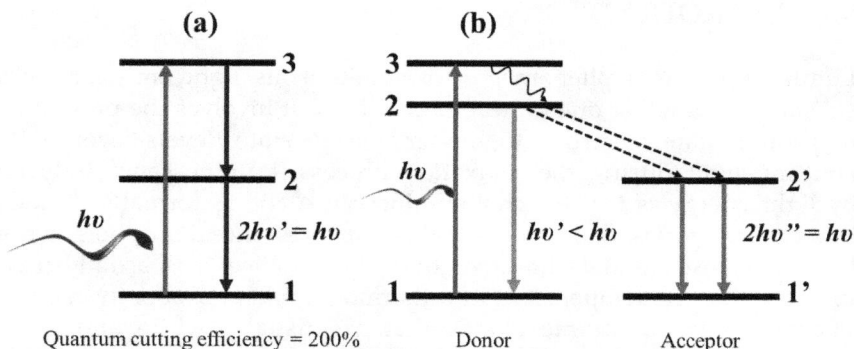

(a)   **(b)**

Quantum cutting efficiency = 200%   Donor   Acceptor

Figure 5.1   Energy level diagrams of quantum-cutting mechanisms for single (a) and double (b) ions [2].

The phosphor doped with a single rare-earth ion is excited with $h\upsilon$ energy, and the ground-state electrons get transferred to a higher (excited state energy level). During the relaxation process, they emit two photons of equal energy. Thus, one high-energy photon is converted successfully into two lower-energy photons during this process, i.e., $2h\upsilon' = h\upsilon$. This process is reported with singly doped Pr^{3+}, Eu^{3+}, and Tm^{3+} host materials [3, 4]. Yu et al. showed that the cross-relaxation between the same ions in $Gd_2O_2S{:}Tm^{3+}$ samples led to the generation of QC emissions of the two, three, and four NIR photons, respectively [5]. Meanwhile, in the two ions' dopant phosphor material, both ions can be excited using the same source or light sources with different wavelengths. Upon excitation, the ground-state electrons get promoted to a higher energy level 3, and during the relaxation process, they arrive at level 2 non-radiatively. From intermediate level 2, the electrons emit a radiative transition of energy lower than the excitation energy, i.e., $h\upsilon' < h\upsilon$. Successful energy transfer through dipole-dipole interaction between donor ions in level 2 and acceptor ions in level 1. This results in the emission of lower-energy photons and subsequently confirms the conversion of one high-energy photon into two lower-energy photons, i.e., $h\upsilon = 2\ h\upsilon'$, making the QC efficiency 200%. This QC mechanism is reported for $Ca_2Al_{14}O_{33}$ doped with Eu^{3+} and Yb^{3+} with a QC efficiency of 199% upon excitation at 394 nm wavelength [6].

5.2.1 Visible Quantum Cutting

Quantum cutting in phosphors occurs when every absorbed photon provides two or more emitted photons. During visible QC, the phosphors utilise vacuum UV (VUV) light and generate visible light with conversion efficiency higher than unity. This help to provide improved energy efficiency in many lighting devices [7]. The energy of the absorbed VUV photon is much higher than the emitted visible photon and thus makes it capable of emitting two visible photons upon every absorbed photon. Wegh et al. explained the energy level diagram for the two types of rare-earth ions (I and II), which shows the down-conversion phenomenon [8]. The concept of obtaining more than 100% quantum efficiencies is based on a combination of two lanthanide dopants. A high QE (close to 200%) can be obtained through partial energy transfer between the co-dopant ions [3].

The process of obtaining higher QE (near 200%) is also known as upconversion [9, 10]. The quantum cutting through energy transfer, i.e., down-conversion, has been shown in Figure 5.2. The concept of down-conversion has been explained with two types of ions (I and II) with a hypothetical energy level scheme. Efficient visible QC obtained from two-photon emission from a high-energy level for a single lanthanide ion is possible theoretically, as shown in Figure 5.2(a) with red lines.

The resulting emissions in the IR and UV regions are shown with black lines. This mechanism is based on one luminescent centre with three energy levels [11].

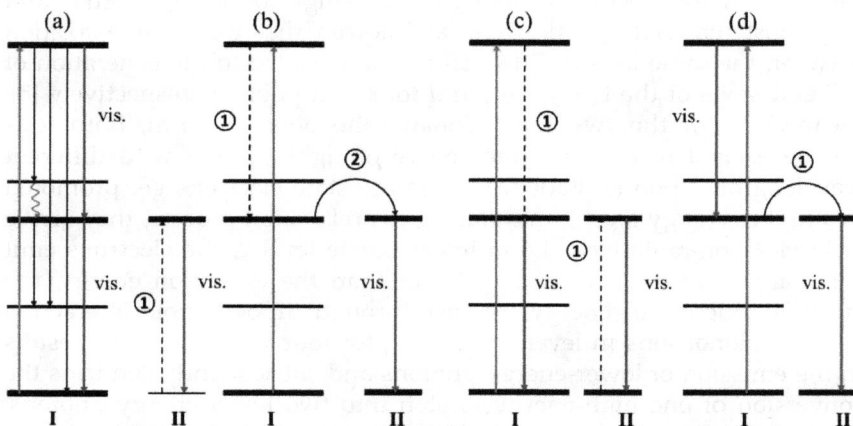

Figure 5.2 Energy level diagram for ions (type I and II) illustrates visible quantum cutting via down conversion. Type I is an ion showing emission from a high-energy level, and Type II emission is dedicated to an activator ion to which energy transfer takes place. (a) Quantum cutting on single ion I by the sequential emission of two visible photons. (b) Quantum cutting by two-step energy transfer. The (c) and (d) involve only one energy transfer step from ion I to ion II. These are the possible mechanisms of obtaining visible quantum cutting if one of two visible photons can be emitted by ion I [8].

5.2.2 Near IR (NIR) Quantum Cutting

Various mechanisms that are possible for NIR quantum cutting have been shown in Figure 5.3. Figure 5.3(a) shows the mechanism based on one luminescent centre with three energy levels. The emission of two consecutive NIR photons is possible, with the optical centre showing a transition to the highest energy level by absorbing one high-energy UV or visible photon. The sequential emission of two NIR photons is possible when the optical centre absorbs one UV or visible photon and shows transition at the highest energy level. The populated intermediate state was also observed when the optical centre returned to its ground state. This phenomenon has been reported for single ions like Ho^{3+}, Tm^{3+}, or Er^{3+} [12–17]. The NIR quantum-cutting mechanism through two luminescent centres has been shown in Figure 5.3(b–e). Here we can see a plausible mechanism via a two-step electron transfer process through ion pairs of physically interacting lanthanide ions, such as Pr^{3+}–Yb^{3+} [18] and Er^{3+}–Yb^{3+} [19], accompanied by the emission of two NIR photons [11]. Figures 5.3(c) and (d) show emitted NIR photons, resulting from quantum cutting due to a one-step energy transfer between two optical centres. This phenomenon was successfully demonstrated by the ion pairs Tm^{3+}–Yb^{3+} [20] and Ho^{3+}–Yb^{3+} [21]. Energy splitting depends on the donor species' population of an intermediate energy level. The relaxation process is dominated by second-order cooperative sensitisation, resulting in the simultaneous excitation of two acceptors with the emission of two NIR photons, as shown in Figure 5.3(e).

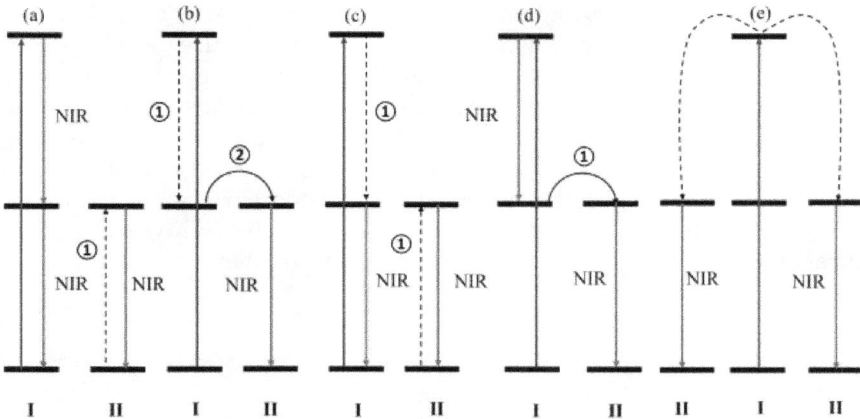

Figure 5.3 Typical mechanisms of NIR quantum cutting for PV applications. NIR quantum cutting is shown through simplified energy level diagrams for ions (type I and II). (a) NIR quantum cutting on a single ion by the successive emission of two NIR photons. (b–d) NIR quantum cutting due to resonant energy transfer from ion I to ion II, and (e) NIR quantum cutting due to cooperative energy transfer from ion I to ion II [11].

5.3 PHOSPHOR THERMOGRAPHY

Thermography uses an infrared camera to look for abnormally hot or cold areas on a given component operating under normal conditions. Phosphor thermometry is an optical method for surface temperature measurement using the phosphor materials' luminescence emitted. Upon stimulation with particular light sources, phosphors emit light of different colours and wavelengths. It emits light change with temperature, including its brightness, colour and afterglow duration, which is most commonly used for temperature measurement. After illumination, the time required for the brightness to decrease to $1/e$ of its original value is known as the decay time indicated by τ. This decay time is a function of temperature T. The intensity, I, of the luminescence decay exponentially, as shown in Equation 5.1:

$$I = I_0 e^{\frac{-t}{\tau}} \qquad (5.1)$$

I_0 is the initial intensity, the t is the time, and τ is the temperature-dependent parameter.

Phosphors show the thermographic property when they display temperature-dependent emission-changing characteristics. During the development of the fluorescent lamp, Nuebert et al. [22] discovered the idea of using phosphors for temperature measurement after they observed the loss of luminescence with increasing temperature. Wickersheim and his team applied a phosphor at the tip of an optical fibre and investigated many phosphors with various applications leading to the commercialisation of fluorescence-based thermometry systems [23]. Cates [24, 25] and co-workers developed remote measurement systems by adhering a phosphor layer over the interested surface instead at the tip of a fibre. It helped the system to have greater flexibility, and remote measurements of moving surfaces were also possible.

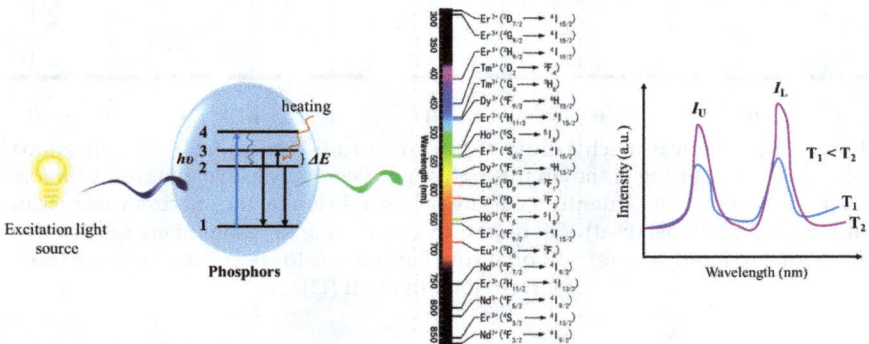

Figure 5.4   Luminescence transitions and mechanism for different rare-earth-doped phosphors used for optical thermometry.

Phosphor thermometry shows a non-contact technique based on the intensity of luminescence signals for the remote temperature measurement. Phosphors are usually composed of inorganic oxides or ceramic materials resistant to high temperatures and harsh chemicals. As shown in Figure 5.4, trivalent lanthanide ions are used as a doping material in inorganic oxides, which are now working as a phosphor material. Trivalent lanthanide ions such as Eu^{3+}, Tm^{3+}, Er^{3+}, Yb^{3+}, Ho^{3+}, Nd^{3+}, and Dy^{3+} are used as luminescence centres or activators for the study of temperature change during thermometric measurements. Different lanthanide ions [26–31] have different luminescence transitions during the emission process, as shown in Figure 5.4. The effect of various shapes of the host material, such as nanorods [29], spherical and tetragonal [32], core-shell [33], hollow nanostructures [34], fibres [35], glass ceramics with fluoride nanocrystals [36], and bulk oxides [37]. Upon excitation with a suitable light source, these phosphors show emitted luminescence in UV, visible, infrared, and NIR regions.

Host matrix doped with rare-earth ions plays a vital role in deciding whether the phosphor material satisfies the thermally coupled energy level required to be used as an optical temperature sensor. Table 5.1 represents host dependent ΔE_f, ΔE_m, and δ in Er^{3+} doped and Er^{3+}–Yb^{3+} co-doped phosphors. The term δ calculates the fluorescence intensity ratio between ΔE_f and ΔE_m. From Table 5.1, we can see that the small δ value is the result of successive excited state absorption, which has overcome the energy transfer process between thermally coupled levels and other energy levels due to irregular ligand fields around the dopant lanthanide ion sites inside the host crystals. The value of δ varies with different host crystals and shows a maximum value of 90.44% for Er^{3+}, Yb^{3+} co-doped $\beta - NaLuF_4$, which is ascribed to the energy of the excited state absorption due to Er^{3+} ion.

Table 5.1   Host dependent ΔE_f, ΔE_m, and δ in Er^{3+} doped and Er^{3+}–Yb^{3+} co-doped phosphors

Samples	ΔE_f (cm^{-1})	ΔE_m (cm^{-1})	δ (%)	References
Er^{3+} doped In – Zn – Sr – Ba glass	861.0	771.8	11.55	[38]
Er^{3+} doped Sr – Ba – Nb – B glass	872.3	748.0	16.62	[39]
Er^{3+} doped $BaTiO_3$ nanocrystals	662.4	729.9	09.25	[40]
Er^{3+} doped Si – B – Ba – Na glass	236.0	511.7	53.89	[41]
Er^{3+}, Yb^{3+} co-doped $\beta - NaLuF_4$	270.6	2830.6	90.44	[42]
Er^{3+}, Yb^{3+} co-doped Y_2SiO_5	781.0	686.1	13.83	[43]
Er^{3+}, Yb^{3+} co-doped $BaMoO_4$	607.0	716.0	15.22	[44]
Er^{3+}, Yb^{3+} co-doped $CaWO_4$	1455.0	1530.0	04.90	[45]
Er^{3+}, Yb^{3+} co-doped $Yb_2Ti_2O_7$	478.6	482.0	0.71	[46]
Er^{3+}, Yb^{3+} co-doped $LiNbO_3$	860.0	686.2	25.33	[47]

In applications with high temperatures, phosphors are chosen according to their high thermal stability and adaptability of dopant lanthanides with the host crystal structure. Yttrium oxide (Y_2O_3) is a popular oxide material for a host crystal due to its wide optical band gap of 5.6 eV [48], high melting point, and high possible adaptation of dopant ions, such as Er^{3+} in the host crystal along with good transparency from UV to IR. Lojpur et al. [42] studied Er^{3+}–Yb^{3+} co-doped Y_2O_3. Nanoparticles for optical temperature sensing properties in the temperature range of 93 K to 613 K. They analysed the temperature-dependent fluorescence intensity ratio of the two green emissions and obtained maximum sensitivity of 528×10^{-4} K^{-1} for 150 K for $Y_{1.97}Yb_{0.02}Er_{0.01}O_3$ nano phosphors. Another case study was done by Dong et al. [49], where they studied the fluorescence intensity ratio of the green upconversion emission at 523 nm and 545 nm in Er^{3+}–Yb^{3+} co-doped Al_2O_3. The study was conducted at 495 K, and the maximum sensitivity was 0.0051 K^{-1}. Furthermore, the study was continued with other oxide materials co-doped with Er^{3+}–Yb^{3+} [45, 50–52].

5.4 THERMAL STABILITY

Thermal stability is one of the critical quality a phosphor that involves the durability, longevity, and consistency of its performance in light emitting devices in the required working temperature at the very least, and ideally at very high temperatures [53]. Figure 5.5 explains the essential parameters that define the thermal stability of a phosphor.

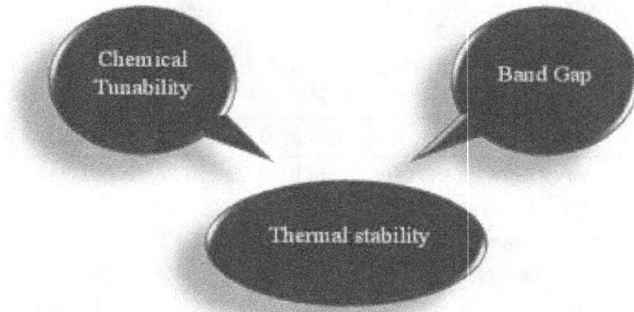

Figure 5.5 Factors affecting thermal stability of phosphors.

The host crystals in the phosphor may have similar structures, such as tetragonal, hexagonal, cubic, etc., but the characteristic emission intensity at an increasing temperature can be assigned to its electronic structure. Thus, the thermal stability of a phosphor is defined by the effective band gap, chemical tunability, and the degree of condensation [53]. Qin et al. [54] mentioned the relation of emission intensity of the $5d – 4f$ transition and temperature by an equation:

$$I(T) = \frac{I(0)}{[1 + (\tau_0/\tau_V)\exp(-\Delta E/k_B T)]} \tag{5.2}$$

where τ_V is the radiative decay rate corresponding to the state of the lanthanide ion, τ_0 is the attempt rate of the thermal quenching process, k is Boltzmann's constant, and ΔE is the energy barrier for the thermal quenching process.

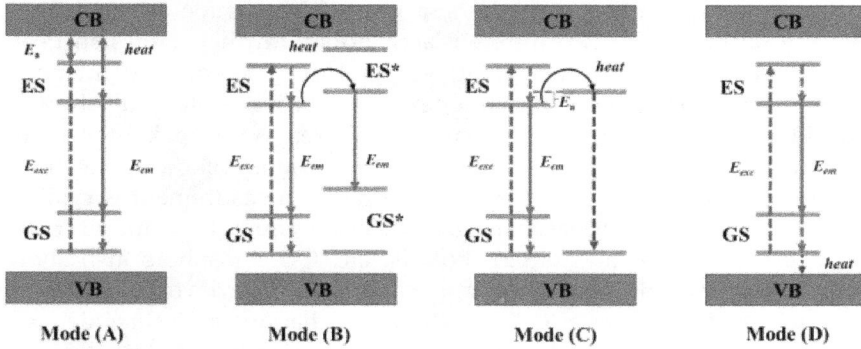

Mode (A)   Mode (B)   Mode (C)   Mode (D)

Figure 5.6 Various modes of the thermal-assisted quenching process [53].

Various modes of the thermal-assisted quenching process have been shown in Figure 5.6. The autoionisation model described by Dorenbos et al. [55] shows that the $5d$ orbital of Eu^{2+} is near the conduction band of the host lattice before the crossing point of the $4f$ and $5d$ orbitals. At high temperatures, the electrons at the $5d$ orbital get thermally activated and transferred to the conduction band. The activated electron in the $4f$ energy level releases its energy through non-radiative processes such as heat loss, traps in host lattice, and defects which can turn into quenching centres and by lattice vibration process. Thus, it was observed that the more significant the gap between the $5d$ orbital to the conduction band, the phosphor shows higher thermal stability [56, 57]; mode (B) shows that different valance states of activators can coexist at similar energy levels positions [58, 59]. During the study of [60] red emitting Eu-doped $SrLiAl_3N_4$ and Ce-doped $SrMg_2Al_2N_4$ with green emission, Leaño Jr. observed that valance states of Eu and Ce show coexistence. Mode (C) and mode (D) are respectively based on the $4f$–$5d$ crossing model and electron holes transfer from the activator's ground state to the valance band [61–63].

5.5 CONCLUSION

It is a well–a known phenomenon that the emission properties of the phosphors are highly affected by temperature. These emission properties include emission intensity, emission and excitation peak wavelength, spectrum shape, decay, and rise time. The temperature of a phosphor can be calculated by measuring one or more of these properties. The phosphors used for temperature measurements are known as thermographic phosphors and are helpful for various applications where temperature measurement is a critical parameter of the system.

In quantum-cutting phosphors, a pair of lanthanide activators are usually used. So, the two different activators with their emission peak show a response to the increasing temperature. The emission intensity of the activator with a higher energy excitation state will show a considerable increase in peak intensity compared to the peak intensity of the lower-energy excitation state. These QC phosphors show applications in many engineering fields where temperature measurement is critical, such as thermal barrier coatings, surface temperature monitoring, fluid flow analysis, and circuit boards, etc. QC phosphors also show their use in various biomedical applications (cellular level temperature measurements) such as tumour monitoring, diagnosing ischaemia, etc.

REFERENCES

[1] Khalid, A.H. and K. Kontis. 2008. Thermographic phosphors for high temperature measurements: principles, current states of the art and recent applications. Sensors 8: 5673–5744.

[2] Yadav, R.S. and R.S. Ningthoujam. 2021. Synthesis and characterization of quantum cutting phosphor materials. *In:* Handbook on Synthesis Strategies for Advanced Materials. Indian Institute of Metal Series. Springer, Singapore.

[3] Wegh, R.T., H. Donker, K.D. Oskam and A. Meijerink. 1999. Visible quantum cutting in $LiGdF_4:Eu^{3+}$ through downconversion. Science 283: 663–666.

[4] Piper, W.W., J.A. DeLuca and F.S. Ham. 1974. Cascade fluorescence decay in Pr^{3+} – doped fluorides: achievement of a quantum yield greater than unity for emission of visible light. J. Lumin. 8: 344–348.

[5] Yu, D.C., R. Martin–Rodriguez, Q.Y. Zhang, A. Meijerink and F.T. Rabouw. 2015. Multi-photon quantum cutting in $Gd_2O_2S:Tm^{3+}$ to enhance the photo response of solar cells. Light Sci. Appl. 4: e344.

[6] Yadav, R.V., R.S. Yadav, A. Bahadur and S.B. Rai. 2016. Down shifting and quantum cutting from Eu^{3+}, Yb^{3+} co-doped $Ca_{12}Al_{14}O_{33}$ phosphor: a dual mode emitting material. RSC Adv. 6: 9049–9056.

[7] Shionoya, S., H. Yamamoto amd W.M. Yen. 2008. Phosphor Handbook, 2nd Ed. CRC. Taylor & Francis Distributor, London.

[8] Wegh, R.T., H. Donker, K.D. Oskam and A. Meijerink. 1999. Visible quantum cutting in Eu^{3+} doped gadolinium fluorides via downconversion. J. Lumin. 82: 93–104.

[9] Blasse, G. and B.C. Grabmaier. 1994. Luminescent Materials. Springer–Verlag, Berlin.

[10] Henderson, B. and G.F. Imbusch. 1989. Optical Spectroscopy in Inorganic Solids. Clarendon, Oxford.

[11]   Huang, X., S. Han, W. Huang and X. Liu. 2013. Enhancing solar cell efficiency: the search for luminescent materials as spectral converters, Chem. Soc. Rev. 42: 173–201.

[12]   Yu, D.C., X.Y. Huang, S. Ye, M.Y. Peng, Q.Y. Zhang and L. Wondraczek. 2011. Three – photon near – infrared quantum splitting in β–NaYF$_4$:Ho^{3+}. Appl. Phys. Lett. 99: 161904.

[13]   Yu, D.C., X.Y. Huang, S. Ye, Q.Y. Zhang and J. Wang. 2011. A sequential two – step near – infrared quantum splitting in Ho^{3+} singly doped NaYF$_4$. AIP Adv. 1: 042161.

[14]   Chen, X.B., J.G. Wu, X.L. Xu, Z.Y. Zhang, N. Sawanobori, C.L. Zhang, et al. 2009. Three photon infrared quantum cutting from single species of rare–earth Er^{3+} ions in Er$_{0.3}$Gd$_{0.7}$VO$_4$ crystalline. Opt. Lett. 34(7): 887–889.

[15]   Miritello, M., R. Lo Savio., P. Cardile and F. Priolo. 2010. Enhanced down conversion of photons emitted by photoexcited Er$_x$Y$_{2-x}$Si$_2$O$_7$ films grown on silicon. Phys. Rev. B: Condens. Matter. 81: 041411.

[16]   Yu, D.C., S. Ye, M.Y. Peng, Q.Y. Zhang and L. Wondraczek. 2012. Sequential three – step three photon near infrared quantum splitting in β–NaYF$_4$:Tm^{3+}. Appl. Phys. Lett. 100: 191911.

[17]   Zhang, W.J., D.C. Yu, J.P. Zhang, Q. Qian, S.H. Xu, Z.M. Yang, et al. 2012. Near infrared quantum splitting in Ho^{3+}:LaF$_3$ nanocrystals embedded germanate glass ceramic. Opt. Mater. Express 2: 636–643.

[18]   van der Ende, B.M., L. Aarts and A. Meijerink. 2009. Near infrared quantum cutting for photovoltaics. Adv. Mater. 21: 3073–3077.

[19]   Eilers, J.J., D. Biner, J.T. van Wijngaarden, K. Kraemer, H.U. Güedel and A. Meijerink. 2010. Efficient visible to infrared quantum cutting through downconversion with the Er^{3+}–Yb^{3+} couple in Cs$_3$Y$_2$Br$_9$. Appl. Phys. Lett. 96: 151106.

[20]   Zheng, W., H. Zhu, R. Li, D. Tu, Y. Liu, W. Luo, et al. 2012. Visible to infrared quantum cutting by phonon assisted energy transfer in YPO$_4$:Tm^{3+}, Yb^{3+} phosphors, Phys. Chem. Chem. Phys. 14: 6974–6980.

[21]   Yu, D.C., X.Y. Huang, S. Ye. and Q.Y. Zhang. 2011. Efficient first order resonant near infrared quantum cutting in β–NaYF$_4$:Ho^{3+}, Yb^{3+}. J. Alloys Compd. 509: 9919–9923.

[22]   Nuebert, P. 1937. Device for indicating the temperature distribution of hot bodies–US Patent No. 2,071,471.

[23]   Wickersheim, K. amd R. Alves. 1979. Recent advances in optical temperature measurement. Ind. Res. Dev. 21: 82

[24]   Cates, M.R. 1984. Applications of pulsed laser techniques and thermographic phosphors to dynamic thermometry of rotating surfaces. ICALEO 84: Proceedings: Inspection, Measurement and Control 45: 4–10.

[25]   Alaruri, S., D. McFarland, A. Brewington, M. Thomas and N. Sallee. 1995. Development of a Fiber-Optic Probe for Thermographic Phosphor Measurements in Turbine Engines. Opt. Lasers Eng. 22: 17–31.

[26] Jadhav, A.P., J.H. Oh, S.W. Park, H.Y. Choi, B.K. Moon, B.C. Choi, et al. 2016. Enhanced down and upconversion emission for Li^+ co–doped Gd_2O_3:Er^{3+} nanostructures. Curr. Appl. Phys. 16: 1374–1381.

[27] Jadhav, A.P., T.D. Thi Dinh, S. Khan, S.Y. Lee, J.K. Park, S.W. Park, et al. 2016. Enhanced photoluminescence due to Bi^{3+}–Eu^{3+} energy transfer and re-precipitation of RE doped homogeneous sized Y_2O_3 nanophosphors. Mater. Res. Bull. 83: 186–192.

[28] Li, D.Y., Y.X. Wang, X.R. Zhang, K. Yang, L. Liu and Y.L. Song. 2012. Optical temperature sensor through infrared excited blue upconversion emission in Tm^{3+}/Yb^{3+} co-doped Y_2O_3. Opt. Commun. 285: 1925–1928.

[29] Zheng, K.Z., Z.Y. Liu, C.J. Lv and W.P. Qin. 2013. Temperature sensor based on the UV upconversion luminescence of Gd^{3+} in Yb^{3+}–Tm^{3+}–Gd^{3+} co-doped $NaLuF_4$ microcrystals. J. Mater. Chem. C 1: 5502–5507.

[30] Xu, W., H. Zhao, Z.G. Zhang and W.W. Cao. 2013. Highly sensitive optical thermometry through thermally enhanced near infrared emissions from Nd^{3+}/Yb^{3+} co-doped oxyfluoride glass ceramic. Sens. Actuators B: Chem. 178: 520–524.

[31] Cao, Z.M., S.S. Zhou, G.C. Jiang, Y.H. Chen, C.K. Duan and M. Yin. 2014. Temperature dependent luminescence of Dy^{3+} doped $BaYF_5$ nanoparticles for optical thermometry. Curr. Appl. Phys. 14: 1067–1071.

[32] Lojpur, V., M. Nikolic, L. Mancic, O. Milosevic and M.D. Dramicanin. 2013. Y_2O_3: Yb, Tm and Y_2O_3: Yb, Ho powders for low–temperature thermometry based on up-conversion fluorescence. Ceram. Int. 39: 1129–1134.

[33] Wang, X.F., J. Zheng, Y. Xuan and X.H. Yan. 2013. Optical temperature sensing of $NaYbF_4$:Tm^{3+} @SiO_2 core–shell micro-particles induced by infrared excitation. Opt. Express 21(18): 21596–21606.

[34] Luis, S.F.L., U.R.R. Mendoza, P.H. González, I.R. Martín and V. Lavín. 2012. Role of the host matrix on the thermal sensitivity of Er^{3+} luminescence in optical temperature sensors, 2012. Sens. Actuators B: Chem. 174: 176–186.

[35] Wade, S.A., S.F. Collins and G.W. Baxter. 2003. Fluorescence intensity ratio technique for optical fiber point temperature sensing. J. Appl. Phys. 94: 4743–4756.

[36] Xu, W., X.Y. Gao, L.J. Zheng, Z.G. Zhang and W.W. Cao. 2012. Short wavelength upconversion emissions in Ho^{3+}/Yb^{3+} co-doped glass ceramic and the optical thermometry behavior. Opt. Express 20(16): 18127–18137.

[37] Dong, B., B.S. Cao, Y.Y. He, Z. Liu, Z.P. Li and Z.Q. Feng. 2012. Temperature sensing and *in vivo* imaging by molybdenum sensitized visible upconversion luminescence of rare earth oxides. Adv. Mater. 24(15): 1987–1993.

[38] González, P.H., S.F.L. Luis, S.G. Pérez and I.R. Martín. 2011. Analysis of Er^{3+} and Ho^{3+} codoped fluoroindate glasses as wide range temperature sensor. Mater. Res. Bull. 46(7): 1051–1054.

[39] González, P.H., I.R. Martín, L.L. Martín, S.F.L. Luis, C.P. Rodríguez and V. Lavín. 2011. Characterization of Er^{3+} and Nd^{3+} doped strontium barium niobate glass ceramic as temperature sensors. Opt. Mater. 33: 742–745.

[40] Alencar, M.A.R.C., G.S. Maciel and C.B. de Araújo. 2004. Er^{3+} doped $BaTiO_3$ nanocrystals for thermometry: influence of nano environment on the sensitivity of a fluorescence–based temperature sensor. Appl. Phys. Lett. 84: 4753–4755.

[41] Li, C.R., B. Dong, C.G. Ming and M.K. Lei. 2007. Application to temperature sensor based on green upconversion of Er^{3+} doped silicate glass, Sensors 7(11): 2652–2659.

[42] Zheng, K.Z., W.Y. Song, G.H. He, Z. Yuan and W.P. Qin. 2015. Five photon UV upconversion emissions of Er^{3+} for temperature sensing. Opt. Express 23(6): 7653–7658.

[43] Rakov, N. and G.S. Maciel. 2012. Three photon upconversion and optical thermometry characterization of Er^{3+}: Yb^{3+} co-doped yttrium silicate powders. Sens. Actuators B: Chem. 164: 96–100.

[44] Soni, A.K., A. Kumari and V.K. Rai. 2015. Optical investigation in shuttle like $BaMoO_4$:Er^{3+} – Yb^{3+} phosphor in display and temperature sensing. Sens. Actuators B: Chem. 216: 64–71.

[45] Xu, W., Z.G. Zhang and W.W. Cao. 2012. Excellent optical thermometry based on short wavelength upconversion emissions in Er^{3+}/Yb^{3+} co-doped $CaWO_4$. Opt. Lett. 37: 4865–4867.

[46] Cao, B.S., Y.Y. He, Z.Q. Feng, Y.S. Li and B. Dong. 2011. Optical temperature sensing behavior of enhanced green upconversion emission from Er–Mo: $Yb_2Ti_2O_7$ nanophosphor. Sens. Actuators B: Chem. 159: 8–11.

[47] Quintanilla, M., E. Cantelar, F. Cussó, M. Villegas and A.C. Caballero. 2011. Temperature sensing with upconverting submicron sized $LiNbO_3$:Er^{3+}/Yb^{3+} particles. Appl. Phys. Express 4: 022601.

[48] Wang, W.C., M. Badylevich, V.V. Afanas'ev, A. Stesmans, C. Adelmann, S. Van Elshocht, et al. 2009. Band alignment and electron traps in Y_2O_3 layers on (100) Si. Appl. Phys. Lett. 95: 132903.

[49] Dong, B. and D.P. Liu. 2007. Optical thermometry through infrared excited green upconversion emissions in Er^{3+}–Yb^{3+} co-doped Al_2O_3. Appl. Phys. Lett. 90, 181117.

[50] Rakov, N. and G.S. Maciel. 2014. Nd^{3+}–Yb^{3+} doped powder for near-infrared optical temperature sensing. Opt. Lett. 39: 3767–3769.

[51] Mahata, M.K., K. Kumar and V.K. Rai. 2015. Er^{3+}–Yb^{3+} doped vanadate nanocrystals: a highly sensitive thermographic phosphor and its optical nano heater behavior. Sens. Actuators B: Chem. 209: 775–780.

[52] Zou, H., J. Li, X.S. Wang, Y.S. Li and X. Yao. 2014. Color tunable upconversion emission and optical temperature sensing behavior in Er – Yb – Mo co-doped $Bi_7Ti_4NbO_{21}$ multifunctional ferroelectric oxide. Opt. Mater. Express 4: 1545–1554.

[53] Leaño Jr., J.L., M.H. Fang and R.S. Liu. 2018. Review—narrow band emission of nitride phosphors for light emitting diodes: perspectives and opportunities. ESC J. Solid State Sci. Technol. 7(1): R3111–R3133.

[54] Qin, X., X. Liu, W. Huang, M. Bertinelli and X. Liu. 2017. Lanthanide activated phosphors based on $4f$–$5d$ optical transitions: theoretical and experimental aspects. Chem. Rev. 117: 4488–4527.

[55] Dorenbos, P. 2005. Thermal quenching of Eu^{2+} $5d$–$4f$ luminescence in inorganic compounds. J. Phys.: Condens. Matter. 17: 8103–8111.

[56] Liu, C., Z. Qi, C.C. Ma, P. Dorenbos, D. Hou, S. Zhan, et al. 2014. High light yield of $Sr_8(Si_4O_{12})$ Cl_8: Eu^{2+} under X–ray excitation and its temperature dependent luminescence characteristics. Chem. Mater. 26: 3709–3715.

[57] Luo, H., L. Ning, Y. Dong, A.J.J. Bos and P. Dorenbos. 2016. Electronic structure and site occupancy of lanthanide doped (Sr, Ca)$_3$(Y, Lu)$_2$Ge$_3$O$_{12}$ garnets: a spectroscopic and first principles study. J. Phys. Chem. C 120: 28743–28752.

[58] Lin, C.C., Z.R. Xia, G.Y. Guo, T.S. Chan and R.S. Liu. 2010. Versatile phosphate phosphors ABPO$_4$ in white light emitting diodes: collocated characteristics analysis and theoretical calculations. J. Am. Chem. Soc. 132: 3020–3028.

[59] Sontakke, A.D., J. Ueda, J. Xu, K. Asami, M. Katayama, Y. Inada, et al. 2016. A comparison on Ce^{3+} luminescence in borate glass and YAG ceramics: understanding the role of host's characteristics. J. Phys. Chem. C 120(31): 17683–17691.

[60] Leaño Jr., J.L., S.Y. Lin, A. Lazarowska, S. Mahlik, M. Grinberg, C. Liang, et al. 2016. Green light excitable Ce-doped nitrodomagnesoalumiate Sr[Mg$_0$Al$_2$N$_4$] Phosphor for white light emitting diodes, Chem. Mater. 28(19): 6822–6825.

[61] Blasse, G. and B.C. Grabmaier. 1994. A general introduction to luminescent materials. pp. 1–9. *In*: Luminescent Materials. Springer, Berlin, Heidelberg.

[62] Ivanovski, K.V., J.M. Ogieglo, A. Zych, C.R. Ronda and A. Meijerink. 2013. Luminescence temperature quenching for Ce^{3+} and Pr^{3+} d–f emission in YAG and LuAG. ECS Solid State Sci. Technol. 2: R3148.

[63] Kramers, M., G.O. Mueller, R.B. Mueller–Mach, H. Bechtel and P.J. Schmidt. 2010. PCT Pat. International Applications. A1. WO201013 1133.

Chapter **6**

A Review of Flexible Sensors

Surendra Maharjan and Ahalapitiya H. Jayatissa*

Nanotechnology and MEMS Laboratory,
Department of Mechanical, Industrial, and
Manufacturing Engineering (MIME),
The University of Toledo, OH 43606, USA.
Email: smaharj7@rockets.utoledo.edu

6.1 INTRODUCTION

A sensor is a device or a unit of an electronic system that perceives external information or stimulus and converts them into electrical signals. The signals are then transferred to the system or other devices. They are built to enable them to interact with the system and surroundings. Thus, it can be seen as a vital component of the perception system in electronic devices [1]. Flexible sensors have been intensively used in numerous applications, including those fields where conventional sensors cannot be imagined due to their flexibility. Advancements in flexible sensing technology can be observed due to the emerging Internet of Things (IoT) and intelligent systems and the massive demand for such systems or devices in human-machine interaction [2].

Recent research and development of flexible sensors are heading towards the innovation of novel materials, ease of mass fabrication, integration into systems, and imparting quality features. The spotlight

*For Correspondence: Ahalapitiya H. Jayatissa (ahalapitiya.jayatissa@utoledo.edu)

of flexible sensors lies in their seamless applications to soft and irregular surfaces, such as textile fabrics, artificial skins, smart tattoos, soft robotics, and wearable devices. Advances in material science and device architecture have enabled large-scale manufacturing of flexible sensors to become a realistic option, simultaneously enhancing their conductivity and deformability [3]. They can be rolled, folded, or twisted without losing functionality [4, 5]. For instance, Samsung and Huawei developed foldable phones using flexible technology [4, 6–8]. Flexible sensors can address conventional sensor issues, such as poor signal transduction, low sensitivity, and rigid nature. The primary importance of flexible sensors can also be visualised in Figure 6.1.

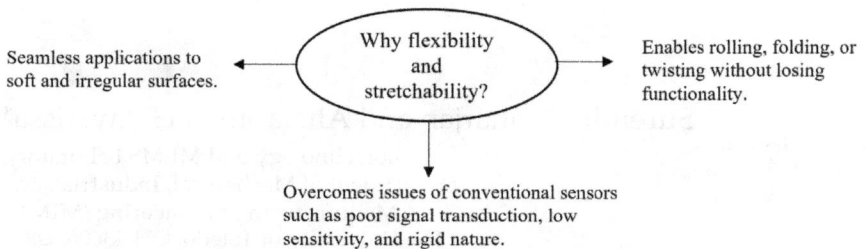

Seamless applications to soft and irregular surfaces.

Why flexibility and stretchability?

Enables rolling, folding, or twisting without losing functionality.

Overcomes issues of conventional sensors such as poor signal transduction, low sensitivity, and rigid nature.

Figure 6.1 Importance of flexible sensors in the real world.

The flexible sensors are made by coupling conductive material with a flexible substrate. Recent advances in semiconductors, such as metal oxides, silicon nanomembranes, and organic materials have enabled high-performance circuits capable of on-site sensor conditioning. Graphene, nanowires, black phosphorus [9], transition metal dichalcogenides (TMDs) [10], and non-transition metal oxides ZnO, SnO_2, In_2O_3, and Ga_2O_3 have been studied for their application in gas and strain sensors. Perovskites have been in the limelight for fabricating ultra-sensitive light sensors. Conductive or active materials are processed as nanoparticles, nanotubes, nanowires, or thin films.

Substrate materials play another significant role along with conductive materials to make feasible for flexible sensors. While selecting substrates, specific properties should be carefully considered, such as chemical resistance, thermal resistance, transparency, and flexibility. In consumer electronics, polyimide (PI), polyethylene (PE), polyethylene terephthalate (PET), polyetheretherketone (PEEK), polyethersulfone (PES), polycarbonate (PC), and polyethylene naphthalene (PEN) are popular substrate materials. Nowadays, flexible sensors are ubiquitous, and the market value is expected to rise from $3.6 billion to $7.6 billion by 2027, which is more than double within a few years [11].

6.2 WORKING MECHANISMS OF FLEXIBLE SENSORS

Flexible sensors are categorised into different groups based on their working mechanism. However, the most common types are piezoresistive, capacitive, and piezoelectric, as shown in Figure 6.2.

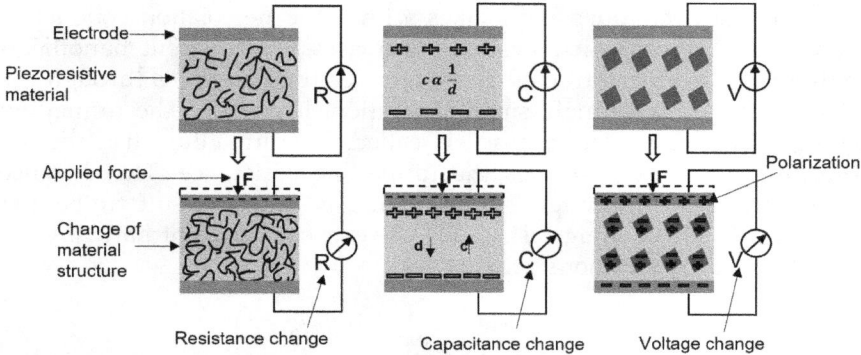

Figure 6.2 Schematics illustration of the different working mechanisms of a flexible sensor [12].

6.2.1 Piezoresistive Type

6.2.1.1 Geometrical Effect

The working mechanism can be explained as the change in resistance of a material caused by mechanical stimulus. The change of resistance can be converted into other types of electrical signals. The factors that result in resistance change depend on the property of the material and its structures, including geometrical effect, structural effect, and disconnection mechanism.

The geometrical effect illustrates the change in resistance due to a change in the geometrical dimensions of a sensor, such as a length (L) and area (A) when pressure or strain is applied. The geometrical changes appear in the form of elongation or contraction. The resistance of a conductor is given by $R = \rho L/A$, where ρ is the electrical resistivity. This effect is minimal compared to other effects for such kinds of sensors.

6.2.1.2 Structural Effect

When pressure or strain is applied to the sensor having semiconductors as an active material, the structural deformation causes a change in interatomic space, which changes the band gap, resulting in the change in resistivity of the material [13]. It illustrates the structural effect. For instance, carbon nanotube (CNT) shows extremely high resistivity change due to their chirality and change in barrier height.

6.2.1.3 *Disconnection Mechanism*

Three situations cause the disconnection between the adjacent nanoflakes: contact area change, tunnelling effect, and crack propagation. The change in the contact area between adjacent nanoflakes is dominant when the applied pressure or strain is slight, and the electrons travel through the overlapped nanoflakes within the percolation conductive network. As the applied pressure increases, the adjacent nanoflakes pull apart, creating a tunnel. However, electrons can pass through the tunnel due to exceedingly small separation. It is called the tunnelling effect, and the separation space is called the tunnelling distance. As the distance grows, so does the tunnelling resistance. The distance no electrons pass through by tunnelling is called cut-off tunnelling distance. The tunnelling resistance between two adjacent particles can be estimated by Simmons's theory [14].

$$R_{tunnel} = \frac{h^2 d}{Ae^2 \sqrt{2m\lambda}} \exp\left(\frac{4\pi d}{h} \sqrt{2m\lambda} \right) \tag{1.1}$$

Here A, e, h, d, m, and λ represent the cross-sectional area of the tunnelling junction, single-electron charge, Plank's constant, the distance between adjacent nanoflakes, the mass of an electron, and the height of energy barrier for insulators, respectively. The third one is crack propagation which occurs when the applied pressure or strain is even higher. Initially, the crack initiates and propagates along with time and pressure conditions. The separation of crack edges critically limits electrical conduction.

The two essential parameters of piezoresistive sensors are gauge factor (GF) and sensitivity, which can be calculated by equations (1.2) [15] and (1.3). The change in resistance relative to the strain determines GF. The sensitivity is defined as the ratio of resistance change to the change in pressure.

$$GF = \frac{\Delta R/R_0}{\text{Strain}(\Delta L/L_0)} \tag{1.2}$$

$$S = \frac{(\Delta R/R_0)}{\Delta P} \tag{1.3}$$

An example of a piezoresistive-based flexible sensor is an airflow sensor made by coating graphene/PVDF nanocomposite on PE material having a sensitivity of 1.21% kPa^{-1} in the pressure range of 0–2.7 kPa [29], as shown in Figure 6.3.

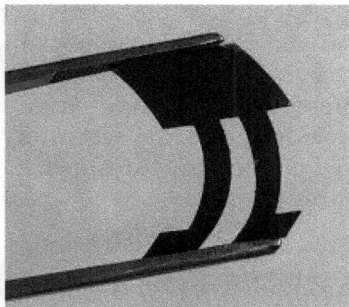

Figure 6.3 A piezoresistive airflow sensor made of graphene/PVDF nanoparticles coated on PE substrate.

6.2.2 Piezoelectric Type

The sensing mechanism of piezoelectric-type sensors is a piezoelectric effect, which means electric charge accumulates in piezoelectric materials when a mechanical stimulus is applied. The change in electrical polarisation inside the material results in a change in surface charge called voltage which can be measured by connecting two surfaces. Piezoelectric materials can be crystals, certain ceramics, and even biological matter. Quartz is the most common natural piezoelectric material, but synthesised piezoelectric materials are more efficient and are primarily ceramics based. Some examples include lead zirconate titanate (PZT), barium titanate ($BaTiO_3$), and lead titanate ($PbTiO_3$). Potassium niobate ($KNbO_3$), lithium niobate ($LiNbO_3$), lithium tantalate ($LiTaO_3$), gallium nitride (GaN), and zinc oxide (ZnO).

6.2.3 Capacitive Type

This type of sensor is based on a change in the capacitance of a capacitor under the mechanical stimulus. The most common type of parallel plate capacitor due to simple configuration and fabrication is presented by $C = kA/d$, where k, A, and d represent the medium's dielectric constant, the electrode's overlap area, and the distance between two plates, respectively. When the parallel plate electrode is stretched, the overlap area will change. When the pressure is applied, the distance between electrodes changes. As any of the above parameters change, the capacitance will change.

Sensitivity is one of the essential parameters for capacitive sensors and parallel plate capacitors. It is expressed by equation (1.4).

$$S = \frac{(\Delta C/C_0)}{\Delta P} \tag{1.4}$$

Here ΔC, C_0, and ΔP represent the relative change of capacitance, initial capacitance, and the change in capacitance when pressure is applied. The main factor that controls sensitivity is the relative change of capacitance under pressure. Many studies have shown that improving the deformation behaviour of the dielectric increases sensitivity, for which they have adopted porous structures such as sponges or foams [16] and some microstructures [17, 18].

6.3   BASIC PARAMETERS OF A FLEXIBLE SENSOR

6.3.1   Sensitivity

The slope of the sensor's response curve determines the sensitivity of a sensor. It is sometimes used as a GF. Sensitivity is a critical factor in determining sensor performance. For weak signals, such as human body signals, to be detected, sensitivity should be higher. The GF demonstrated by the flexible strain sensor based on $ZnSnO_3$ nanowires is up to 3,740, which is 19 times higher than the silicon-based strain sensor [1]. The sensitivity of flexible pressure sensors could be achieved up to 133.1 kPa^{-1} by designing hollow-sphere structured conducting polymers [19]. Another approach to increase sensitivity is using unique electrical designs such as a charge-coupled device (CCD). The pH sensor based on CCD could have sensitivity up to 240 mV pH^{-1}, which is four times higher than the value obtained by the Nernst equation at room temperature (59 mV pH^{-1}) [20].

6.3.2   Linearity

Linearity is one of the essential parameters the sensor should display the response while detecting the external stimulus to integrate with the system quickly. If the response curve is not linear, then extra signal processing circuits are needed for calibration, dramatically increasing design complexity and compensation [21]. An example of a strain sensor made by CNT thin film and PDMS as active material and dielectric, respectively, the sensor could respond near-perfect linear under strain.

6.3.3   Selectivity

Sensors should be capable of detecting selective things and giving an accurate signal. For instance, the SiO_2 sensing layer in the pH sensor interacts with hydrogen ions (H^+), which are present in hydroxyl groups (OH) when it is in contact with a certain kind of solution. Sodium (Na^+) and potassium (K^+) ions in the solution would not react with the

OH groups; thus, the measured signals from the pH sensor would not be affected by the presence of those ions. Selective detection of other biomarkers, such as glucose, would be beneficial.

6.3.4 Resolution

The sensor's capability to detect very subtle changes depends upon its resolution. Many physiological indexes vary over a limited range, and a slight change may contain much information about the physiological status. For instance, the average human body temperature is typically between 36.5 and 37.5 °C, and changes occur due to physical activity, fatigue, hormone level, etc. Minimising interference from the measurement environment and suppressing noise level could promote the sensor's resolution. For example, graphene having low thermal noise as an active material could achieve high resolution. The temperature sensor made up of a mixture of SnO_2 nanoparticles and single-walled carbon nanotubes and a graphene-based sensor (low thermal noise) could achieve higher resolution [22].

6.3.5 Detection Limit

The detection limit can be the lowest quantity of a substance the sensor can detect to meet the target requirements. For example, a flexible glucose sensor fabricated based on In_2O_3 nanoribbon FETs can achieve an ultralow detection limit of 10 nmol of glucose in biofluids on a human body surface, such as tears and sweat.

6.3.6 Durability

Durability shows a sensor's ability to perform typically for a long time without excessive repair or maintenance. It is usually measured by cyclic stability (endurance to periodic loading and unloading cycles). The sensing material or substrates are prone to buckle, fracture, and even strip after enough cycles, resulting in the cyclic unstable problem. One of the ways to promote durability is to prepare a sensor enabling self-healing ability.

6.3.7 Hysteresis and Response Time

Hysteresis is the dependence of performance on the history of the sensor, which should be reduced or avoided. The interfacial bonding between sensing material and substrate significantly affects the optimisation of hysteresis. Capacitive sensors have lower hysteresis than piezoresistive sensors as they respond immediately to the variation of overlapped areas. Response time means the speed of achieving a steady response to the external stimulus. Piezoresistive sensors display a significant response time than others due to the re-establishment of the percolation network in resistive composites.

6.4 MATERIALS AND FABRICATION TECHNIQUES

Materials used in flexible sensors are usually divided into four groups depending upon their typical roles: conductors, semiconductors, insulators, and substrates. Conductive materials are discussed based on their conductivity and transparency. Semiconductors play a role in field-effect mobility and stability while bending, insulators, or dielectrics are selected based on their dielectric constant and breakdown voltage. In contrast, substrates are highlighted based on flexibility, stretchability, surface quality, transparency, and thermal and chemical stability [2]. Figure 6.4 presents the overview of everyday materials and respective fabrication methods used in flexible sensing technology.

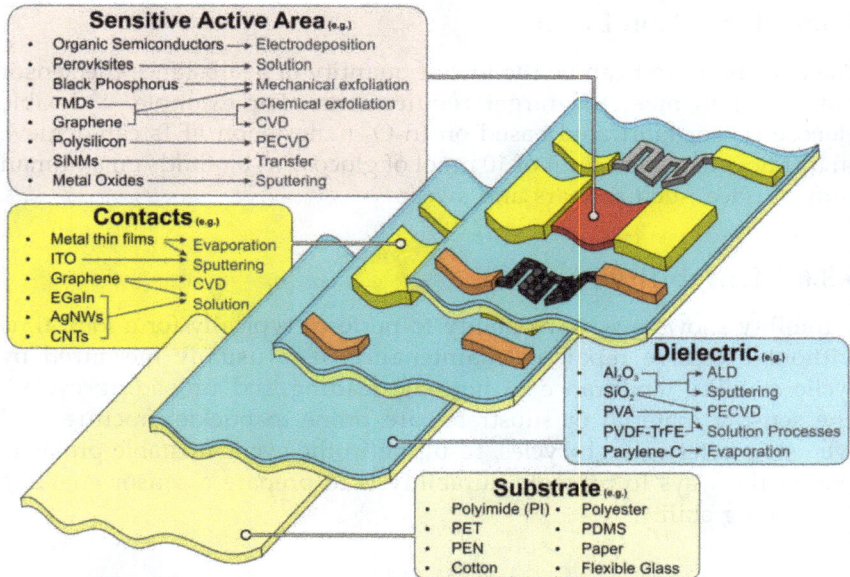

Sensitive Active Area (e.g.)
- Organic Semiconductors → Electrodeposition
- Perovksites → Solution
- Black Phosphorus → Mechanical exfoliation
- TMDs → Chemical exfoliation
- Graphene → CVD
- Polysilicon → PECVD
- SiNMs → Transfer
- Metal Oxides → Sputtering

Contacts (e.g.)
- Metal thin films → Evaporation
- ITO → Sputtering
- Graphene → CVD
- EGaIn → Solution
- AgNWs
- CNTs

Dielectric (e.g.)
- Al₂O₃ → ALD
- SiO₂ → Sputtering
- PVA → PECVD
- PVDF-TrFE → Solution Processes
- Parylene-C → Evaporation

Substrate (e.g.)
- Polyimide (PI)
- PET
- PEN
- Cotton
- Polyester
- PDMS
- Paper
- Flexible Glass

Figure 6.4 Standard materials and respective fabrication methods are used to fabricate flexible sensors [3].

6.4.1   Conductors

Usually, conductors are used as contact materials in flexible sensors. However, nanoparticles and nanowires are best suited for conductors. Some conductors provide transparency such as AgNWs. The general overview of conductors used in flexible sensors is presented in Figure 6.5.

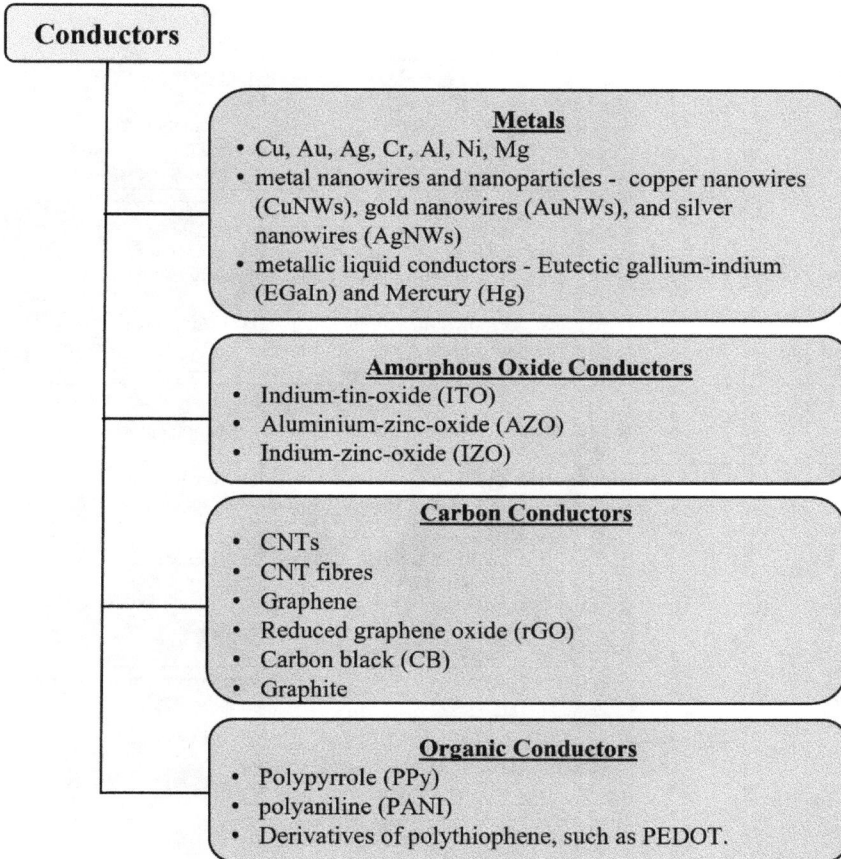

Conductors

Metals
- Cu, Au, Ag, Cr, Al, Ni, Mg
- metal nanowires and nanoparticles - copper nanowires (CuNWs), gold nanowires (AuNWs), and silver nanowires (AgNWs)
- metallic liquid conductors - Eutectic gallium-indium (EGaIn) and Mercury (Hg)

Amorphous Oxide Conductors
- Indium-tin-oxide (ITO)
- Aluminium-zinc-oxide (AZO)
- Indium-zinc-oxide (IZO)

Carbon Conductors
- CNTs
- CNT fibres
- Graphene
- Reduced graphene oxide (rGO)
- Carbon black (CB)
- Graphite

Organic Conductors
- Polypyrrole (PPy)
- polyaniline (PANI)
- Derivatives of polythiophene, such as PEDOT.

Figure 6.5   Classification of conductors used in flexible sensors.

6.4.2   Semiconductors

Semiconductors are used as sensitive active areas or sensing material. The conductivity of semiconductors lies between metal and insulators, and the transport of charge carriers depends upon covalent bonds. Figure 6.6 depicts diverse types of semiconductors used in flexible sensors. Silicon (Si) and germanium (Ge) are the most used semiconductors. However, they are doped with other elements to obtain desirable properties for sensor applications. Typically, semiconductors are compared in terms of carrier mobility, compatibility with large-area fabrication, and processing temperature. Due to their rigid nature, they are mostly preferred as thin-film transistor (TFT) channels or as sensitive layers.

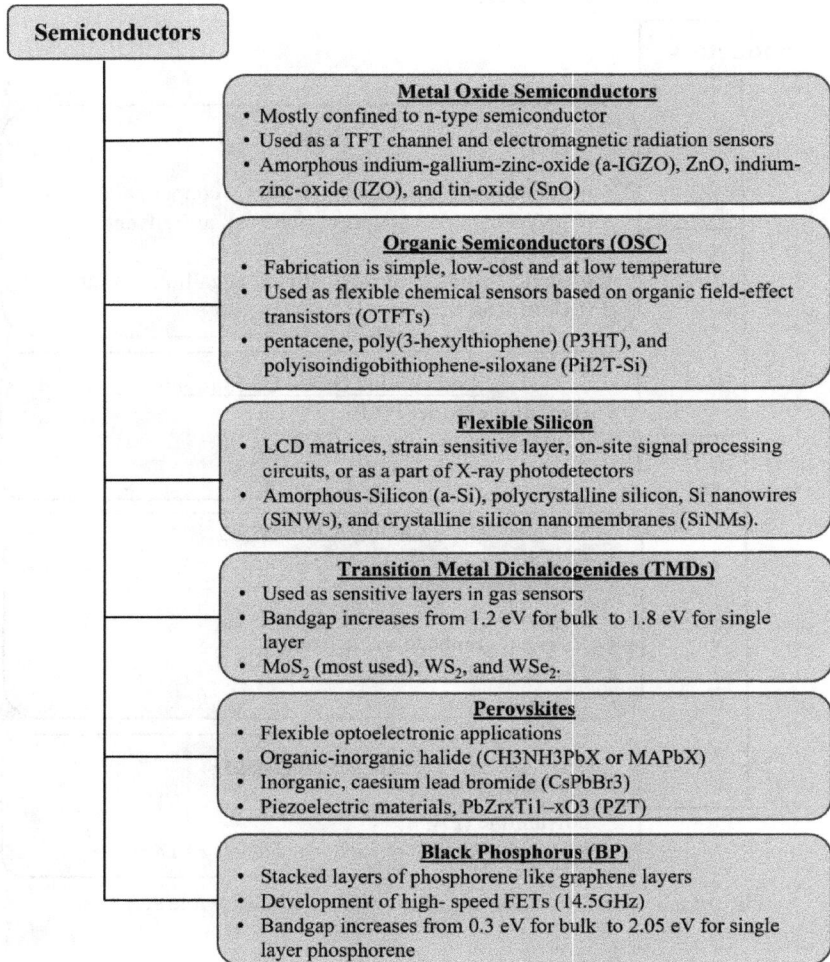

Semiconductors

Metal Oxide Semiconductors
• Mostly confined to n-type semiconductor
• Used as a TFT channel and electromagnetic radiation sensors
• Amorphous indium-gallium-zinc-oxide (a-IGZO), ZnO, indium-zinc-oxide (IZO), and tin-oxide (SnO)

Organic Semiconductors (OSC)
• Fabrication is simple, low-cost and at low temperature
• Used as flexible chemical sensors based on organic field-effect transistors (OTFTs)
• pentacene, poly(3-hexylthiophene) (P3HT), and polyisoindigobithiophene-siloxane (PiI2T-Si)

Flexible Silicon
• LCD matrices, strain sensitive layer, on-site signal processing circuits, or as a part of X-ray photodetectors
• Amorphous-Silicon (a-Si), polycrystalline silicon, Si nanowires (SiNWs), and crystalline silicon nanomembranes (SiNMs).

Transition Metal Dichalcogenides (TMDs)
• Used as sensitive layers in gas sensors
• Bandgap increases from 1.2 eV for bulk to 1.8 eV for single layer
• MoS_2 (most used), WS_2, and WSe_2.

Perovskites
• Flexible optoelectronic applications
• Organic-inorganic halide (CH3NH3PbX or MAPbX)
• Inorganic, caesium lead bromide (CsPbBr3)
• Piezoelectric materials, PbZrxTi1–xO3 (PZT)

Black Phosphorus (BP)
• Stacked layers of phosphorene like graphene layers
• Development of high- speed FETs (14.5GHz)
• Bandgap increases from 0.3 eV for bulk to 2.05 eV for single layer phosphorene

Figure 6.6 Classification of semiconductors used in flexible sensors.

6.4.3 Insulators/Dielectrics

Dielectrics used in flexible sensors are either organic materials such as polyvinyl phenol (PVP), polyvinylpyrrolidone, poly(perfluorobut enylvinylether), PDMS, PLA, polyvinylidene fluoride (PVDF), PVDF-Trifluoroethylene (PVDF-TrFE), or inorganic materials such as Al_2O_3, SiO_2, HfO_2, and TiO_2. Such dielectric materials are just part of field-effect transistors or capacitive sensors. Al_2O_3 and TiO_2 have higher dielectric constants, 9.7 and 27, respectively, when spin coating, followed by an annealing step, obtains thin films. SiO_2 presents a dielectric constant of 3.9, which is lower than some organic materials. Due to well-established properties and fabrication techniques, inorganic materials are more widely used as dielectrics than organic materials. PDMS is used as stretchable dielectrics. PVDF-TrFE presents a higher dielectric constant than Al_2O_3 when fabricated by spin coating and processed at 140°C. PVDF possesses good dielectric and piezoelectric properties.

6.4.4 Substrates

Substrates are desired to conform to soft and irregular surfaces and hold properties such as thermal and chemical resistance, extensive area compatibility, transparency, low surface roughness, and process-compatible glass transition temperature to be compatible with flexible sensor applications. Polyimide (PI), polyethylene terephthalate (PET), and polyethylene naphtholate (PEN) substrates dominate in flexible applications due to their excellent properties, whereas polydimethylsiloxane (PDMS) is widely used in the stretchable sensor. Other substrates include polyethylene (PE), polycarbonate (PC), polyvinyl alcohol (PVA), polylactic acid (PLA), polysulfide (PSU), and polyetheretherketone (PEEK).

6.5 TYPES OF FLEXIBLE SENSORS AND THEIR APPLICATIONS

Flexible sensors are categorised into strain, pressure, temperature, humidity, magnetic, chemical, electromagnetic radiation, multi-modal, and orientation.

6.5.1 Strain Sensors

Strain sensors convert a mechanical stimulus into an electrical signal. A strain sensor is characterised by its GF (sensitivity), stretchability, hysteresis, and response time. Strain sensors are categorised into resistive, capacitive, piezoelectric, and triboelectric based on their working mechanism. Materials such as ZnONWs, $ZnSnO_3$, liquid metals, metal nanowires/nanostructures, carbon black, CNT, and graphene have been used to fabricate strain sensors. These sensors are used to detect the physical dimension changes in the body or structures.

6.5.2 Pressure Sensors

Pressure sensors are among the most used sensors that detect the pressure of the human body, machinery device, or surrounding environment. The critical parameters to the performance of pressure sensors include sensitivity, detection range, and response time. Different approaches have been adopted to fabricate pressure sensors, such as the resistive, capacitive, piezoelectric, field-effect transistor, and peizocapacitive. Some examples of pressure sensors include ZnO-microparticle pressure sensors, laser-scribed graphene (LSG) pressure sensors with foam-like structures, multi-functional P(VDF-TrFE) field-effect organic transistors using a microstructured dielectric layer, and AgNWs/PDMS pressure sensor array using a stacked structure.

6.5.3 Shear Stress Sensors

Shear stress sensors are essential for monitoring and characterising the dynamic properties of the fluid. In real life, many things continuously interact with a fluid, such as an aircraft, automobiles, and fluid flow in a pipe. By determining the shear stress, the design of an object can be optimised for better performance. The critical parameters for the wall shear stress are fluid viscosity and the velocity gradient. Rigid shear stress sensors are found to be challenging to apply on uneven or curved surfaces, which the flexible type can mitigate. Hasegawa et al. [23] developed a smooth-surfaced flexible thermal shear stress sensor from thin parylene film on a micromachined PDMS structure. It was tested by placing it on the curved streamlined surface of a vehicle to measure shear stress with high accuracy.

6.5.4 Temperature Sensors

Flexible temperature sensors belong to the electrical contact sensors whose electrical properties change to temperature. They are categorised into resistive, capacitive, thermoelectric, pyroelectric, transistors, or diodes. The critical factors of a temperature sensor are temperature sensitivity, temperature range, hysteresis, and response time. Resistive thin-film temperature sensors, developed by integrating gold resistance temperature detectors (RTDs) on flexible PI substrates, are used to detect the temperature of the metal. Other materials fabricating these types of sensors include lead zirconate titanate (PZT), ZnO, PVDF, graphene, CNTs, etc.

6.5.5 Humidity Sensors

Humidity sensors measure relative humidity (RH) levels in the air. RH is the ratio of water vapour in the air to the air needed for saturation at a given temperature and pressure. The most critical parameters are sensitivity to RH, humidity range, response time, and hysteresis, and bendability, stretchability, and transparency are further investigated. Humidity sensors are categorised into resistive, piezoelectric, capacitive, TFT, and fibre optic. Some examples of humidity sensors include a humidity-reactive PTFE sensor applied on a curved surface and performance comparison before and after NaOH treatment and a GO hydrophilic quartz crystal microbalance humidity sensor.

6.5.6 Magnetic Sensors

The performance of a flexible magnetic sensor is assessed with sensitivity, magnetic field direction, and bendability. These sensors are developed by depositing stacked thin films on a flexible substrate. The approaches that are adopted while developing magnetic sensors include giant magnetoresistance (GMR), anisotropic magnetoresistance (AMR), tunnelling magnetoresistance (TMR), magnetoimpedance (MI), and hall sensor. For example, an AMR sensor, fabricated by sequentially depositing permalloy, Au, and Pt layers on a PI substrate [24] is used to detect geomagnetic fields (40–60 μT).

6.5.7 Chemical Sensors

The most essential parameters for chemical sensors are sensitivity, chemical concentration, detection range, and response time. Chemical sensors have taken multiple approaches for development including resistive chemical, electrochemical, FET chemical, and optical. Some examples of chemical sensors are a photo-induced room temperature gas sensor using an IGZO thin film transistor [25], SnO_2 flexible sensor, and CNTs field-effect transistor for enzymatic acetylcholinesterase detection [26].

6.5.8 Electromagnetic Radiation Sensors

These sensors use the photoelectric properties of semiconductors to detect electromagnetic radiation. When light hits on these materials, electrical properties will change. The critical parameters required for these sensors are sensitivity and electromagnetic response. ZnO and perovskites have dominated this field. An X-ray sensor is an example of this type of sensor.

6.5.9 Multi-modal Sensors

Multi-modal sensors can sense independent phenomena with an individual sensing component. A sensor made by PVDF with attached tetrapod ZnO (T-ZnO) nanostructures on a PET substrate [27] displays this type of sensing behaviour. This sensor can detect bending. Meantime, it is suitable for O_2 gas sensing and humidity detection.

6.5.10 Electropotential Sensors

Electropotential sensors consist of electrode and signal acquisition electronics that measure electric potentials and fields. Signal acquisition electrodes are the critical factor for the performance of electropotential sensors. These sensors are categorised into resistively coupled electrodes (direct conductive contact with the skin) and capacitively coupled electrodes (not direct conductive contact with the skin). Flexible resistive coupled electrodes are fabricated by Ag/AgCl conductive inks, Au, or graphene. Copper electrodes fabricate a capacitively coupled electrode on a PI substrate, and the textile is used as the dielectric [28]; for example, a washable textile electrode for ECG monitoring.

6.5.11 Orientation Sensors

Orientation sensors measure the tilt angle of an object in 3D space. The critical parameters for the better performance of these sensors are determined by resolution and hysteresis. The conductive microspheres create an electrical connection between the contact and ground pads based on the sensor orientation. Such sensors are used in gaming platforms.

6.5.12 Ultrasonic Sensors

These sensors perform non-destructive imaging of 3D objects in the free space by detecting reflected pulse-echo ultrasound waves. Flexible electronics contribute to solidly adhering to irregular and non-planar surfaces, improving test results' reliability. The resonant frequency is the critical parameter to determine the performance of such sensors. A piezoelectric, stretchable ultrasonic array of anisotropic PZT elements and flexible capacitive micromachined ultrasonic transducers (CMUTs) are some materials used for these applications.

6.6 SUMMARY

Flexible sensors are widely used in various sectors, such as artificial intelligence, soft robotics, biomedical devices, human-machine interface, etc., due to their advantages over conventional sensors. The implementation on soft and irregular surfaces and the capability to twist, turn and roll without change in functionality made humans invent new things that seemed impossible before. Most of the sensors work on piezoresistive, piezoelectric, and capacitive principles. While developing the sensors, the basic parameters of the sensors, such as sensitivity, linearity, selectivity, resolution, durability, detection limit, hysteresis, and response time, should be analysed appropriately to make them viable for commercial use. The four main types of materials play a significant role in designing and developing a sound sensor. For instance, conductors are used as contact material. Semiconductors are used as sensitive active areas. Dielectrics are used to isolate substrate with conductive materials, whereas substrates provide flexibility to the sensor. Flexible substrates are conductive by pasting nanoparticles, nanowires, or nanotubes either by coating or 3D printing. Some flexible sensors work based on physical parameter change, some on chemical parameters, and some on optical. Extensive research is going on in this field to reduce cost, size, and weight and find new possibilities.

REFERENCES

[1] Wu, J.M., C.Y. Chen, Y. Zhang, K.H. Chen, Y. Yang, Y. Hu, et al. 2012. Ultrahigh sensitive piezotronic strain sensors Based on a $ZnSnO_3$ nanowire/microwire. ACS Nano 6: 4369–4374.

[2] Li, X. and Y. Chai. 2021. Design and applications of graphene-based flexible and wearable physical sensing devices. 2D Mater. 8: 022001.

[3] Costa, J.C., F. Spina, P. Lugoda, L. Garcia-Garcia, D. Roggen and N. Münzenrieder. 2019. Flexible sensors—from materials to applications. Technologies 7: 35.

[4] Weiser, M. 1999. The computer for the 21st century. ACM SIGMOBILE Mobile Computing and Communications Review 3: 3–11. Accessed: (2022).

[5] Bauer, S., S. Bauer-Gogonea, I. Graz, M. Kaltenbrunner, C. Keplinger and R. Schwödiauer. 2014. 25th anniversary article: a soft future: from robots and sensor skin to energy harvesters. Adv. Mater. 26: 149–162.

[6] Myny, K. 2018. The development of flexible integrated circuits based on thin-film transistors. Nat. Electron. 1: 30–39.

[7] Nathan, A., A. Ahnood, M.T. Cole, S. Lee, Y. Suzuki, P. Hiralal, et al. 2012. Flexible electronics: the next ubiquitous platform. Proc. IEEE, Vol. 100, no. Special Centennial Issue: 1486–1517.

[8] Featherstone, D.J., R.J. Werner, C.A. Camarce and S.E. Cullen. 2014. Flexible display patent landscape and implications from the america invents act. Nanotech. L. & Bus. 11: 181–194.

[9] Coleman, J.N., M. Lotya, A. O'Neill, S.D. Bergin, P.J. King, U. Khan, et al. 2011. Two-dimensional nanosheets produced by liquid exfoliation of layered materials. Science. 331: (80) 568–571.

[10] Zhang, X., H. Xie, Z. Liu, C. Tan, Z. Luo, H. Li, et al. 2015. Black phosphorus quantum dots. Angew. Chemie. Int. Ed. 54: 3653–3657.

[11] Das, R. and X. He. 2018. Flexible, Printed and organic electronics 2019-2029: forecasts, players & opportunities. IDTechEx.

[12] Liu, D. and G. Hong. 2019. Wearable electromechanical sensors and its applications, in wearable devices—the big wave of innovation. Intech. Open 13.

[13] Yin, B., X. Liu, H. Gao, T. Fu and J. Yao. 2018. Bioinspired and bristled microparticles for ultrasensitive pressure and strain sensors. Nat. Commun. 9: 5161.

[14] Amjadi, M., A. Pichitpajongkit, S. Lee, S. Ryu and I. Park. 2014. Highly stretchable and sensitive strain sensor based on silver nanowire–elastomer nanocomposite. ACS Nano 8: 5154–5163.

[15] Boland, C.S. 2019. Stumbling through the research wilderness, standard methods to shine light on electrically conductive nanocomposites for future healthcare monitoring. ACS Nano 13: 13627–13636.

[16] Yang, Y.F., L.Q. Tao, Y. Pang, H. Tian, Z.Y. Ju, X.M. Wu, et al. 2018. An ultrasensitive strain sensor with a wide strain range based on graphene armour scales. Nanoscale 10: 11524–11530.

[17] Cai, Y., J. Shen, G. Ge, Y. Zhang, W. Jin, W. Huang, et al. 2018. Stretchable $Ti_3C_2T_x$ MXene/Carbon nanotube composite based strain sensor with ultrahigh sensitivity and tunable sensing range. ACS Nano 12: 56–62.

[18] Ma, Y., Y. Yue, H. Zhang, F. Cheng, W. Zhao, J. Rao, et al. 2018. 3D synergistical mxene/reduced graphene oxide aerogel for a piezoresistive sensor. ACS Nano 12: 3209–3216.

[19] Pan, L., A. Chortos, G. Yu, Y. Wang, S. Isaacson, R. Allen, et al. 2014. An ultra-sensitive resistive pressure sensor based on hollow-sphere microstructure induced elasticity in conducting polymer film. Nat. Commun. 5: 3002.

[20] Nakata, S., M. Shiomi, Y. Fujita, T. Arie, S. Akita and K. Takei. 2018. A wearable pH sensor with high sensitivity based on a flexible charge-coupled device. Nat. Electron. 1: 596–603.

[21] Hille, P., R. Höhler and H. Strack. 1994. A linearisation and compensation method for integrated sensors. Sens. Actuators, A: Phys. 44: 95–102.

[22] Cheng, Z., Q. Li, Z. Li, Q. Zhou and Y. Fang. 2010. Suspended graphene sensors with improved signal and reduced noise. Nano Lett. 10: 1864–1868.

[23] Hasegawa, Y., C. Okihara, T. Yamada, K. Komatsubara, K. Iwano, Y. Sakai, et al. 2017. Smooth-surfaced flexible wall shear stress sensor fabricated by film transfer technology. Sens. Actuators, A: Phys. 265: 86–93.

[24] Yu, J., X. Tang, H. Su and Z. Zhong. 2020. A self-biased linear anisotropic magnetoresistance sensor realised by exchange biased bilayers. J. Magn. Magn. Mater. 493: 165695.

[25] Knobelspies, S., B. Bierer, A. Daus, A. Takabayashi, G.A. Salvatore, G. Cantarella, et al. 2018. Photo-induced room-temperature gas sensing with a-IGZO based thin-film transistors fabricated on flexible plastic foil. Sensors 18: 358.

[26] Bhatt, V., S. Joshi, M. Becherer and P. Lugli. 2017. Flexible, low-cost sensor based on electrolyte gated carbon nanotube field effect transistor for organo-phosphate detection. Sensors 17: 1147.

[27] He, H., Y. Fu, W. Zang, Q. Wang, L. Xing, Y. Zhang, et al. 2017. A flexible self-powered T-ZnO/PVDF/fabric electronic-skin with multi-functions of tactile-perception, atmosphere-detection and self-clean. Nano Energy 31: 37–48.

[28] Lee, S.M., K.S. Sim, K.K. Kim, Y.G. Lim and K.S. Park. 2010. Thin and flexible active electrodes with shield for capacitive electrocardiogram measurement. Med. Biol. Eng. Comput. 48: 447–457.

Chapter 7

The Transition from Pb- to Pb-Free Halide-Based Perovskite Inks for Optoelectronic Applications

Sonali Mehra[1,2], A.K. Srivastava[1,3] and
Shailesh Narain Sharma[1,2]*

[1]CSIR- National Physical Laboratory,
Dr. KS Krishnan Marg, New Delhi - 110012, India.

[2]Academy of Scientific and Innovative Research (AcSIR),
Ghaziabad - 201002, India.

[3]CSIR- Advanced Materials and Processes Research Institute,
Bhopal - 462026, India.

7.1 INTRODUCTION

OIMH perovskites are highly emphasized instead of their characteristic properties of wide absorption-coefficient, band gap tunability, high carrier mobility, longer carrier diffusion length, high charge transport properties, and weak exciton binding. OIMH perovskites are composed of the general formula ABX_3, which is composed of three different species, where A is an organic and inorganic monovalent cation [for

*For Correspondence: Shailesh Narain Sharma (shailesh@nplindia.org)

instance, methylammonium ($CH_3NH_3^+$), formamidinium ($CH_3(NH_3)_2^+$) and cesium (Cs^+)], B is a divalent cation [for instance lead (Pb^{2+}), tin (Sn^{2+}), germanium (Ge^{2+}), and bismuth (Bi^{2+})], and X is a monovalent halide anion (for instance, Cl^-, Br^-, I^-) [1]. These halide perovskites are known for their diverse applications in solar harvesting, lasing, quantum dots, light emission, water splitting, and thin-film electronics [2]. The properties of these perovskite materials can vary by modifying A, B, or X site in the ABX_3 perovskite structural lattice. Generally, perovskites possess a cubic structure consisting of a closely-packed sub-lattice of AX_3 with divalent B-site cations in the six-fold coordinated cavity. B-X bonding of perovskite rules over the electronic behavior of perovskite semiconductors, while A cation has no direct rule for the electronic properties and its size leads to symmetry disruption of the material [3]. Perovskites also possess cubic, tetragonal, and orthorhombic phases. Even organic-inorganic perovskite materials show different optical energy gaps than their bulk counterpart because of the quantum-confinement effect [4].

Other than the properties mentioned above, perovskite materials are easily solution-processable, making them a potential candidate for optoelectronic applications with enhanced power conversion efficiency. Perovskites emit light very strongly that can be tuned in visible to IR region, which makes them ideal for making LEDs and other optoelectronic applications. Perovskite inks have a significant role in displays and solid-state lightning with improved efficiency and advantages, such as color gamut and low material cost [5].

Despite the high performance, OMH perovskites are facing stability issues. These OMH perovskites possess very low thermal decomposition temperatures due to unstable organic monovalent cations, such as $MAPbI_3$ materials that undergo poor water, oxygen, and thermal stability and decompose rapidly within 30 minutes at 150 °C in the air. At the same time, α-$FAPbI_3$ is comparatively more stable, and its black phase is stable only at a temperature of more than 160 °C, while it forms a yellow non-perovskite δ phase at room temperature or below the phase transition temperature [6].

The mixed-halide (Br^- and I^-) perovskites show a high degree of segregation under the illumination of the full sun, which leads to poor light stability, but this issue can be minimized by incorporating small inorganic cesium (Cs) ions. Other than decomposition temperature, moisture [7], light-induced [8] trap-state formation, and halide segregation are also considered essential degradation factors for the OMH perovskites. While inorganic materials possess higher stability than organic materials, specifically at higher temperatures, thus substituting the inorganic cations in place of organic cations in the perovskite lattice is the best way to increase the stability of perovskites.

However, due to high toxicity and low stability, the inorganic lead (Pb) cation is getting replaced by another cation of equivalent electronic configuration. Here, the main disadvantage of Pb perovskites is their degradation in the presence of water leading to the formation of soluble PbI_2. Soluble PbI_2 is highly toxic to humans and the natural environment [9] because it gets consumed into the human blood, causing harmful impacts on the central nervous system, cardiovascular system, kidneys, and immune system. Due to high levels of Pb-toxicity, nowadays, Pb-free perovskites are an emerging field in optoelectronics [10–12]. Thus, the need to develop alternate Pb-free perovskites arises [13, 14], such as $CsSnI_3$, $MASnI_3$, $FABiI_3$, $FABiICl_2$, $FABiClBr_2$, etc. [15]. The non-toxic and environmental-friendly perovskites can be synthesized by replacing Pb with non-toxic elements [16, 17], which can be achieved using the following two techniques:

1. Homovalent substitution
2. Heterovalent substitution

Homovalent Substitution

Homovalent elements with +2 stable oxidation-states are the best alternative for Pb-free perovskite materials. For instance, Sn^{2+} and Ge^{2+} also belong to group 14 as of Pb^{2+}, and they are the best alternate for Pb-substitution. Besides, transition metals, such as Cd^{2+}, Mn^{2+}, Fe^{2+}, Cu^{2+}, and Zn^{2+}, alkaline-earth metals, such as Ba^{2+}, Sr^{2+}, and Ca^{2+}, and rare-earth elements, such as Eu^{2+} and Yb^{2+}, could be advised for Pb-free perovskites [16], while following tolerance factor calculations Pb-substituents were perfect for perovskite structure. However, some elements such as Ba^{2+}, Sr^{2+}, and Ca^{2+} have a larger band gap, which makes them unsuitable for semiconducting materials. In contrast, working with Cu^{2+} and Zn^{2+} in ambient conditions per the perovskite requirement is not easy. Thus, Sn^{2+} and Mn^{2+} are the most favorable candidate for homovalent substitution in perovskite quantum dots.

Heterovalent Substitution

This is a second feasible substitution method for Pb-free perovskite materials. Here, the Pb^{2+} can be replaced with a cation in different valence states, e.g., mono-, tri, or tetra-valent cation as well as the direct substitution with heterovalent-cations is viable due to their existence in different valence states. Thus, two approaches for heterovalent substitution can be given:

(i)   Mixed-valence approach, where the equal number of mono- and trivalent cations give an average valence state of +2; for instance, Pb^{2+}, thallium, and gold-halide perovskites.

(ii) Heterovalent substitution of Pb^{2+} with trivalent cations like Sb^{3+} and Bi^{3+}. It is also accompanied by the appreciable variation in perovskite structure from ABX_3-type to $A_3B_2X_9$-type to maintain the charge-neutrality [14, 18].

In spite, organic halide perovskites (OHP) also faced photo and thermal instability issues due to the presence of organic groups, which were sensitive to oxygen and moisture and sustained the particular environment restrictions for storage, fabrication, and device operation using these perovskite inks. Thus, the environmental stability and the stability of all these OHP in electrical and optical properties can be improved by substituting the organic methylammonium (MA) group with inorganic cations, like cesium (Cs^+), and formamidinium (FA^+) by surface modification with moisture tolerant molecules, by integrating OHP with polymers, and by substituting organic carrier-transport layers with inorganic oxides. However, replacing these organic cations with inorganic cations leads to the formation of inorganic halide perovskites (IHP) [19, 20]. However, various methods were used to improve efficiency. However, still, the low stability of organic groups in metal-halide perovskites (MHP) hinders the commercialization of perovskite LED devices, as shown in Figure 7.1. So, the low stability of perovskite LEDs is attributed to the intrinsic instability of the operation of PeLEDs as well as of MHP materials. Thus, the inorganic materials in these IHP have a higher melting point (>500 °C) and improved photo-stability, which makes them a promising candidate for optoelectronic applications in LEDs. According to previous literature, these perovskite inks possess enhanced photoluminescence quantum yield (PLQY), narrow line width, wide color gamut, and PL emission in the broad spectral region. However, as in the case of $CsSnI_3$, Sn^{2+} can be easily oxidized into Sn^{4+} [21] and its maximum reported PLQY is below 1%, but Cs_2SnX_6 perovskite possesses good stability compared with other IHP but still suffers from high PLQY. However, for IHP, Mn^{2+} can be doped into Pb-based perovskites, but until Pb^{2+} is replaced entirely, it can only attain half PLQY in comparison to other Pb-based perovskites [22–25]. Similarly, bismuth (Bi^{2+}) and lead (Pb^{2+}) are adjacent periodic elements of period-6, and Bi is not as toxic as Pb. Here, Bi ($6s^2$ and $6p^3$) has a similar state of lone pair $6s^2$ as that of Pb ($6s^2$ and $6p^2$) and exhibits the same electronic properties and comparable energy levels with Pb. Bi-based perovskites show a different configuration than Pb-based as $A_3Bi_2X_9$ configuration to balance the charge, and Bi^{3+} produces a layered form of vacancy-ordered perovskite unit cell with $2/3^{rd}$ of octahedral position fully occupied. It results in the variation of 3D to 2D crystal dimensions, while OIH Bi-based perovskites $MA_3Bi_2X_9$ QDs have a 2D structure and exhibit good ethanol stability and PLQY of 12% with emission at 423 nm in the blue region. On the other hand, air-stable inorganic

Bi-based $Cs_3Bi_2Br_9$ perovskites exhibit the emission in blue region at 410 nm with PLQY of 19.4%, but still, this PLQY value in Bi-based perovskites is lower as compared to Pb-based perovskites [25]. Thus, it is observed that substituting lead with Sn^{2+}, Mn^{2+}, Zn^{2+}, and Bi^{2+} homovalent cations reduces the toxicity of perovskite material and helps enhance the optical performance of LEDs. Pb-free perovskite quantum dots (PeQDs) based LEDs were useful for environment-friendly and highly performing Pe ink-based LEDs. Table 7.1 gives the comparison between inorganic metal-halide perovskites in respect of their shape and emission spectra.

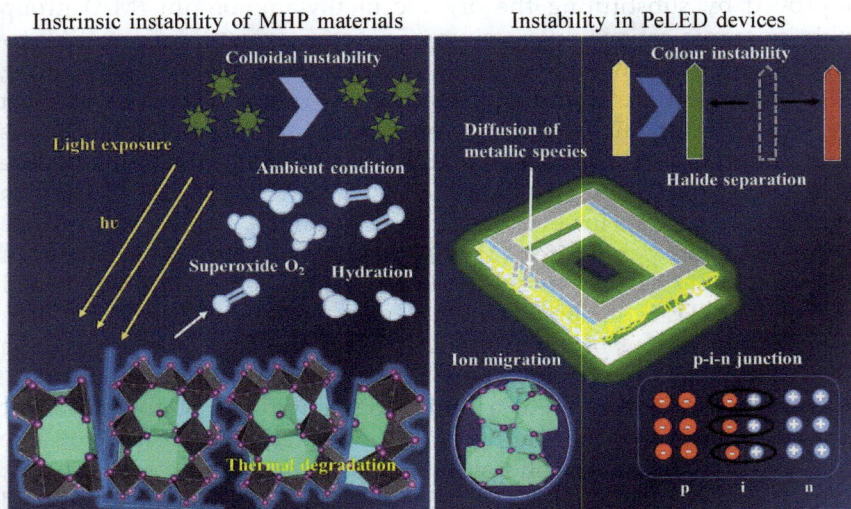

Figure 7.1 Diagrammatic representation of the intrinsic stability of inorganic-organic metal-halide perovskite materials and instability in Pe-based devices [4].

It was observed that these hybrid Pb-free OIMH perovskite inks are a far better option than conventional Pb-based perovskite quantum nanocrystals [26]. Due to these unique optical properties, these perovskite inks are observed to be a suitable candidate for advanced PL-based light-emitting devices with high efficiency, high color purity, and broad wavelength tunability. Generally, all compounds with ABX_3 geometry cannot be defined as a perovskite; the criteria for the perovskite structure are as follows:

1. Charge-neutrality: Here, cations and anions charge would be equal. For ABX_3 structure, A and B are monovalent (+1) and divalent (+2) cations, respectively (total= +3), while anion-X has a −3 charge, which implies that both the cations and anions have equal charge [27].

2. Goldschmidt tolerance factor, t and Octahedral factor, μ [28].

Here, t and μ are important factors that decide the existence and structure of perovskite material [29, 30]. t and μ are defined according to the given equation:

$$t = \frac{R_A + R_X}{2^{\frac{1}{2}}(R_X + R_B)} \text{ and } \mu = \frac{R_B}{R_X}$$

R_A, R_B, and R_X denote the ionic radii of A, B, and X ions in the above equation [28, 31, 32].

To maintain the symmetry of the perovskite structure, the tolerance factor and octahedral factor values should lie between the given limit: $0.8 < t < 1.1$ and $0.442 < \mu < 0.895$ [33, 34]. Or else, the cubic structure of perovskite will be exhausted, and as the Cs^+ ion complies well with the requisites of t; thus, it is the most suitable alternate for replacing MA^+ (inorganic metal halide) with the inorganic monovalent cation [35]. Therefore, the size of A, B, and X ions play a significant part in the perovskites material, and the t factor is also essential to determine the Pb-free perovskites material depending on the ionic radius of the implicated ions [36–39].

7.2 SYNTHESIS METHODS

Generally, the hybrid OIMH perovskites were synthesized using a colloidal or solution-reaction process that exhibits high crystallinity and quantum-confinement effect [40]. These colloidal processes generally used for the synthesis of perovskites include:

7.2.1 Ligand-Assisted Reprecipitation Method (LARP)

In the LARP technique, two precursor solutions were mixed, which induces supersaturated precipitation at room temperature and is poured into a non-polar or wrong solvent, such as toluene (as shown in Fig. 7.2); the resulting precipitate will not get dispersed into the non-polar solvent [41]. For instance, to synthesize $MAPbBr_3$ QDs [42], the precursors lead bromide, methylammonium bromide, n-octylamine (OM), and oleic acid (OA) were mixed and dispersed in dimethylformamide (DMF) to obtain a clear precursor solution; an optimized quantity of this solution was further added to toluene with vigorous stirring, thus obtaining a yellowish-green colloidal solution which indicated the formation of nanoparticles. However, as-synthesized $MAPbBr_3$ perovskites have a longer recombination lifetime [43], which is associated with surface-passivation using octyl ammonium bromide capping-ligand and reduced defect density in the nanoscale single crystals. Even $MAPbBr_3$ perovskites became resistant to moisture using long-chain capping ligands, which helps reduce the probability of water-induced degradation. We can also synthesize colloidal $MAPbX_3$, $CsPbX_3$ and $FAPbX_3$ using the LARP technique [26, 44] but after certain modifications in a process. Here, we work to avoid using organic solvents, like oleic acid [45] and octadecene (ODE), because excessive use of these organic solvents will cause difficulties in charge transport mechanism during device fabrication. Thus, the LARP method is suitable for introducing good solvents into the wrong solvent and obtaining clear solutions. This LARP technique is easily process-able, producing QDs with high luminescence and QY at room temperature. But, the QDs synthesized by the LARP method were not easy to separate from the colloidal solution because of their non-stability in various non-polar solvents like DMF, ethanol, methanol, etc., that show lower QY and also causes hindrance in the performance of the photoelectric devices [45].

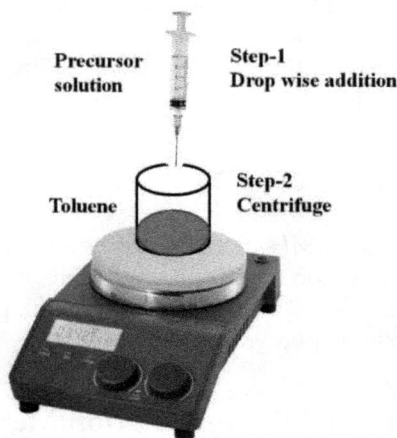

Figure 7.2 Ligand-assisted reprecipitation method (LARP).

7.2.2 Hot-Injection Method with Centrifugation or Solvothermal Synthesis

In the hot-injection method, Pb-based and Pb-free perovskite QDs were synthesized with a high degree of compositional band gap engineering. Figure 7.3 shows the schematic for Hot-injection method, where the Cs-oleate precursor was injected into the precursor solution of PbX_2 containing the hot, high boiling point solvents under an optimized temperature and inert-gas atmosphere using OA and OM as the capping ligands to dissolve the Pb-halide sources and to stabilize the QDs [3, 46, 47]. Then, the colloidal inorganic perovskite QDs can be synthesized by cooling the solution in an ice-water bath. In this rapid process, the reaction temperature and halide elements are major controlling factors for tuning the emissive color and size of QDs. These QDs crystallize in the cubic phase rather than the tetragonal or orthorhombic phase at high temperatures. After synthesizing perovskite QDs, these QDs were purified by centrifugation at 8,000 rpm for 10 minutes to remove the unreacted precursors. Further, these precipitates were re-dispersed in hexane and again centrifuged at a speed of 3,000 rpm to eliminate larger particles [48]. These QDs show bright PL with higher QY and long-term stability.

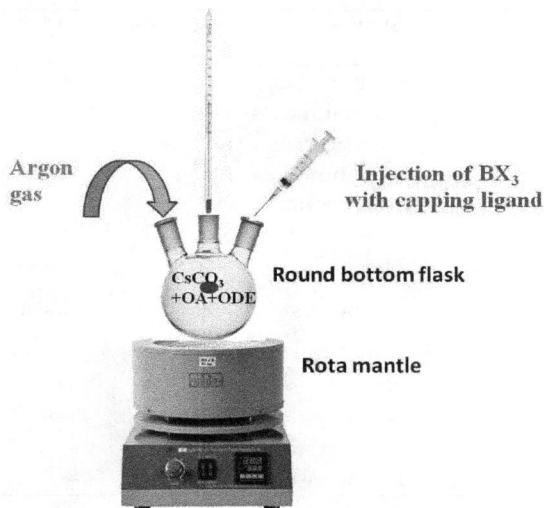

Figure 7.3 Hot-Injection Method.

Similarly, in a solvothermal reaction, cesium carbonate, lead halide, octadecene, oleic acid and octyl-amine were added in an autoclave and put in the oven for 30 minutes to complete the reaction process. The obtained product was washed with hexane and collected by centrifugation for further characterization. This method is known for controlling the size of QDs because this method is characterized by long reaction times [49]. So, to obtain the homogeneous QDs, the particles were purified and segregated to remove the large particles, which generated a tremendous amount of waste-product in this technique. Due to various characteristics, such as easy-setup, controlled morphology and composition, high uniformity, and high crystallinity of products over other methods, this method is widely used for the preparation of perovskites NCs [50].

It can be concluded that these techniques can synthesize perovskite QDs to increase mass production quickly as LARP and emulsion routes operate at room temperature owing to the easy synthesis technique and small reaction-time. Although in the LARP method, due to low temperature, QDs synthesized by this method are of low crystallinity and stability. On the other hand, QDs synthesized using the hot-injection method possess comparatively high crystallinity because of high reaction temperature and uniform particle size. Therefore, ignoring the complicated sequence, the hot-injection technique is known as the most specific and reliable method for synthesizing perovskite QDs.

Table 7.1 Comparison between different inorganic metal-halide perovskites to their shapes and emission spectra [4]

S.No.	Materials	Shapes	PL Peaks (nm)	PLQY (%)	References
1.	$CsPbX_3$	Nanocrystals	410–700	90	[51]
2.	$CsPbX_3$	Quantum dots	410–700	72	[34]
3.	$CsPbBr_3$–$CsPb(Br/I)_3$	Nanocrystals	520–580	75	[52]
4.	$CsPbX_3$	Quantum dots	Blue-red	70–95	[53]
5.	$CsSnX_3$	Nanocrystals	607–696	~	[54]
6.	$CsPbBr_3$/CdS	Quantum dots	514	88	[55]
7.	$CsPbX_3$	Nanorods	505	34	[56]
8.	$CsPbX_3$	Nanowires	Entire visible spectrum	20–80	[49]
9.	$CsPbBr_3$	Nanowires	442	30	[34]
10.	$CsPbBr_3$	Nanoplates	452	33	[57]
11.	$CsPbBr_3$	Nanocrystals	390–660	1–78	[58]
12.	$CsPbI_3$	Nanocrystals	648–692	~	[59]
13.	$CsPbCl_3$	Nanowires	410–460	~	[60]

7.3   A BRIEF REVIEW OF THE WORK ALREADY BEING DONE

Pal et al. [61] reported the colloidal synthesis of Pb-free $Cs_3Sb_2I_9$ nanoplatelets (NPLs) and $Rb_3Sb_2I_9$ nanorods (NRs) for the first time along with the characterization of their crystal structure, morphological characteristics, capping ligands, and thermal stability for the optimization of their properties. He observed that these Pb-free perovskites have similar optical and optoelectronic properties as that of colloidal $CsPbX_3$ nanocrystals (NCs). Nevertheless, Sb-halide perovskite NCs are more prone to defects than Pb-halide-based perovskite NCs (PeNCs) within the band gap. Thus, the defect chemistry of Sb-halide-based PeNCs should be controlled for comparison with Pb-halide-based PeNCs.

Protesescu et al. [51] studied the synthesis of monodisperse $CsPbX_3$ nanocrystals in the size range of 4–15 nm possessing a cubic shape of the perovskite crystal structure. These nanocrystals exhibit size-tunability of band gap in the visible spectral range from 410–700 nm because the exciton Bohr diameter ranges up to 12 nm. Here, the size of these nanocrystals can be varied from 4 nm to 15 nm using a reaction temperature of 140–200°C instead of the growth-time. As $(Cl/Br)_3$ and $(Br/I)_3$ mixed-halide perovskites were easily synthesized using the optimized amount of PbX_2 salts, while $(Cl/I)_3$ perovskites cannot synthesize with much ease due to the significant difference in the ionic radius of Cl^- and I^- ions that also correlates with phase diagram of bulk materials. $CsPbX_3$ materials are ionic, stoichiometric and highly-arranged due to the difference in the size and charge of Cs^+ and Pb^{2+} ions. These obtained chalcogenide QDs are bright (50–90%), stable, spectrally-narrow, and broadly-tunable photoluminescent for optoelectronic applications.

Jellicoe et al. [54] stated the synthesis of perovskite $CsSnX_3$ (X = I, Br, Cl, Br/I, and Cl/Br) NCs and studied the tunability of optical band gap in NIR and visible spectral region using quantum-confinement effect and varying the halide composition in perovskite lattice. This research elaborated on the various techniques to produce mixed-halide perovskites, either by synthesizing them using different halide sources or by using a post-synthesis anion exchange approach through mixing as-synthesized pure halide perovskite particles. However, the band gap of Sn-containing PeNCs shows the red-shift compared to lead-based perovskites due to higher electronegativity of Sn-ion in place of Pb-ion at B-site in ABX_3 perovskite structure.

Shen et al. [62] studied the synthesis of Pb-free $FA_3Bi_2Br_9$ PeQDs via the ligand-assisted reprecipitation (LARP) technique. These PeQDs exhibit bright blue-emission at 437 nm with PLQY of 52%, while Pb-free Sn and Bi-halide-based ($Cs_3Bi_2Br_9$ and $MA_3Bi_2Br_9$) PeQDs suffer from low PLQY. This configuration of Bi-based perovskites maintains charge balance because Bi^{3+} produces a layered form of vacancy-ordered perovskite unit cell with 2/3rd of octahedral position fully occupied, which leads to the reduction from 3D to 2D crystal dimensions. Due to the Br-rich component, these QDs possess good air stability and exceptional ethanol stability, although the PLQY value of $FA_3Bi_2Br_9$ PeQDs is highest compared to the other Pb-free ($Cs_3Bi_2Br_9$ and $MA_3Bi_2Br_9$) and Pb-based perovskites in the blue-emitting region. The primary reasons for the high PLQY are the very low defect density of $FA_3Bi_2Br_9$ QDs because of ligand surface passivation that leads to the suppression of non-radiative-recombination channels. Moreover, high exciton binding energy (BE) of materials enable efficient radiative recombination of excitons leading to enhanced PLQY. These QDs show a PL emission in the 399 to 526 nm range and a direct band gap of 2.84 eV that coincides nicely with the experimental data. These $FA_3Bi_2Br_9$ QDs with bright blue color lead to the fabrication of LED with promising applications in blue light emission.

Song et al. [63] reported the synthesis of all-inorganic $CsPbX_3$ (X = I, Br, Cl) PeQDs with sharp emission wavelength (FWHM < 30 nm). In these high-quality PeQDs, the luminescence wavelength and color of QDs from blue to orange can be tuned with different sized QDs and varying the halide composition (I, Br, Cl). They can easily dissolve in different non-polar solvents (toluene, octane, and hexane) that serve as ink for solution-based optoelectronic devices. The inorganic PeQDs can be stored for a more extended period of >60 days, which exhibits higher stability than OI $MAPbX_3$ PeQDs. These all-inorganic PeQDs exhibit single-crystal structure, good dispersity, and high-ink stability, making them suitable for low-cost, solution-processed, and flexible optoelectronics. Thus, due to high PL efficiency and the whole visible wavelength range, unlike organic-inorganic QDs, all-inorganic PeQDs are a potential candidate for LED devices.

In summary, Table 7.2 gives a detailed view about the metal-halide perovskites along with their synthesis method, principal solvent and conclusive findings.

7.4   GLIMPSE ON DEVICE FABRICATION

The high performance of halide perovskite (HP) LEDs is assigned to the intrinsic properties of HP inks, such as low defect density, high crystallinity, high absorption, high PLQY, and efficient charge transport. In device fabrication using these perovskite inks, the critical parameters that determine the performance of LEDs are external quantum efficiency (EQE), current efficiency (CE), turn-on voltage (V_{on}), maximum luminance (L_{max}), and stability. Here, these parameters can be calculated as follows:

$$EQE = IQE.\eta$$

where, IQE is the internal quantum efficiency.

H is the fraction of photons emitted to free space.

$$PE = \frac{P}{IV}$$

where, P = Power emitted into free space

$$CE = \frac{L}{J}$$

where, L = Luminance of LEDs

J = Current density

It also helps to feature the effect of dimensional evolution on the optical and electrical properties of HP materials and their performance in LEDs. The robustness of highly luminescent properties and effective

Table 7.2 Metal-halide perovskites along with their synthesis method, principal solvent, and conclusive findings

S.No.	Pb-Based Perovskite	Pb-Free Perovskite	Technique Used	Principal Solvent	Results/Conclusions	References
1.	—	$Cs_3Sb_2I_9$ $Rb_3Sb_2I_9$	Hot-Injection	Octadecene, Oleylamine, and Octanoic acid	• Have potential for optical and optoelectronic perovskites. • Improved defect chemistry is required for Sb-halide PeNCs in comparison to Pb-halide PeNCs.	[61]
2.	$CsPbX_3$ (X = I^-, Br^-, Cl^-, Cl^-/Br^- or Br^-/I^-)	—	Hot-Injection	Oleic acid, Octadecene, and tri-octyl-phosphine	• The combination of optical properties and chemical robustness makes $CsPbX_3$ NCs the best candidate for optoelectronic applications in the green and blue spectral region (410–530 nm).	[51]
3.	—	$CsSnX_3$ (X = I^-, Br^-, Cl^-, $Cl_{0.5}Br_{0.5}$ $Br_{0.5}I_{0.5}$)	Hot-Injection	Oleic acid, Octadecene, and Tri-octyl-phosphine	• Here, the replacement of Pb with Sn reveals their spectral tenability via the quantum-confinement effect and varying the halide ions.	[54]
4.	—	$FA_3Bi_2Br_9$	Ligand-Assisted Reprecipitation (LARP)	Toluene, N, N-DMF	• As-synthesized $FA_3Bi_2Br_9$ QDs exhibit high PLQY of over 50% in the blue-emitting region, and these quantum-confined QDs exhibit high exciton BE leading to the generation of excitons and their high-rate recombination.	[62]
5.	MAPbI$_3$/ MAPbBr$_3$	—	Hot-Injection	Oleic acid and Oleylamine	• Phase transitions lead to the variation in the band-width and peak positions of the MA-cage vibrations and some bands associated with the NH_3^+ group.	[64]

(Contd.)

Table 7.2 Metal-halide perovskites along with their synthesis method, principal solvent, and conclusive findings (*Contd.*)

S.No.	Pb-Based Perovskite	Pb-Free Perovskite	Technique Used	Principal Solvent	Results/Conclusions	References
6.	$CsPbX_3$ (X = Cl⁻, Br⁻, I⁻)	—	Hot-Injection	Oleic acid, Oleylamine, and Octadecene	• These all-inorganic PeQDs can be stored for more than two months and exhibit higher stability than $MAPbX_3$PeQDs, and when they disperse in various non-polar solvents can serve as ink for solution-based optoelectronic devices.	[63]
7.	$MAPb(I_xBr_{1-x})_3$	—	Thin-film growth method	N, N-Dimethyl-formamide	• Mixed-halide perovskites are more critical than pure-phase perovskites because band gaps of mixed lead-halide perovskites are tuned by halide composition, which is the most optimum characteristic for optoelectronic applications.	[65]
8.	$MAPbX_3$ (X= Br⁻, Cl⁻/Br⁻)	—	Aqueous method	Phenylalanine (PLLA) and Di-dodecyl-dimethyl-ammonium bromide (DDAB)	• $MAPbX_3$ Perovskite nanocrystals have been synthesized using an aqueous method, which identifies the influence of the Pb-halide complex, pH-value, and ligands on the formation of these nanocrystals.	[66]
9.	—	$Cs_3Bi_2Br_9/Cs_3Sb_2Br_9/$ $(CH_3NH_3)_3Bi_2X_9$	Ligand-Assisted Reprecipitation (LARP) method	Dimethylformamide or Dimethyl-sulfoxide and octane with oleic acid	• $Cs_3Sb_2Br_9$ exhibits a relatively high PLQY of 46% where it was shelled with a Br-rich surface, and high exciton BE contributes to enhanced quantum yield of material. While PLQY of $(CH_3NH_3)_3Bi_2X_9$ nanocrystals improved and reported values of 0.03–15% and 0.018–26.4%.	[13, 67, 68]
10.	—	$(CH_3NH_3)_2MnCl_4$	Single-Crystal Growth Method	Dimethyl-formamide and Dimethyl-sulfoxide, Hydrochloric acid	• $(CH_3NH_3)_2MnCl_4$ single crystal shows enhanced luminescence properties leading to red light emission upon excitation at 417 nm based on the purple light absorption for white light illumination.	[69]

electrical injection and transportation features of QD films were assumed to achieve high exciton recombination efficiency in devices. Here, the device fabrication consists of multiple layers arranged in the following sequence: ITO, PEDOT-PSS (40 nm), PTAA (40 nm), perovskite QDs ink (40 nm), TPBi (40 nm), and LiF/Al (1/100 nm). The HTL layer of PTAA used exhibits high hole mobility [70, 71], which helps to achieve high efficiency. The electron transport mechanism and device structure in perovskite-based devices is shown in Fig. 7.4

ITO-coated glass substrates were cleaned following a sequence using ultra-sonication in acetone and IPA for 10 minutes each. Firstly, the HTL layer of PEDOT: PSS is deposited on a glass substrate via spin-coating at 5,000 rpm for 40 seconds. It was followed by annealing the substrate for 15 minutes @ 130°C. Then, the active layer of perovskite NCs dispersed in toluene (20–40 mg/ml) was deposited by spin-coating on PEDOT: PSS at 1,000 rpm for 30 seconds in the glovebox and annealed at 50°C for 2 minutes. Further, the ETL SPB-02T dispersed in chlorobenzene (0.4 wt.%) was spin-coated at 2,000 rpm for 45 seconds. Lastly, thermal evaporation deposited LiF (1 nm) and silver (100 nm).

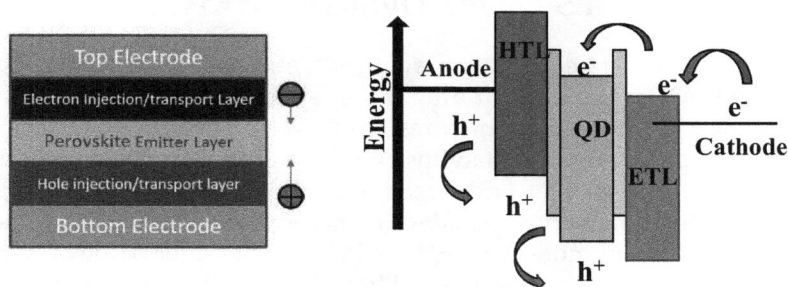

Figure 7.4   Electron transport mechanism in Perovskite-based devices.

For instance, in a device fabrication of $CsPbBr_3$ and FA-doped $CsPbBr_3$, the current-voltage (I–V) curve of the device was measured where the controlled devices (without FA cation) showed higher current [72]. After introducing the FA^+ dopant, the devices' current reduces slightly, leading to the suppression of charge in balance. The voltage-dependent luminance of QLEDs shows a minor increase in turn-on-voltage after incorporation of FA^+, which ascribes to the deeper valence band of FA-doped $CsPbBr_3$ (5.96 eV) compared to $CsPbBr_3$ (5.67 eV). However, the QLEDs based on FA-doped $CsPbBr_3$ QDs possess high luminance compared to pure $CsPbBr_3$ [34, 73]. For the device, internal quantum efficiency (IQE) can be calculated as follows:

$$IQE = 2n^2EQE$$

Here, EQE is external quantum efficiency, and n is the refractive index of the glass substrate.

An excess amount of FA$^+$ cation mixed in equal stoichiometry with CsPbBr$_3$ shows decreased LED performance which attributes to the poor morphology of QD films originating from poor inks, and these devices exhibit high reproducibility for future practical applications [74].

Thus, these CsPbBr$_3$ QD inks can be referred to as ideal emitters with single radiative decay in the intrinsic channel synthesized through synergistic ligands [75, 76] at room temperature in the open air. Although, QD films with few non-radiative-recombination centers show improved PL efficiency after introducing a small amount of FA$^+$ dopant. It also demonstrates that the CsPbBr$_3$ QDs-based electroluminescent LEDs show a peak EQE of 11.6% and peak CE of 45.4 Cd/A, which is the most significant value for perovskite QLEDs [77, 78]. The proposed simple and easily accessible synthesis method is used to scale up the QDs production without any noticeable change in material properties on the device's performance, which holds excellent commercialized industrial applications [74].

7.5 PROPERTIES OF PEROVSKITE MATERIALS

Swarnkar et al. stated [79] the synthesis of stable cubic phase QDs of CsPbI$_3$ QDs by transforming their synthesized cubic phase to orthorhombic. Core-shell structures are the best alternative to enhance the optical properties and stability of all-inorganic MHP nanomaterials (NMs) [13]. Luo et al. synthesized [80] the CsPbBr$_3$/Cs$_4$PbBr$_6$ core-shell structure nanocrystals using a microchannel reactor and these nanocrystals retain more than 90% of PL intensity after 84 days. While Tang et al. stated the synthesis of CsPbBr$_3$/CdS core-shell structured QDs by hot-injection method [55]. Chen et al. adopted an ion-doping strategy for improving the stability and efficiency of as-synthesized CsPbI$_2$Br and related semiconductor devices [81]. Zhao et al. [59] synthesized CsPbI$_3$ nanocrystals with variation in size distribution and also studied the perovskite structure and confinement property of these PeNCs. Thus, it is summarized that the perovskite NC structure can more effectively improve stability as compared to thin-film perovskite materials. It is also important to discuss the excellent optical properties of AIMHP concerning bulk materials, including the following three characteristics:

7.5.1 Strong Quantum-Confinement Effect

The principal property of nanomaterials is that quantum conversion efficiency (QCE) was taken under consideration among other optical properties of materials when the size of an AIMHP is tiny as compared with Bohr radius of excitons where excitons were confined in three spatial dimensions leading to a transition from continuous to discrete energy levels. For instance, the exciton Bohr radius of inorganic $CsPbBr_3$ perovskite nanocrystals (PeNCs) is approx. 7 nm theoretically, and the QCE is an essential factor in these PeNCs when their size is compared with the exciton Bohr radius [82]. So, if the emission band gap of $CsPbBr_3$ perovskite NCs is tuned from 2.7 to 2.4 eV with a change in particle size from 4 to 12 nm, theoretically, thus it shows well-agreement with the quantum-confinement effect [58, 82, 83].

7.5.2 A Wider Range of Optical Properties

Here, the band gap and optical properties can be tuned to the size of AIMHP nanomaterials. The characteristic property of AIMHP is its visible range emitting tunable wavelength [73] by controlling the material composition, dimensions, and structure. As the emission range of all-inorganic perovskite ($CsPbX_3$ and other mixtures) QDs, quantum wells (QWs), and nanoplatelets (NPLs) lies in the whole visible range of 400–700 nm (blue to red) that can also be tuned to NIR or UV range [84–86] via particle replacement method and new device structure for display applications [87].

7.5.3 High Quantum Efficiency

Quantum efficiency can be defined as the ratio of the number of converted photons to absorbed photons, and it is a well-known characteristic property of light emitters [80, 88–96]. High quantum efficiency denotes that more absorbed photons were converted via the radiative-recombination process instead of the non-radiative-recombination process. Perovskites are also known as exceptional light emitters because of their high absorption coefficient and quantum efficiency [93, 95]. The high quantum efficiency in perovskites is due to a clear band gap with negligible charge-trapping states promoting the exciton radiative-recombination efficiency [97, 98], although these unique properties of AIMHP nanomaterials still require some essential aspects for improving the performance of the perovskite-based devices:

(i) Understanding physics in low-dimensional perovskites: Generally, low-dimensional perovskites are not very popular, and there is minimal research on these perovskites. However, this low-dimensionality of perovskites could bring restricted charge transport in an appropriate direction, which forms the basis for field-effect devices showing various unique electrical properties, such as 2D electron gas, etc. However, these studies are lagging for various reasons and will be under discussion in the upcoming years [99–101].

(ii) Precisely control in size and dimensions: To date, the optimistic control on size and dimension for band gaps is not attracting much attention but still, some synthesis methods of perovskites with low dimensions were tried for the development of perovskites. At the same time, size control and low-dimensionality are critical factors in inducing the morphology and applications of as-synthesized perovskites [102, 103].

(iii) Lead-free HP nanocrystals: The toxicity problem of lead (Pb) arises from the need to replace Pb with other earth-abundant germanium (Ge), tin (Sn), bismuth (Bi), and antimony (Sb) metal ions. Even the development of these environment-friendly Pb-free halide perovskites with improved optical and electronic properties as well as excellent environmental stability makes these Pb-free perovskites (for instance, $CsSnX_3$, $Cs_3Bi_2X_9$, $FA_3Bi_2X_9$, $MASnX_3$, etc.) promising candidates for display applications [104, 105].

7.6 APPLICATIONS OF METAL-HALIDE PEROVSKITES

In general, inorganic-organic metal-halide (IOMH) perovskite nanomaterials are a topic of discussion among researchers because of their enhanced device performance and higher thermal stability compared with organic metal-halide perovskite (OMHP). These NCs and NWs utilized for nanoscale photonic, electronic, and optoelectronic devices, such as semiconductors, LEDs, lasers, photodetectors, single-quantum-photon sources (SPQS), etc., some of the significant applications are discussed below [4].

7.6.1 Solar Cells

High-efficiency perovskite semiconductors (PSCs) were realized by tuning the band gap and stabilizing the black perovskite phase of $CsPbI_2Br$ at lower temperatures. However, the power conversion efficiency (PCE) of $CsPbX_3$-based PSCs has enhanced rapidly from 2.9% to 23.7% with a tremendous increase in stability [106–108], and here the device also realized 9.8% PCE and over 5% stabilized power output [83]. Based on the slow-photon effect of carbon-QD sensitized $CsPbBr_3$, Zhou et al. [109] synthesized inverse opal-perovskite semiconductors (SCs), but as compared to planar $CsPbBr_3$, the perovskite SCs exhibit power conversion efficiency (PCE) up to 8.29% and incident-photon-to-electron conversion efficiency up to 76.9%. Further, researchers have adapted the various device optimization techniques like interface engineering (self-passivation and electron-transporting layer) and defect engineering to improve the stability as well as QCE of PSCs [107, 110–115]. Nanocrystalline-controlled synthesis methods were mainly utilized for improving the interfacial energy, optoelectronic and recombination properties of devices and materials [116–118].

7.6.2   Light-Emitting Diodes (LEDs)

The external quantum efficiencies (EQE) of the blue, orange and green LEDs were 0.07%, 0.09%, and 0.12%, respectively. Thus, inorganic perovskites can be used as LED-device emitters instead of low EQE [119–122]. According to the literature, the highest EQE of the inorganic LEDs is greater than 20% compared with organic LEDs [123]. At the same time, the operational stability of $CsPbBr_3$ QD-based LED devices was found to be 36 times higher due to the lower surface-ligand density of corresponding QDs [20, 124–126]. Thus, it is advisable to utilize the materials and devices optimization techniques for improving the characteristics of LEDs [74, 89, 127–129]. Further, Shan et al. [130] synthesized perovskite $CsPbX_3$ (X = Br, Cl) QDs with halide-ion pair ligands and using this perovskite, green and blue LEDs were fabricated, which exhibits the higher EQE in comparison with untreated QDs. In contrast, Song et al. [74] used a tetra butyl-ammonium bromide to reduce non-radiative defects of $CsPbBr_3$, promoting the electroluminescence performance of LEDs. However, various device engineering technology was utilized to improve the efficiency and performance of LEDs [127–130]. Therefore, various approaches like improved hole injection, enhanced electron injection, increase in charge balance and decrease in charge leakage, etc., were adopted to enhance the efficiency [131–133] and stability of perovskite LEDs.

7.6.3   Lasing

The emission-wavelength tunability illustrated in perovskite materials was an essential property for nanoscale lasers. Generally, the lasing can be maintained for over 1 hour on persisting on exposure to the surrounding atmosphere signifying their improved performance compared to hybrid perovskite NW lasers [134, 135]. For instance, Wang et al. [136] stated the multi-photon pumped lasing from $CsPbX_3$ nanorods with a threshold value of ~0.6 and 1.7 mJ cm^{-2} under an excitation wavelength of 800 nm and 1,200 nm with 80 fs and 1 kHz as shown in Fig. 7.5. AILHP-based temperature-dependent nanostructured lasers were studied due to their temperature sensitivity [55, 137–139]. Thus researchers have also adopted the new material structure or device engineering to improve the detector properties [140–145]. Pushkarev et al. [143] reported a new method for synthesizing high-quality $CsPbBr_3$ nano-lasers using the rapid precipitation from DMSO solution sprayed onto hydrophobic substrates at ambient conditions. However, in-plane self-assembly $CsPbBr_3$ NWs and $CsPbBr_3$ QDs, when incorporated in a silica sphere, was the best way to improve the performance and stability of $CsPbBr_3$ nano-lasers [146, 147]. On the other hand, mainly inorganic and hybrid perovskites were utilized to enhance the external luminescence efficiency of lasers [140, 142, 145].

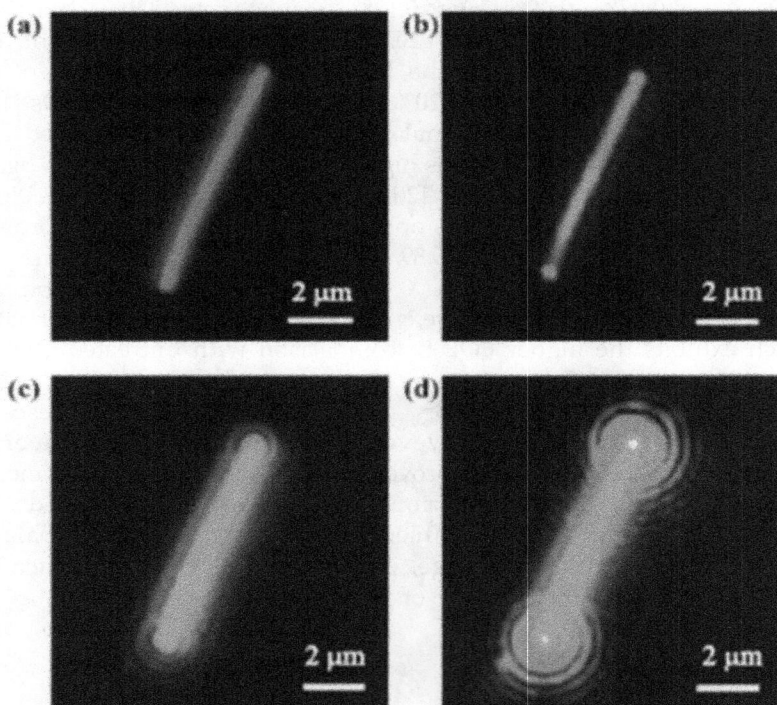

Figure 7.5 Lasing in single-crystal $CsPbBr_3$ NWs were (a) dark-field images of $CsPbBr_3$ NWs (b-d) under excitation of NWs from femtosecond pulsed laser with increasing excitation fluence [60].

7.6.4 Photodetectors

An ideal photodetector (PD) must have high speed, spectral selectivity, signal-to-noise ratio, and sensitivity. PDs have diversified applications such as imaging, optical communication, remote control, chemical/ biological sensing, etc. [139]. However, Inorganic scattered $CsPbBr_3$ nanoplatelets-based low-voltage PDs with high-responsivity were fabricated [148, 149]. In addition to 0D and 1D $CsPbBr_3$ nanomaterials, 1D AILHP were also used to fabricate optical detectors with high stability and detection rate [139, 150–155]. Thus, PDs based on 1D $CsPbI_3$ nanorods were developed, exhibiting 2.92×10^3 AW^{-1} responsivity, 0.05 ms ultrafast response time, and detectivity approaching up to 5.17×10^3 jones [155]. Then, Sn-doped $CsPbI_3$ nanobelts were synthesized, which holds stability for 15 days under air conditions. These as-synthesized PDs of $CsPb_{0.922}Sn_{0.078}I_3$ nanobelts show ultrahigh selectivity of up to 6.43×10^{13} jones [151]. Indeed, SPPDs with no external power source, high sensitivity, and fast-response speed an essential type of PDs [156–160]. Various researchers have reported the synthesis of PDs with characteristic properties [157]. However, they have still accepted the material and its device optimization techniques like the core-shell method, vertical Schottky junction, inverse structure, and fast interfacial charge-transfer layer for improving the PD properties [150, 161–163]. Thus, the improved performance of PDs is assigned to improved crystalline and fewer surface defects of AIMHP NMs.

7.6.5   In Opto-electronic Device

AIMHP nanomaterials were also used in various optoelectronic devices, unlike SCs, LEDs, lasers, and PDs. For instance, the development of polarization-sensitive and $CsPbBr_3$ and $CsPbCl_3$ nanocrystals-based light sources [164]. Similarly, $CsPbBr_3$ thin-film-based hybrid phototransistors were reported with a responsivity of 4.9×10^6 AW^{-1}, fast-response of 0.45 s/0.55 s, and long-term stability of 200 h in ambient conditions [165]. On the other hand, field-effect transistors were developed using the $CsPbBr_3$ nanocrystals [166–168] that exhibit unipolar transport characteristics in a p-type mode featured well-aligned linear and saturation regimes.

7.7   CONCLUSION

This chapter gives a very conceptual summary of the transition from Pb-based to Pb-free halide perovskite inks for optoelectronic applications. Moreover, this chapter discusses the various synthesis methods, characteristic properties, and applications of the Pb-based and Pb-free perovskite inks. It also gives an insight into the organic and inorganic constituting ions of the perovskite inks and the advantages of the hot-injection synthesis technique over other methods for the perovskite inks. This chapter gives a brief idea about the device fabrication using the specific HTL and ETL for the Pb-based/Pb-free perovskite active layer depending on the thickness of each layer required for the optoelectronic devices using these perovskite inks. However, here we have also discussed various important comparable properties and properties required to enhance the performance of IOMH perovskite inks-based devices. Thus, this chapter generally gives a detailed description of the Pb-based and Pb-free halide perovskite inks, along with the advantages and challenges ahead in the transition from Pb-based to Pb-free perovskite inks for commercial and industrial applications.

7.8 CHALLENGES AND FUTURE SCOPE

Researchers are struggling with the stability issues of metal-halide perovskite inks. Here, incorporating large organic cations was suggested to improve the stability of these perovskite inks. However, by tuning the dimensionality and crystal structure, some halide perovskite inks have to pertain to the stability of optoelectronic applications. Moreover, concerning the issue of lead toxicity, the development of Pb-free halide-based perovskite inks will be the priority for researchers. It raises the possibility of wholly or partly substituting the Pb with Sn in the perovskite lattice because of several parameters, such as suitable ionic radii and existence in the unstable +2 state. This substitution has various advantages of band gap narrowing on the perovskite inks-based materials and devices. Despite its advantages, there are still pertinent issues that need to be addressed regarding the fabrication of stable and high-performance Sn-based devices compared to the stability characteristics of Pb-based devices, although working on the stability of Sn-halide perovskite inks is a more troublesome task than Pb-halide-based perovskite inks due to the conversion of Sn from +2 to +4 state in ambient conditions. Indeed, there are various disadvantages of using Sn over Pb for perovskite device fabrication due to the higher cost and restricted production of Sn than that of Pb. In the case of Sn, many optimizations are required for the fabrication process of Sn-based devices. Pb-based analogues have shown remarkable properties with optimum PCE but still work for the devices' stability. Therefore, along with PCE, we will work further on the stability of Pb-based and Sn-based halide perovskite inks-based devices for optoelectronic applications.

ACKNOWLEDGMENT

The authors sincerely thank the Director, CSIR-NPL, for his constant support and encouragement. Author SM gratefully acknowledges the Department of Science and Technology for providing a WOS-A project grant (WOS-A/CS-132/2018).

REFERENCES

[1] Li, J., J. Duan, X. Yang, Y. Duan, P. Yang and Q Tang. 2020. Review on recent progress of lead-free halide perovskites in optoelectronic applications. Nano Energy 80: 105526. https://doi.org/10.1016/j.nanoen.2020.105526

[2] Jian, W., R. Jia, H.-X. Zhang and F.-Q Bai. 2020. Arranging strategies for A-site cations: impact on the stability and carrier migration of hybrid perovskite materials. Inorg. Chem. Front. 7: 1741–1749. https://doi.org/10.1039/D0QI00102C

[3] Jesper Jacobsson, T., J.-P. Correa-Baena, M. Pazoki, M. Saliba, K. Schenk, M. Grätzel, et al. 2016. Exploration of the compositional space for mixed lead halogen perovskites for high efficiency solar cells. Energy Environ. Sci. 9: 1706–1724.. https://doi.org/10.1039/C6EE00030D

[4] Cao, L., X. Liu, Y. Li, X. Li, L. Du, S. Chen, et al. 2021. Recent progress in all-inorganic metal halide nanostructured perovskites: materials design, optical properties, and application. Front. Phys. 16: 33201. https://doi.org/10.1007/s11467-020-1026-9

[5] Chen, P., Y. Bai, M. Lyu, J.-H. Yun, M. Hao and L. Wang. 2018. Progress and perspective in low-dimensional metal halide perovskites for optoelectronic applications. Sol. RRL 2: 1700186. https://doi.org/10.1002/solr.201700186

[6] Saidaminov, M.I., J. Kim, A. Jain, R. Quintero-Bermudez, H. Tan, G. Long, et al. 2018. Suppression of atomic vacancies via incorporation of isovalent small ions to increase the stability of halide perovskite solar cells in ambient air. Nat Energy 3: 648–654. https://doi.org/10.1038/s41560-018-0192-2

[7] Yang, J., B.D. Siempelkamp, D. Liu and T.L. Kelly. 2015. Investigation of $CH_3NH_3PbI_3$ degradation rates and mechanisms in controlled humidity environments using *in situ* techniques. ACS Nano 9: 1955–1963. https://doi.org/10.1021/nn506864k

[8] Conings, B., J. Drijkoningen, N. Gauquelin, A. Babayigit, J. D'Haen, L. D'Oliieslaeger, et al. 2015. Intrinsic thermal instability of methylammonium lead trihalide perovskite. Adv. Energy Mater 5: 1500477. https://doi.org/10.1002/aenm.201500477

[9] Lyu, M., J. Yun, P. Chen, M. Hao and L. Wang. 2017. Addressing toxicity of lead: progress and applications of low-toxic metal halide perovskites and their derivatives. Adv. Energy Mater. 7: 1602512. https://doi.org/10.1002/aenm.201602512

[10] Chen, Q., N. De Marco, Y. Yang, (Michael), T.-B. Song, C.-C. Chen, H. Zhao, et al. 2015. Under the spotlight: the organic–inorganic hybrid halide perovskite for optoelectronic applications. Nano Today 10: 355–396. https://doi.org/10.1016/j.nantod.2015.04.009

[11] Marshall, K.P., R.I. Walton and R.A. Hatton. 2015. Tin perovskite/fullerene planar layer photovoltaics: improving the efficiency and stability of lead-free devices. J. Mater. Chem. A 3: 11631–11640. https://doi.org/10.1039/C5TA02950C

[12] Park, B.-W., B. Philippe, X., Zhang, H. Rensmo, G. Boschloo and E.M.J. Johansson. 2015. Bismuth based hybrid perovskites $A_3Bi_2I_9$ (A: Methylammonium or Cesium) for solar cell application. Adv. Mater. 27: 6806–6813. https://doi.org/10.1002/adma.201501978

[13] Sun, J., J. Yang, J.I. Lee, J.H. Cho and M.S. Kang. 2018. Lead-free perovskite nanocrystals for light-emitting devices. J. Phys. Chem. Lett. 9: 1573–1583. https://doi.org/10.1021/acs.jpclett.8b00301

[14] Fan, Q., G.V. Biesold-McGee, J. Ma, Q. Xu, S. Pan, J. Peng, et al. 2020. Lead-free halide perovskite nanocrystals: crystal structures, synthesis, stabilities, and optical properties. Angewandte Chemie Intl. Edit. 59: 1030–1046. https://doi.org/10.1002/anie.201904862

[15] Dimesso, L., C. Das, M. Stöhr, T. Mayer and W. Jaegermann. 2017. Properties of cesium tin iodide (Cs-Sn-I) systems after annealing under different atmospheres. Mater. Chem. Phy. 197: 27–35. https://doi.org/10.1016/j.matchemphys.2017.05.018

[16] Ning, W. and F. Gao. 2019. Structural and functional diversity in lead-free halide perovskite materials. Adv. Mater. 31: 1900326. https://doi.org/10.1002/adma.201900326

[17] Xiao, Z., Z. Song and Y. Yan. 2019. From lead halide perovskites to lead-free metal halide perovskites and perovskite derivatives. Adv. Mater. 31: 1803792. https://doi.org/10.1002/adma.201803792

[18] Hoefler, S.F., G. Trimmel and T. Rath. 2017. Progress on lead-free metal halide perovskites for photovoltaic applications: a review. Monatsh Chem. 148: 795–826. https://doi.org/10.1007/s00706-017-1933-9

[19] Chang, S., Z. Bai and H. Zhong. 2018. *In Situ* Fabricated perovskite nanocrystals: a revolution in optical materials. Adv. Opt. Mater. 6: 1800380 https://doi.org/10.1002/adom.201800380

[20] Wang, Y., S. Chang, X. Chen, Y. Ren, L. Shi, Y. Liu, et al. 2019. Rapid growth of halide perovskite single crystals: from methods to optimization control. Chin. J. Chem. 37: 616–629. https://doi.org/10.1002/cjoc.201900071

[21] Kwoka, M., L. Ottaviano, M. Passacantando, S. Santucci, G. Czempik and J. Szuber. 2005. XPS study of the surface chemistry of L-CVD SnO_2 thin films after oxidation. Thin Solid Films 490: 36–42. https://doi.org/10.1016/j.tsf.2005.04.014

[22] Takahashi, Y., R. Obara, Z.-Z. Lin, Y. Takahashi, T. Naito, T. Inabe, et al. 2011. Charge-transport in tin-iodide perovskite $CH_3NH_3SnI_3$: origin of high conductivity. Dalton Trans. 40: 5563. https://doi.org/10.1039/c0dt01601b

[23] Takahashi, Y., H. Hasegawa, Y. Takahashi and T. Inabe. 2013. Hall mobility in tin iodide perovskite $CH_3NH_3SnI_3$: evidence for a doped semiconductor. J. Solid State Chem. 205: 39–43. https://doi.org/10.1016/j.jssc.2013.07.008

[24] Kumar, M.H., S. Dharani, W.L. Leong, P.P. Boix, R.R. Prabhakar, T. Baikie, et al. 2014. Lead-free halide perovskite solar cells with high photocurrents realized through vacancy modulation. Adv. Mater. 26: 7122–7127. https://doi.org/10.1002/adma.201401991

[25] Noel, N.K., S.D. Stranks, A. Abate, C. Wehrenfennig, S. Guarnera, A.-A. Haghighirad, et al. 2014. Lead-free organic–inorganic tin halide perovskites for photovoltaic applications. Energy Environ. Sci. 7: 3061–3068. https://doi.org/10.1039/C4EE01076K

[26] Li, Q., H. Li, H. Shen, F. Wang, F. Zhao, F. Li, et al. 2017. Solid ligand-assisted storage of air-stable formamidinium lead halide quantum dots via restraining the highly dynamic surface toward brightly luminescent light-emitting diodes. ACS Photonics 4: 2504–2512. https://doi.org/10.1021/acsphotonics.7b00743

[27] Qian, J., B. Xu and W. Tian. 2016. A comprehensive theoretical study of halide perovskites ABX_3. Org. Electron. 37: 61–73. https://doi.org/10.1016/j.orgel.2016.05.046

[28] Goldschmidt, V.M. 1926. Die gesetze der krystallochemie. Naturwissenschaften 14: 477–485. https://doi.org/10.1007/BF01507527

[29] Zhou, L., J. Liao, Z. Huang, J. Wei, X. Wang, W. Li, et al. 2019. A highly red-emissive lead-free indium-based perovskite single crystal for sensitive water detection. Angew. Chem. 131: 5331–5335. https://doi.org/10.1002/ange.201814564

[30] Zhou, C., Y. Tian, M. Wang, A. Rose, T. Besara, N.K. Doyle, et al. 2017. Low-dimensional organic tin bromide perovskites and their photoinduced structural transformation. Angew. Chem. Int. Ed. 56: 9018–9022. https://doi.org/10.1002/anie.201702825

[31] Huang, T.J., Z.X. Thiang, X. Yin, C. Tang, G. Qi and H. Gong. 2016. $(CH_3NH_3)_2 PdCl_4$: A compound with two-dimensional organic-inorganic layered perovskite structure. Chem. Eur. J. 22: 2146–2152. https://doi.org/10.1002/chem.201503680

[32] Travis, W., E.N.K. Glover, H. Bronstein, D.O. Scanlon and R.G. Palgrave. 2016. On the application of the tolerance factor to inorganic and hybrid halide perovskites: a revised system. Chem. Sci. 7: 4548–4556. https://doi.org/10.1039/C5SC04845A

[33] Li, C., X. Lu, W. Ding, L. Feng, Y. Gao and Z. Guo. 2008. Formability of ABX_3 (X = F, Cl, Br, I) halide perovskites. Acta Crystallogr B Struct Sci. 64: 702–707. https://doi.org/10.1107/S0108768108032734

[34] Li, Z., M. Yang, J.-S. Park, S.-H. Wei, J.J. Berry and K. Zhu. 2016. Stabilizing perovskite structures by tuning tolerance factor: formation of formamidinium and cesium lead iodide solid-state alloys. Chem. Mater. 28: 284–292. https://doi.org/10.1021/acs.chemmater.5b04107

[35] Kieslich, G., S. Sun and A.K. Cheetham. 2015. An extended tolerance factor approach for organic-inorganic perovskites. Chem. Sci. 6: 3430–3433. https://doi.org/10.1039/C5SC00961H

[36] Uribe, J.I., D. Ramirez, J.M. Osorio-Guillén, J. Osorio and F. Jaramillo. 2016. $CH_3NH_3CaI_3$ Perovskite: synthesis, characterization, and first-principles studies. J. Phys. Chem. C 120: 16393–16398. https://doi.org/10.1021/acs.jpcc.6b04207

[37] Filip, M.R. and F. Giustino. 2016. Computational screening of homovalent lead substitution in organic–inorganic halide perovskites. J. Phys. Chem. C 120: 166–173. https://doi.org/10.1021/acs.jpcc.5b11845

[38] Becker, M., T. Klüner and M. Wark. 2017. Formation of hybrid ABX_3 perovskite compounds for solar cell application: first-principles calculations of effective ionic radii and determination of tolerance factors. Dalton Trans. 46: 3500–3509. https://doi.org/10.1039/C6DT04796C

[39] Kieslich, G., S. Sun and A.K. Cheetham. 2014. Solid-state principles applied to organic–inorganic perovskites: new tricks for an old dog. Chem. Sci. 5: 4712–4715. https://doi.org/10.1039/C4SC02211D

[40] Deng, W., H. Fang, X. Jin, X. Zhang, X. Zhang and J. Jie. 2018. Organic–inorganic hybrid perovskite quantum dots for light-emitting diodes. J. Mater. Chem. C. 6: 4831–4841. https://doi.org/10.1039/C8TC01214H

[41] Zhang, F., H. Zhong, C. Chen, X. Wu, X. Hu, H. Huang, et al. 2015. Brightly luminescent and color-tunable colloidal $CH_3NH_3PbX_3$ (X = Br, I, Cl) Quantum dots: potential alternatives for display technology. ACS Nano 9: 4533–4542. https://doi.org/10.1021/acsnano.5b01154

[42] Dong, H. and W. Hu. 2013. Organic nanomaterials. pp. 905–940. *In:* R. Vajtai (ed.). Springer Handbook of Nanomaterials. Springer Berlin Heidelberg, Berlin, Heidelberg.

[43] Huang, H., A.S. Susha, S.V. Kershaw, T.F. Hung and A.L. Rogach. 2015. Control of emission color of high quantum yield $CH_3NH_3PbBr_3$ perovskite quantum dots by precipitation temperature. Adv. Sci. 2: 1500194. https://doi.org/10.1002/advs.201500194

[44] Levchuk, I., A. Osvet, X. Tang, M. Brandl, J.D. Perea, F. Hoegl, et al. 2017. Brightly luminescent and color-tunable formamidinium lead halide perovskite $FAPbX_3$ (X = Cl, Br, I) colloidal nanocrystals. Nano Lett. 17: 2765–2770. https://doi.org/10.1021/acs.nanolett.6b04781

[45] Kumar, S., J. Jagielski, S. Yakunin, P. Rice, Y.-C. Chiu, M. Wang, et al. 2016. Efficient blue electroluminescence using quantum-confined two-dimensional perovskites. ACS Nano 10: 9720–9729. https://doi.org/10.1021/acsnano.6b05775

[46] Gonzalez-Carrero, S., R.E. Galian and J. Pérez-Prieto. 2015. Maximizing the emissive properties of $CH_3 NH_3 PbBr_3$ perovskite nanoparticles. J. Mater. Chem. A 3: 9187–9193. https://doi.org/10.1039/C4TA05878J

[47] Meng, L., E. Yao, Z. Hong, H. Chen, P. Sun, Z. Yang, et al. 2017. Pure formamidinium-based perovskite light-emitting diodes with high efficiency and low driving voltage. Adv. Mater. 29: 1603826. https://doi.org/10.1002/adma.201603826

[48] Protesescu, L., S. Yakunin, M.I. Bodnarchuk, F. Bertolotti, N. Masciocchi, A. Guagliardi, et al. 2016. Monodisperse formamidinium lead bromide nanocrystals with bright and stable green photoluminescence. J. Am. Chem. Soc. 138: 14202–14205. https://doi.org/10.1021/jacs.6b08900

[49] Zhang, D., Y. Yu, Y. Bekenstein, A.B. Wong, A.P. Alivisatos and P. Yang. 2016. Ultrathin colloidal cesium lead halide perovskite nanowires. J. Am. Chem. Soc. 138: 13155–13158. https://doi.org/10.1021/jacs.6b08373

[50] Chen, M., Y. Zou, L. Wu, Q. Pan, D. Yang, H. Hu, et al. 2017. Solvothermal synthesis of high-quality all-inorganic cesium lead halide perovskite nanocrystals: from nanocube to ultrathin nanowire. Adv. Funct. Mater. 27: 1701121. https://doi.org/10.1002/adfm.201701121

[51] Protesescu, L., S. Yakunin, M.I. Bodnarchuk, F. Krieg, R. Caputo, C.H. Hendon, et al. 2015. Nanocrystals of cesium lead halide perovskites ($CsPbX_3$, X = Cl, Br, and I): novel optoelectronic materials showing bright emission with wide color gamut. Nano Lett. 15: 3692–3696. https://doi.org/10.1021/nl5048779

[52] Swarnkar, A., V.K. Ravi and A. Nag. 2017. Beyond colloidal cesium lead halide perovskite nanocrystals: analogous metal halides and doping. ACS Energy Lett. 2: 1089–1098. https://doi.org/10.1021/acsenergylett.7b00191

[53] Li, J., X. Shan, S.G.R. Bade, T. Geske, Q. Jiang, X. Yang, et al. 2016. Single-layer halide perovskite light-emitting diodes with sub-band gap turn-on voltage and high brightness. J. Phys. Chem. Lett. 7: 4059–4066. https://doi.org/10.1021/acs.jpclett.6b01942

[54] Jellicoe, T.C., J.M. Richter, H.F.J. Glass, M. Tabachnyk, R. Brady, S.E. Dutton, et al. 2016. Synthesis and optical properties of lead-free cesium tin halide perovskite nanocrystals. J. Am. Chem. Soc. 138: 2941–2944. https://doi.org/10.1021/jacs.5b13470

[55] Tang, X., Y. Bian, Z. Liu, J. Du, M. Li, Z. Hu, et al. 2019. Room-temperature up-conversion random lasing from $CsPbBr_3$ quantum dots with TiO_2 nanotubes. Opt. Lett. 44: 4706. https://doi.org/10.1364/OL.44.004706

[56] Seth, S. and A. Samanta. 2016. A facile methodology for engineering the morphology of $CsPbX_3$ perovskite nanocrystals under ambient condition. Sci. Rep. 6: 37693. https://doi.org/10.1038/srep37693

[57] Shamsi, J., Z. Dang, P. Bianchini, C. Canale, F. Di Stasio, R. Brescia, et al. 2016. Colloidal Synthesis of quantum confined single crystal $CsPbBr_3$ nanosheets with lateral size control up to the micrometer range. J. Am. Chem. Soc. 138: 7240–7243. https://doi.org/10.1021/jacs.6b03166

[58] Akkerman, Q.A., V. D'Innocenzo, S. Accornero, A. Scarpellini, A. Petrozza, M. Prato, et al. 2015. Tuning the optical properties of cesium lead halide perovskite nanocrystals by anion exchange reactions. J. Am. Chem. Soc. 137: 10276–10281. https://doi.org/10.1021/jacs.5b05602

[59] Zhao, Q., A. Hazarika, L.T. Schelhas, J. Liu, E.A. Gaulding, G. Li, et al. 2020. Size-dependent lattice structure and confinement properties in $CsPbi_3$ perovskite nanocrystals: negative surface energy for stabilization. ACS Energy Lett. 5: 238–247. https://doi.org/10.1021/acsenergylett.9b02395

[60] Eaton, S.W., M. Lai, N.A. Gibson, A.B. Wong, L. Dou, J. Ma, et al. 2016. Lasing in robust cesium lead halide perovskite nanowires. Proc. Natl. Acad. Sci. 113: 1993–1998. https://doi.org/10.1073/pnas.1600789113

[61] Pal, J., S. Manna, A. Mondal, S. Das, K.V. Adarsh and A. Nag. 2017. Colloidal synthesis and photophysics of $M_3Sb_2I_9$(M=Cs and Rb) nanocrystals: lead-free perovskites. Angew. Chem. Int. Ed. 56: 14187–14191. https://doi.org/10.1002/anie.201709040

[62] Shen, Y., J. Yin, B. Cai, Z. Wang, Y. Dong, X. Xu, et al. 2020. Lead-free, stable, high-efficiency (52%) blue luminescent $FA_3Bi_2Br_9$ perovskite quantum dots. Nanoscale Horiz. 5: 580–585. https://doi.org/10.1039/C9NH00685K

[63] Song, J., J. Li, X. Li, L. Xu, Y. Dong and H. Zeng. 2015. Quantum dot light-emitting diodes based on inorganic perovskite cesium lead halides ($CsPbX_3$). Adv. Mater. 27: 7162–7167. https://doi.org/10.1002/adma.201502567

[64] Nakada, K., Y. Matsumoto, Y. Shimoi, K. Yamada and Y. Furukawa. 2019. Temperature-dependent evolution of raman spectra of methylammonium lead halide perovskites, $CH_3NH_3PbX_3$(X = I, Br). Molecules. 24: 626. https://doi.org/10.3390/molecules24030626

[65] Gan, Z., Z. Yu, M. Meng, W. Xia and X. Zhang. 2019. Hydration of mixed halide perovskites investigated by Fourier transform infrared spectroscopy. APL Materials 7: 031107. https://doi.org/10.1063/1.5087914

[66] Geng, C., S. Xu, H. Zhong, A.L. Rogach and W. Bi. 2018. Aqueous synthesis of methylammonium lead halide perovskite nanocrystals. Angew. Chem. Int. Ed. 57: 9650–9654. https://doi.org/10.1002/anie.201802670

[67] Leng, M., Z. Chen, Y. Yang, Z. Li, K. Zeng, K. Li, et al. 2016. Lead-free, blue emitting bismuth halide perovskite quantum dots. Angew. Chem. Int. Ed. 55: 15012–15016. https://doi.org/10.1002/anie.201608160

[68] Leng, M., Y. Yang, K. Zeng, Z. Chen, Z. Tan, S. Li, et al. 2018. All-inorganic bismuth-based perovskite quantum dots with bright blue photoluminescence and excellent stability. Adv. Funct. Mater. 28: 1704446. https://doi.org/10.1002/adfm.201704446

[69] Cheng, X., L. Jing, Y. Yuan, S. Du, Q. Yao, J. Zhang, et al. 2019. Centimeter-size square 2D layered Pb-free hybrid perovskite single crystal $(CH_3NH_3)_2$ $MnCl_4$ for red photoluminescence. Cryst. Eng. Comm. 21: 4085–4091. https://doi.org/10.1039/C9CE00591A

[70] Zhang, X., Y. Zhang, Y. Wang, S. Kalytchuk, S.V. Kershaw, Y. Wang, et al. 2013. Color-switchable electroluminescence of carbon dot light-emitting diodes. ACS Nano. 7: 11234–11241. https://doi.org/10.1021/nn405017q

[71] Zielke, D., A.C. Hübler, U. Hahn, N. Brandt, M. Bartzsch, U. Fügmann, et al. 2005. Polymer-based organic field-effect transistor using offset printed source/drain structures. Appl. Phys. Lett. 87: 123508. https://doi.org/10.1063/1.2056579

[72] Lee, J.-W., D.-H. Kim, H.-S. Kim, S.-W. Seo, S.M. Cho and N.-G. Park. 2015. Formamidinium and cesium hybridization for photo- and moisture-stable perovskite solar cell. Adv. Energy Mater. 5: 1501310. https://doi.org/10.1002/aenm.201501310

[73] Schelhas, L.T., Z. Li, J.A. Christians, A. Goyal, P. Kairys, S.P. Harvey, et al. 2019. Insights into operational stability and processing of halide perovskite active layers. Energy Environ. Sci. 12: 1341–1348. https://doi.org/10.1039/C8EE03051K

[74] Song, L., X. Guo, Y. Hu, Y. Lv, J. Lin, Y. Fan, et al. 2018. Improved performance of $CsPbBr_3$ perovskite light-emitting devices by both boundary and interface defects passivation. Nanoscale. 10: 18315–18322. https://doi.org/10.1039/C8NR06311G

[75] Yassitepe, E., Z. Yang, O. Voznyy, Y. Kim, G. Walters, J.A. Castañeda, et al. 2016. Amine-free synthesis of cesium lead halide perovskite quantum dots for efficient light-emitting diodes. Adv. Funct. Mater. 26: 8757–8763. https://doi.org/10.1002/adfm.201604580

[76] Huang, H., F. Zhao, L. Liu, F. Zhang, X. Wu, L. Shi, et al. 2015. Emulsion synthesis of size-tunable $CH_3NH_3PbBr_3$ quantum dots: an alternative route toward efficient light-emitting diodes. ACS Appl. Mater. Interfaces 7: 28128–28133. https://doi.org/10.1021/acsami.5b10373

[77] Wei, S., Y. Yang, X. Kang, L. Wang, L. Huang and D. Pan. 2017. Homogeneous synthesis and electroluminescence device of highly luminescent $CsPbBr_3$ perovskite nanocrystals. Inorg. Chem. 56: 2596–2601. https://doi.org/10.1021/acs.inorgchem.6b02763

[78] Di Stasio, F., S. Christodoulou, N. Huo and G. Konstantatos. 2017. Near-unity photoluminescence quantum yield in $CsPbBr_3$ nanocrystal solid-state films via postsynthesis treatment with lead bromide. Chem. Mater. 29: 7663–7667. https://doi.org/10.1021/acs.chemmater.7b02834

[79] Swarnkar, A., A.R. Marshall, E.M. Sanehira, B.D. Chernomordik, D.T. Moore, J.A. Christians, et al. 2016. Quantum dot–induced phase stabilization of α-$CsPbI_3$ perovskite for high-efficiency photovoltaics. Science 354: 92–95. https://doi.org/10.1126/science.aag2700

[80] Luo, S.-Q., J.-F. Wang, B. Yang and Y.-B. Yuan. 2019. Recent advances in controlling the crystallization of two-dimensional perovskites for optoelectronic device. Front. Phys. 14: 53401. https://doi.org/10.1007/s11467-019-0901-8

[81] Chen, C., L. Zhang, T. Shi, G. Liao and Z. Tang. 2019. Controllable synthesis of all inorganic lead halide perovskite nanocrystals with various appearances in multiligand reaction system. Nanomater. 9: 1751. https://doi.org/10.3390/nano9121751

[82] Pan, A., B. He, X. Fan, Z. Liu, J.J. Urban, A.P. Alivisatos, et al. 2016. Insight into the ligand-mediated synthesis of colloidal $CsPbBr_3$ perovskite nanocrystals: the role of organic acid, base, and cesium precursors. ACS Nano 10: 7943–7954. https://doi.org/10.1021/acsnano.6b03863

[83] Li, X., F. Cao, D. Yu, J. Chen, Z. Sun, Y. Shen, et al. 2017. All inorganic halide perovskites nanosystem: synthesis, structural features, optical properties and optoelectronic applications. Small 13: 1603996. https://doi.org/10.1002/smll.201603996

[84] Lao, X., X. Li, H. Ågren and G. Chen. 2019. Highly controllable synthesis and DFT calculations of double/triple-halide $CsPbX_3$(X = Cl, Br, I) perovskite quantum dots: application to light-emitting diodes. Nanomater. 9: 172. https://doi.org/10.3390/nano9020172

[85] Ha, S.-T., R. Su, J. Xing, Q. Zhang and Q. Xiong. 2017. Metal halide perovskite nanomaterials: synthesis and applications. Chem. Sci. 8: 2522–2536. https://doi.org/10.1039/C6SC04474C

[86] Kovalenko, M.V., L. Protesescu and M.I. Bodnarchuk. 2017. Properties and potential optoelectronic applications of lead halide perovskite nano-crystals. Science 358: 745–750. https://doi.org/10.1126/science.aam7093

[87] Oksenberg, E., A. Merdasa, L. Houben, I. Kaplan-Ashiri, A. Rothman, I.G. Scheblykin, et al. 2020. Large lattice distortions and size-dependent bandgap modulation in epitaxial halide perovskite nanowires. Nat. Commun. 11: 489. https://doi.org/10.1038/s41467-020-14365-2

[88] Dasgupta, N.P. and P. Yang. 2014. Semiconductor nanowires for photovoltaic and photoelectrochemical energy conversion. Front. Phys. 9: 289–302. https://doi.org/10.1007/s11467-013-0305-0

[89] Shan, G.-C., Z.-Q. Yin, C.H. Shek and W. Huang. 2014. Single photon sources with single semiconductor quantum dots. Front. Phys. 9: 170–193. https://doi.org/10.1007/s11467-013-0360-6

[90] Lei, J.-C., X. Zhang and Z. Zhou. 2015. Recent advances in MXene: preparation, properties, and applications. Front. Phys. 10: 276–286. https://doi.org/10.1007/s11467-015-0493-x

[91] Yan, Z.-Z., Z.-H. Jiang, J.-P. Lu and Z.-H. Ni. 2018. Interfacial charge transfer in WS_2 monolayer/$CsPbBr_3$ microplate heterostructure. Front. Phys. 13: 138115. https://doi.org/10.1007/s11467-018-0785-z

[92] Novoselov, K.S., D.V. Andreeva, W. Ren and G. Shan. 2019. Graphene and other two-dimensional materials. Front. Phys. 14: 13301. https://doi.org/10.1007/s11467-018-0835-6

[93] Zhang, B., Y. Wang, S. Chou, H. Liu and S. Dou. 2019. Fabrication of superior single-atom catalysts toward diverse electrochemical reactions. Small Methods 3: 1800497. https://doi.org/10.1002/smtd.201800497

[94] Zhang, B., T. Sheng, Y. Wang, S. Chou, K. Davey, S. Dou, et al. 2019. Long-life room-temperature sodium–sulfur batteries by virtue of transition-metal-nanocluster–sulfur interactions. Angew. Chem. Int. Ed. 58: 1484–1488. https://doi.org/10.1002/anie.201811080

[95] Chen, Z., Y. Zhang, S. Chu, R. Sun, J. Wang, J. Chen, et al. 2020. Grain boundary induced ultralow threshold random laser in a single GaTe flake. ACS Appl. Mater. Interfaces 12: 23323–23329. https://doi.org/10.1021/acsami.0c03419

[96] Wang, J., X. Li, B. Wei, R. Sun, W. Yu, H.Y. Hoh, et al. 2020. Activating Basal planes of $NiPS_3$ for hydrogen evolution by nonmetal heteroatom doping. Adv. Funct. Mater. 30: 1908708. https://doi.org/10.1002/adfm.201908708

[97] Xin, B., Y. Pak, S. Mitra, D. Almalawi, N. Alwadai, Y. Zhang, et al. 2019. Self-patterned $CsPbBr_3$ nanocrystals for high-performance optoelectronics. ACS Appl. Mater. Interfaces 11: 5223–5231. https://doi.org/10.1021/acsami.8b17249

[98] Liu, J., K. Chen, S.A. Khan, B. Shabbir, Y. Zhang, Q. Khan, et al. 2020. Synthesis and optical applications of low dimensional metal-halide perovskites. Nanotechnol. 31: 152002. https://doi.org/10.1088/1361-6528/ab5a19

[99] Akkerman, Q.A., G. Rainò, M.V. Kovalenko and L. Manna. 2018. Genesis, challenges and opportunities for colloidal lead halide perovskite nanocrystals. Nature Mater. 17: 394–405. https://doi.org/10.1038/s41563-018-0018-4

[100] Hassanabadi, E., M. Latifi, Andrés. F. Gualdrón-Reyes, S. Masi, S.J. Yoon, M. Poyatos, et al. 2020: Ligand and band gap engineering: tailoring the protocol synthesis for achieving high-quality $CsPbI_3$ quantum dots. Nanoscale 12: 14194–14203. https://doi.org/10.1039/D0NR03180A

[101] Zhang, R., Y. Yuan, J. Li, Z. Qin, Q. Zhang, B. Xiong, et al. 2020. Ni and K ion doped $CsPbX_3$ NCs for the improvement of luminescence properties by a facile synthesis method in ambient air. J. Lumin. 221: 117044. https://doi.org/10.1016/j.jlumin.2020.117044

[102] Wang, R., Y. Muhammad, X. Xu, M. Ran, Q. Zhang, J. Zhong, et al. 2020. Facilitating all-inorganic halide perovskites fabrication in confined-space deposition. Small Methods 4: 2000102. https://doi.org/10.1002/smtd.202000102

[103] Thesika, K. and A. Vadivel Murugan. 2020. Microwave-enhanced chemistry at solid–liquid interfaces: synthesis of all-inorganic $CsPbX_3$ nanocrystals and unveiling the anion-induced evolution of structural and optical properties. Inorg. Chem. 59: 6161–6175. https://doi.org/10.1021/acs.inorgchem.0c00294

[104] Bekenstein, Y., J.C. Dahl, J. Huang, W.T. Osowiecki, J.K. Swabeck, E.M. Chan, et al. 2018. The making and breaking of lead-free double perovskite nanocrystals of cesium silver–bismuth halide compositions. Nano Lett. 18: 3502–3508. https://doi.org/10.1021/acs.nanolett.8b00560

[105] Huang, J., T. Lei, M. Siron, Y. Zhang, S. Yu, F. Seeler, et al. 2020. Lead-free cesium europium halide perovskite nanocrystals. Nano Lett. 20: 3734–3739. https://doi.org/10.1021/acs.nanolett.0c00692

[106] Sutton, R.J., G.E. Eperon, L. Miranda, E.S. Parrott, B.A. Kamino, J.B. Patel, et al. 2016. Bandgap-tunable cesium lead halide perovskites with high thermal stability for efficient solar cells. Adv. Energy Mater. 6: 1502458. https://doi.org/10.1002/aenm.201502458

[107] Liang, J., C. Wang, Y. Wang, Z. Xu, Z. Lu, Y. Ma, et al. 2016. all-inorganic perovskite solar cells. J. Am. Chem. Soc. 138: 15829–15832. https://doi.org/10.1021/jacs.6b10227

[108] Kulbak, M., D. Cahen and G. Hodes. 2015. How important is the organic part of lead halide perovskite photovoltaic cells? Efficient $CsPbBr_3$ Cells. J. Phys. Chem. Lett. 6: 2452–2456. https://doi.org/10.1021/acs.jpclett.5b00968

[109] Zhou, C., Y. Tian, M. Wang, A. Rose, T. Besara, N.K. Doyle, et al. 2017. Low-dimensional organic tin bromide perovskites and their photoinduced structural transformation. Angew. Chem. Int. Ed. 56: 9018–9022. https://doi.org/10.1002/anie.201702825

[110] Li, H., G. Tong, T. Chen, H. Zhu, G. Li, Y. Chang, et al. 2018. Interface engineering using a perovskite derivative phase for efficient and stable $CsPbBr_3$ solar cells. J. Mater. Chem. A 6: 14255–14261. https://doi.org/10.1039/C8TA03811B

[111] Jiang, Y., J. Yuan, Y. Ni, J. Yang, Y. Wang, T. Jiu, et al. 2018. Reduced-dimensional α-$CsPbX_3$ perovskites for efficient and stable photovoltaics. Joule 2: 1356–1368. https://doi.org/10.1016/j.joule.2018.05.004

[112] Qian, C.-X., Z.-Y. Deng, K. Yang, J. Feng, M.-Z. Wang, Z. Yang, et al. 2018. Interface engineering of $CsPbBr_3/TiO_2$ heterostructure with enhanced optoelectronic properties for all-inorganic perovskite solar cells. Appl. Phys. Lett. 112: 093901. https://doi.org/10.1063/1.5019608

[113] Yan, L., Q. Xue, M. Liu, Z. Zhu, J. Tian, Z. Li, et al. 2018. Interface engineering for all-inorganic $CsPbI_2Br$ perovskite solar cells with efficiency over 14%. Adv. Mater. 30: 1802509. https://doi.org/10.1002/adma.201802509

[114] Li, J., J. Duan, X. Yang, Y. Duan, P. Yang and Q. Tang. 2021. Review on recent progress of lead-free halide perovskites in optoelectronic applications. Nano Energy 80: 105526. https://doi.org/10.1016/j.nanoen.2020.105526

[115] Bian, H., H. Wang, Z. Li, F. Zhou, Y. Xu, H. Zhang, et al. 2020. Unveiling the effects of hydrolysis-derived $DMAI/DMAPbI_x$ intermediate compound on the performance of $CsPbI_3$ solar cells. Adv. Sci. 7: 1902868. https://doi.org/10.1002/advs.201902868

[116] Wang, L., B. Fan, B. Zheng, Z. Yang, P. Yin and L. Huo. 2020. Organic functional materials: recent advances in all-inorganic perovskite solar cells. Sustainable Energy Fuels 4: 2134–2148. https://doi.org/10.1039/D0SE00214C

[117] Zeng, Q., X. Zhang, C. Liu, T. Feng, Z. Chen, W. Zhang, et al. 2019. inorganic $CsPbI_2Br$ perovskite solar cells: The progress and perspective. Sol. RRL 3: 1800239. https://doi.org/10.1002/solr.201800239

[118] Li, J., J. Xia, Y. Liu, S. Zhang, C. Teng, X. Zhang, et al. 2020. Ultrasensitive organic-modulated $CsPbBr_3$ quantum dot photodetectors via fast interfacial charge transfer. Adv. Mater. Interfaces 7: 1901741. https://doi.org/10.1002/admi.201901741

[119] Swarnkar, A., R. Chulliyil, V.K. Ravi, M. Irfanullah, A. Chowdhury and A. Nag. 2015. Colloidal $CsPbBr_3$ perovskite nanocrystals: luminescence beyond traditional quantum dots. Angew. Chem. 127: 15644–15648. https://doi.org/10.1002/ange.201508276

[120] Yettapu, G.R., D. Talukdar, S. Sarkar, A. Swarnkar, A. Nag, P. Ghosh, et al. 2016. Terahertz conductivity within colloidal $CsPbBr_3$ perovskite nanocrystals: remarkably high carrier mobilities and large diffusion lengths. Nano Lett. 16: 4838–4848. https://doi.org/10.1021/acs.nanolett.6b01168

[121] Wei, S., Y. Yang, X. Kang, L. Wang, L. Huang and D. Pan, et al. 2017. Homogeneous synthesis and electroluminescence device of highly luminescent $CsPbBr_3$ perovskite nanocrystals. Inorg. Chem. 56: 2596–2601. https://doi.org/10.1021/acs.inorgchem.6b02763

[122] Zhihai, W., W. Jiao, S. Yanni, W. Jun, H. Yafei, W. Pan. 2019. Air-stable all-inorganic perovskite quantum dot inks for multicolor patterns and white LEDs. J. Mater. Sci. 54: 6917–6929. https://doi.org/10.1007/s10853-019-03382-2

[123] Chiba, T., Y. Hayashi, H. Ebe, K. Hoshi, J. Sato, S. Sato, et al. 2018. Anion-exchange red perovskite quantum dots with ammonium iodine salts for highly efficient light-emitting devices. Nature Photon. 12: 681–687. https://doi.org/10.1038/s41566-018-0260-y

[124] Wu, S., S. Zhao, Z. Xu, D. Song, B. Qiao, H. Yue, et al. 2018. Highly bright and stable all-inorganic perovskite light-emitting diodes with methoxypolyethylene glycols modified $CsPbBr_3$ emission layer. Appl. Phys. Lett. 113: 213501. https://doi.org/10.1063/1.5054367

[125] Sasaki, H., N. Kamata, Z. Honda and T. Yasuda. 2019. Improved thermal stability of $CsPbBr_3$ quantum dots by ligand exchange and their application to light-emitting diodes. Appl. Phys. Express 12: 035004. https://doi.org/10.7567/1882-0786/ab0019

[126] Wang, K.-H., B.-S. Zhu, J.-S. Yao and H.-B. Yao. 2018. Chemical regulation of metal halide perovskite nanomaterials for efficient light-emitting diodes. Sci. China Chem. 61: 1047–1061. https://doi.org/10.1007/s11426-018-9325-7

[127] Xu, L., J. Li, B. Cai, J. Song, F. Zhang, T. Fang, et al. 2020. A bilateral interfacial passivation strategy promoting efficiency and stability of perovskite quantum dot light-emitting diodes. Nat. Commun. 11: 3902. https://doi.org/10.1038/s41467-020-17633-3

[128] Lozano, G. 2018. The role of metal halide perovskites in next-Generation lighting devices. J. Phys. Chem. Lett. 9: 3987–3997. https://doi.org/10.1021/acs.jpclett.8b01417

[129] Luo, D., Q. Chen, Y. Qiu, M. Zhang and B. Liu. 2019. Device engineering for all-inorganic perovskite light-emitting diodes. Nanomaterials 9: 1007. https://doi.org/10.3390/nano9071007

[130] Shan, Q., J. Song, Y. Zou, J. Li, L. Xu, J. Xue, et al. 2017. High performance metal halide perovskite light-emitting diode: from material design to device optimization. Small 13: 1701770. https://doi.org/10.1002/smll.201701770

[131] Wang, X., H. Zhou, S. Yuan, W. Zheng, Y. Jiang, X. Zhuang, et al. 2017. Cesium lead halide perovskite triangular nanorods as high-gain medium and effective cavities for multiphoton-pumped lasing. Nano Res. 10: 3385–3395. https://doi.org/10.1007/s12274-017-1551-1

[132] Park, J.H., A. Lee, J.C. Yu, Y.S. Nam, Y. Choi, J. Park, et al. 2019. Surface ligand engineering for efficient perovskite nanocrystal-based light-emitting diodes. ACS Appl. Mater. Interfaces 11: 8428–8435. https://doi.org/10.1021/acsami.8b20808

[133] Wu, H., Y. Zhang, M. Lu, X. Zhang, C. Sun, T. Zhang, et al. 2018. Surface ligand modification of cesium lead bromide nanocrystals for improved light-emitting performance. Nanoscale 10: 4173–4178. https://doi.org/10.1039/C7NR09126E

[134] Fu, Y., H. Zhu, C.C. Stoumpos, Q. Ding, J. Wang, M.G. Kanatzidis, et al. 2016. Broad wavelength tunable robust lasing from single-crystal nanowires of cesium lead halide perovskites ($CsPbX_3$, X = Cl, Br, I). ACS Nano 10: 7963–7972. https://doi.org/10.1021/acsnano.6b03916

[135] Evans, T.J.S., A. Schlaus, Y. Fu, X. Zhong, T.L. Atallah, M.S. Spencer, et al. 2018. Continuous-wave lasing in cesium lead bromide perovskite nanowires. Adv. Opt. Mater. 6: 1700982. https://doi.org/10.1002/adom.201700982

[136] Wang, Z., Z. Luo, C. Zhao, Q. Guo, Y. Wang, F. Wang, et al. 2017. Efficient and stable pure green all-inorganic perovskite $CsPbBr_3$ light-emitting diodes with a solution-processed NiO_x interlayer. J. Phys. Chem. C. 121: 28132–28138. https://doi.org/10.1021/acs.jpcc.7b11518

[137] Liu, Z., J. Yang, J. Du, Z. Hu, T. Shi, Z. Zhang, et al. 2018. Robust Subwavelength single-mode perovskite nanocuboid laser. ACS Nano 12: 5923–5931. https://doi.org/10.1021/acsnano.8b02143

[138] Liu, Z., Q. Shang, C. Li, L. Zhao, Y. Gao, Q. Li, et al. 2019. Temperature-dependent photoluminescence and lasing properties of $CsPbBr_3$ nanowires. Appl. Phys. Lett. 114: 101902. https://doi.org/10.1063/1.5082759

[139] Yang, L., Z. Li, C. Liu, X. Yao, H. Li, X. Liu, et al. 2019. Temperature-dependent lasing of $CsPbI_3$ triangular pyramid. J. Phys. Chem. Lett. 10: 7056–7061. https://doi.org/10.1021/acs.jpclett.9b02703

[140] Bao, C., J. Yang, S. Bai, W. Xu, Z. Yan, Q. Xu, et al. 2018. High performance and stable all-inorganic metal halide perovskite-based photodetectors for optical communication applications. Adv. Mater. 30: 1803422. https://doi.org/10.1002/adma.201803422

[141] Wang, Y., X. Li, V. Nalla, H. Zeng and H. Sun. 2017. Solution-processed low threshold vertical cavity surface emitting lasers from all-inorganic perovskite nanocrystals. Adv. Funct. Mater. 27: 1605088. https://doi.org/10.1002/adfm.201605088

[142] Stylianakis, M., T. Maksudov, A. Panagiotopoulos, G. Kakavelakis and K. Petridis. 2019. Inorganic and hybrid perovskite based laser devices: a review. Materials 12: 859. https://doi.org/10.3390/ma12060859

[143] Pushkarev, A.P., V.I. Korolev, D.I. Markina, F.E. Komissarenko, A. Naujokaitis, A. Drabavičius, et al. 2019. A few-minute synthesis of $CsPbBr_3$ Nanolasers with a high quality factor by spraying at ambient conditions. ACS Appl. Mater. Interfaces 11: 1040–1048 (2019). https://doi.org/10.1021/acsami.8b17396

[144] Liu, Z., Z. Hu, T. Shi, J. Du, J. Yang, Z. Zhang, et al. 2019. Stable and enhanced frequency up-converted lasing from $CsPbBr_3$ quantum dots embedded in silica sphere. Opt. Express 27: 9459. https://doi.org/10.1364/OE.27.009459

[145] Liu, Z., S. Huang, J. Du, C. Wang and Y. Leng. 2020. Advances in inorganic and hybrid perovskites for miniaturized lasers. Nanophotonics 9: 2251–2272. https://doi.org/10.1515/nanoph-2019-0572

[146] Xu, G., Y. Li, J. Yan, X. Lv, Y. Liu and B. Cai. 2019. In-plane self-assembly and lasing performance of cesium lead halide perovskite nanowires. Mater. Res. Lett. 7: 203–209. https://doi.org/10.1080/21663831.2019.1576797

[147] Yan, F., S.T. Tan, X. Li and H.V. Demir. 2019. Light generation in lead halide perovskite nanocrystals: LEDs, color converters, lasers, and other applications. Small 15: 1902079. https://doi.org/10.1002/smll.201902079

[148] Li, X., D. Yu, J. Chen, Y. Wang, F. Cao, Y. Wei, et al. 2017. Constructing fast carrier tracks into flexible perovskite photodetectors to greatly improve responsivity. ACS Nano 11: 2015–2023. https://doi.org/10.1021/acsnano.6b08194

[149] Liu, X., D. Yu, F. Cao, X. Li, J. Ji, J. Chen, et al. 2017. Low-voltage photodetectors with high responsivity based on solution-processed micrometer-scale all-inorganic perovskite nanoplatelets. Small 13: 1700364. https://doi.org/10.1002/smll.201700364

[150] Gui, P., Z. Chen, B. Li, F. Yao, X. Zheng, Q. Lin, et al. 2018. High-performance photodetectors based on single all-inorganic $CsPbBr_3$ perovskite microwire. ACS Photonics 5: 2113–2119. https://doi.org/10.1021/acsphotonics.7b01567

[151] Du, Z., D. Fu, T. Yang, Z. Fang, W. Liu, F. Gao, et al. 2018. Photodetectors with ultra-high detectivity based on stabilized all-inorganic perovskite $CsPb_{0.922} Sn_{0.078} I_3$ nanobelts. J. Mater. Chem. C 6: 6287–6296. https://doi.org/10.1039/C8TC01837E

[152] Zhai, W., J. Lin, C. Li, S. Hu, Y. Huang, C. Yu, et al. 2018. Solvothermal synthesis of cesium lead halide perovskite nanowires with ultra-high aspect ratios for high-performance photodetectors. Nanoscale 10: 21451–21458. https://doi.org/10.1039/C8NR05683H

[153] Han, M., J. Sun, M. Peng, N. Han, Z. Chen, D. Liu, et al. 2019. Controllable growth of lead-free all-inorganic perovskite nanowire array with fast and stable near-infrared photodetection. J. Phys. Chem. C 123: 17566–17573. https://doi.org/10.1021/acs.jpcc.9b03289

[154] Zhang, Z.-X., C. Li, Y. Lu, X.-W. Tong, F.-X. Liang, X.-Y. Zhao, et al. 2019. Sensitive deep ultraviolet photodetector and image sensor composed of inorganic lead-free $Cs_3Cu_2I_5$ perovskite with wide bandgap. J. Phys. Chem. Lett. 10: 5343–5350. https://doi.org/10.1021/acs.jpclett.9b02390

[155] Yang, T., Y. Zheng, Z. Du, W. Liu, Z. Yang, F. Gao, et al. 2018. Superior photodetectors based on all-inorganic perovskite $CsPbI_3$ nanorods with ultrafast response and high stability. ACS Nano 12: 1611–1617. https://doi.org/10.1021/acsnano.7b08201

[156] Xue, M., H. Zhou, G. Ma, L. Yang, Z. Song, J. Zhang, et al. 2018. Investigation of the stability for self-powered $CsPbBr_3$ perovskite photodetector with an all-inorganic structure. Sol. Energy Mater. Sol. Cells. 187: 69–75. https://doi.org/10.1016/j.solmat.2018.07.023

[157] Cen, G., Y. Liu, C. Zhao, G. Wang, Y. Fu, G. Yan, et al. 2019. Atomic-layer deposition-assisted double-side interfacial engineering for high-performance flexible and stable $CsPbBr_3$ perovskite photodetectors toward visible light communication applications. Small. 15: 1902135. https://doi.org/10.1002/smll.201902135

[158] Yang, Y., H. Dai, F. Yang, Y. Zhang, D. Luo, X. Zhang, et al. 2019. All-perovskite photodetector with fast response. Nanoscale Res. Lett. 14: 291. https://doi.org/10.1186/s11671-019-3082-z

[159] Tian, C., F. Wang, Y. Wang, Z. Yang, X. Chen, J. Mei, et al. 2019. Chemical vapor deposition method grown all-inorganic perovskite microcrystals for self-powered photodetectors. ACS Appl. Mater. Interfaces 11: 15804–15812. https://doi.org/10.1021/acsami.9b03551

[160] Du, Z., D. Fu, J. Teng, L. Wang, F. Gao, W. Yang, et al. 2019. $CsPbI_3$ Nanotube photodetectors with high detectivity. Small 15: 1905253. https://doi.org/10.1002/smll.201905253

[161] Ji, Z., Y. Liu, W. Li, C. Zhao and W. Mai. 2020. Reducing current fluctuation of $Cs_3Bi_2Br_9$ perovskite photodetectors for diffuse reflection imaging with wide dynamic range. Sci.e Bull. 65: 1371–1379. (2020). https://doi.org/10.1016/j.scib.2020.04.018

[162] Saleem, M.I., S. Yang, R. Zhi, M. Sulaman, P.V. Chandrasekar, Y. Jiang, et al. 2020. Surface engineering of all-inorganic perovskite quantum dots with quasi core–shell technique for high-performance photodetectors. Adv. Mater. Interfaces 7: 2000360. https://doi.org/10.1002/admi.202000360

[163] Tong, G., M. Jiang, D.-Y. Son, L. Qiu, Z. Liu, L.K. Ono, et al. 2020. Inverse growth of large-grain-size and stable inorganic perovskite micronanowire photodetectors. ACS Appl. Mater. Interfaces 12: 14185–14194. https://doi.org/10.1021/acsami.0c01056

[164] Gao, Y., L. Zhao, Q. Shang, Y. Zhong, Z. Liu, J. Chen, et al. 2018. Ultrathin $CsPbX_3$ nanowire arrays with strong emission anisotropy. Adv. Mater. 30: 1801805. https://doi.org/10.1002/adma.201801805

[165] Hou, Y., L. Wang, X. Zou, D. Wan, C. Liu, G. Li, et al. 2020. Substantially improving device performance of all-inorganic perovskite-based phototransistors via indium tin oxide nanowire incorporation. Small 16: 1905609. https://doi.org/10.1002/smll.201905609

[166] Kim, D.-K., D. Choi, M. Park, K.S. Jeong and J.-H. Choi. 2020. Cesium lead bromide quantum dot light-emitting field-effect transistors. ACS Appl. Mater. Interfaces 12: 21944–21951. https://doi.org/10.1021/acsami.0c06904

[167] Rainò, G., G. Nedelcu, L. Protesescu, M.I. Bodnarchuk, M.V. Kovalenko, R.F. Mahrt, et al. 2016. Single cesium lead halide perovskite nanocrystals at low temperature: fast single-photon emission, reduced blinking, and exciton fine structure. ACS Nano 10: 2485–2490. https://doi.org/10.1021/acsnano.5b07328

[168] Zhou, S., G. Zhou, Y. Li, X. Xu, Y.-J. Hsu, J. Xu, et al. 2020. Understanding charge transport in all-inorganic halide perovskite nanocrystal thin-film field effect transistors. ACS Energy Lett. 5: 2614–2623. https://doi.org/10.1021/acsenergylett.0c01295

Chapter **8**

Impacts of Working Electrode Parameters on Dye-Sensitised Solar Cell Performance

V. Sasirekha[1*], J. Mayandi[2], J. Vinodhini[1], R. Selvapriya[1],
P. Jayabal[3], V. Ragavendran[2] and J.M. Pearce[4]

[1]Department of Physics, Avinashilingam Institute for Home Science and Higher Education for Women, Coimbatore – 641043, Tamil Nadu, India.

[2]Department of Materials Science, School of Chemistry, Madurai Kamaraj University, Madurai – 625 021, Tamil Nadu, India.

[3]Department of Physics, Gobi Arts and Science College, Gobichettipalayam – 638453, Tamil Nadu, India.

[4]Department of Electrical and Computer Engineering, Western University, London, ON, Canada.

8.1   INTRODUCTION

In the world of booming artificial intelligence (AI) and indoor Internet of Things (IoTs) applications, there is a need for uninterrupted low-power supplies for sensors and transducers. To prevent climate destabilisation demands, sustainable renewable energy sources must be opted for powering them. Although solar photovoltaic (PV) power generation

* For Correspondence: V. Sasirekha (sasirekha_phy@avinuty.ac.in)

is the least expensive globally, the capital costs of transitioning the entire global energy system to sustainable energy are challenging. One approach directly provides electricity for low-power appliances with small-scale PV systems [1]. Historically, the PV industry was dominated by first-generation silicon-based single-crystalline and polycrystalline wafers. The second generation thin films were developed that included amorphous silicon, chalcogenide [CdTe, $Cu(In,Ga)(S,Se)_2$, (CIGSSe), $Cu_2ZnSn(S,Se)_4$, and (CZTSSe)], semiconductors, and multijunction III–V solar cells [2–4]. Emerging third-generation solar cells have the potential to be even more accessible, which include dye-sensitised solar cells (DSSCs), quantum dots, perovskite, organic, inorganic, and tandem or multijunction low-cost PV material-based solar cells. Of these new types of solar cells, DSSCs show particular promise to utilise indoor light sources, compared with other conventional solar cells, with added advantages of flexibility and, if needed, some degrees of transparency [5]. The history of DSSCs started with zinc oxide (ZnO), using chlorophyll and rose bengal dyes as sensitisers, in 1971 with an efficiency of 1% [6]. By 1991, Michael Grätzel's research group had attained an efficiency of 7.12% and 7.9% in AM 1.5 solar simulator (750 W/m^2) and with low light levels (83 W/m^2). They achieved this with a 10-micrometre titanium dioxide (TiO$_2$) cubic particle active layer made from a colloidal solution and enabled charge transfer with ruthenium (Ru) complex as a sensitiser. The same DSSC device showed an increase in efficiency of 12% in diffuse daylight and proved the ability of DSSCs better performance under shaded conditions with current densities in the order of 13 mA cm^{-2} [7]. This discovery caused a flurry of activity in the PV community because DSSCs could be researched and manufactured for minimal capital investments. It led to rapid technological progress. The U.S. National Renewable Energy Lab (NREL) maintains a chart to the maximum certified efficiency of 13.0% in DSSC, as shown in Figure 8.1(a) [8]. On the laboratory scale, 14.7% efficiency has been reported [9].

The architecture of DSSC, as shown in Figure 8.1(b), includes a conductive substrate (FTO) to transport charge carriers to the load, a compact layer, a mesoporous active layer (TiO$_2$), a scattering layer, sensitiser, electrolyte, and counter electrode with electrocatalyst (FTO/Pt). The photoanode should possess a high surface area for anchoring sensitiser to harvest the maximum number of photons per unit area to enhance photon-to-electron conversion. The generated carriers should be efficiently collected from the sensitiser to reduce back-electron transfer. The collected electrons should be transported to the load to avoid recombination, and the overall power conversion efficiency can be increased by introducing a compact layer with high electron mobility.

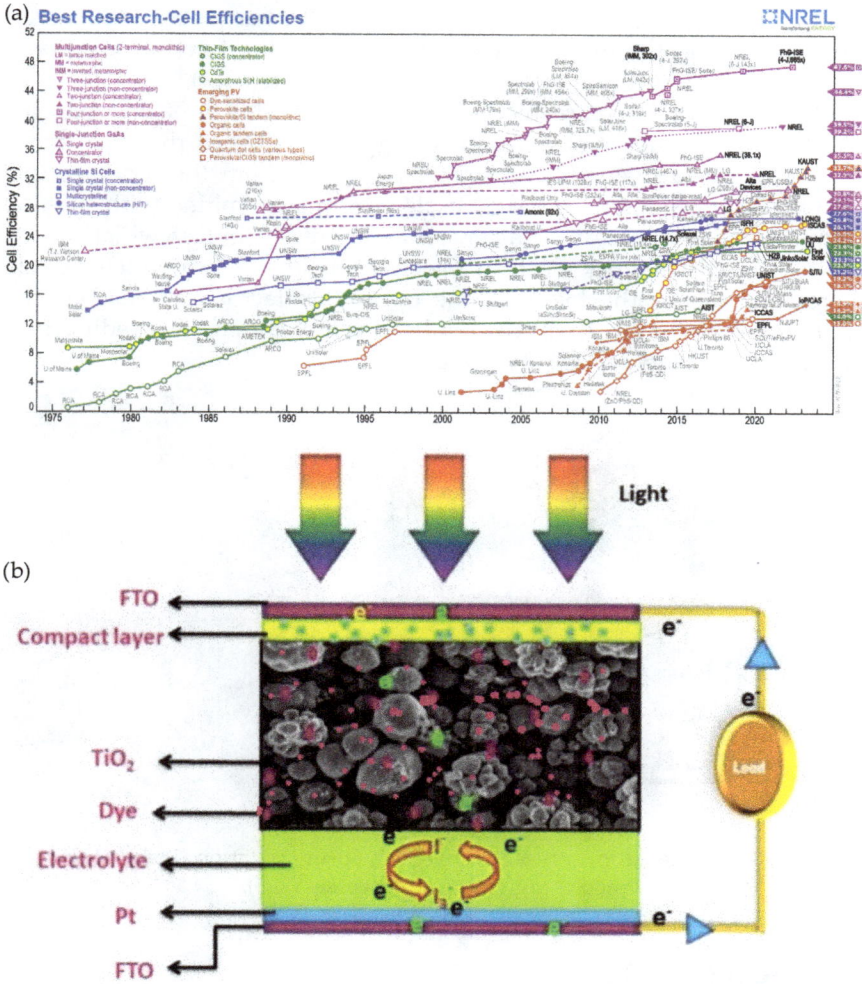

Figure 8.1 (a) Classification of solar cells and the highest efficiency reported in each category from 1976 to till date. The emerging solar cells are highlighted (b) Architecture of DSSC device.

8.2 WORKING PRINCIPLE

In a working DSSC, the incident photons pass through the transparent conductive substrate and sensitisers (S), which are adsorbed on the active layer. Sensitisers trap the photons and utilise the energy to move electrons into an excited state (S*), represented as (1) excited energy carriers are injected into the conduction band (CB) of active layers and (2) return to the ground state with a loss of energy (S$^+$).

When the electrons are injected into the CB band of the semiconductor (3), the oxidised dye molecules must regenerate to start the cycle again so that oxidised dye molecules get an electron from the redox couple (4). On the other hand, the excited electron may combine with a redox couple via back-electron transfer, and the energy is lost which is not desired. Higher energy electron injected into the semiconductor oxide materials reaches the load via the current collector and reaches the counter electrode (5). Then, the oxidised redox couple gets reduced (6) and completes the cycle (Figure 8.2).

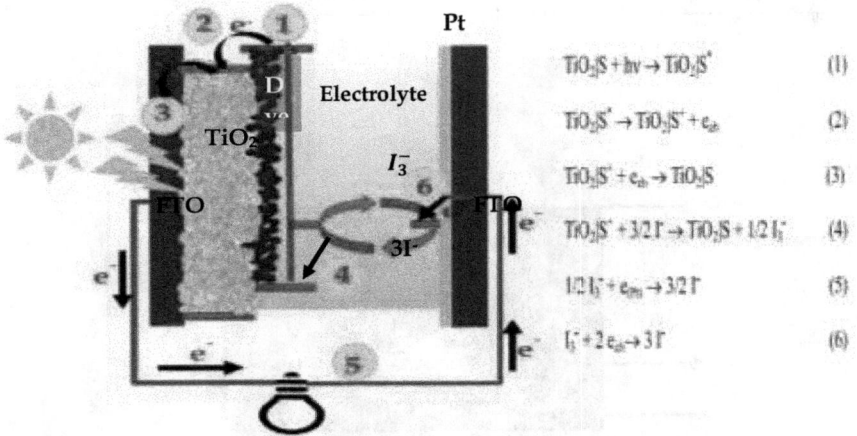

$$TiO_2|S + h\nu \rightarrow TiO_2|S^* \qquad (1)$$

$$TiO_2|S^* \rightarrow TiO_2|S^+ + e_{cb} \qquad (2)$$

$$TiO_2|S^+ + e_{cb} \rightarrow TiO_2|S \qquad (3)$$

$$TiO_2|S^+ + 3/2\,I^- \rightarrow TiO_2|S + 1/2\,I_3^- \qquad (4)$$

$$1/2\,I_3^- + e_{cm} \rightarrow 3/2\,I^- \qquad (5)$$

$$I_3^- + 2e_{cd} \rightarrow 3I^- \qquad (6)$$

Figure 8.2   Schematic representation of sensitiser excitation (1) and step-by-step charge transfer processes (2–5).

8.3   SUBSTRATE

One crucial parameter that determines DSSC's performance is the nature of the conductive substrate and its resistivity. In general, n-type semiconductor materials are usually used as substrates due to their wide band gap, high free electron concentration, and high photons transmittance. Usually, DSSCs are constructed with transparent conducting oxides (TCO) that act as a substrate for preparing the working and counter electrodes and as current collectors. Fluorine-doped tin oxide (FTO) and indium-doped tin oxide (ITO) are the two commonly used TCO substrates. In addition, aluminium-doped zinc oxide glass substrates are also used to prepare photoanodes. The research community mostly chooses FTO or three material properties:

(i) high thermal stability (up to 600°C);

(ii) good conductivity;

(iii) good transparency for visible light [10].

Instead of using FTO and ITO, polymer substrates have recently surged in popularity due to their flexibility and lightweight. Flexible substrates can be made from common thermoplastics like polyethylene terephthalate (PET), polyethylene naphthalate (PEN), nickel-polyethylene, and polyether sulfone (PES) [11, 12]. Even though there is an increase in sheet resistance, which is in the order of 40–100 Ohm/cm^2 compared to 7 Ohm/cm^2 in FTO, flexible substrates allow DSSCs to be formed into various shapes and enable applications that demand malleabilities like clothes and bags as wearable solar cells [13, 14]. However, the high-temperature instability of polymer substrates restricts the use of polymer-based substrates in DSSC. Generally, the TCO substrate used in DSSC fabrication should possess more than 90% transparency to allow the maximum photons to enter the cell and offer higher electrical conductivity for efficient charge transport [15]. In addition, metal and metal wire-based substrates were also investigated as working and counter electrodes. Class I (Ti, stainless steel, W, and Zn) metal substrates were widely used instead of class II (Al, Ni, Co, and Pt) metals because the latter becomes an insulating layer during annealing, which prevents electron transfer from TiO$_2$ to the conductive metal substrate. Class I metal-based working electrodes are sandwiched against glass-FTO-Pt based counter electrodes or PET/PEN/ITO/Pt (PEDOT/SWCNTs). In these configurations, illumination was done at the counter electrode side (i.e. back illumination). Similarly, stainless steel (Ni, carbon steel)/Pt, and Ti metal-based counter electrodes against the glass (PET)/FTO/TiO$_2$ working electrodes were used for DSSCs [16].

8.4 COMPACT LAYER AND BLOCKING LAYER

The primary function of the compact layer is to avoid direct contact between TCO and electrolyte to prevent the recombination of charges. This compact layer should also offer good adhesion for the photoactive mesoporous layer on the TCO substrate. Typically, TiO$_2$ compact layers made from the hydrolysis of the different molar concentration of TiCl$_4$ (pre-treatment with annealing) is known to improve the efficiency of DSSCs. The thickness of the compact layer also plays a vital role in the performance of DSSC. Generally, thicknesses of the compact layers at or below 20 nm offer higher power conversion efficiency, while a thick compact layer film causes a reduction in the optical transmittance, which in turn decreases the power conversion efficiency. If the layer is too thin, it does not shield it from recombination and the loss of carriers. For instance, in a recent study, TiO_2 compact layers with different thicknesses of 13, 19, 25, 38, and 50 nm were prepared using hydrolysis of 40 mM $TiCl_4$ as a function of dipping time, and the maximum efficiency was reported for 25 nm thickness compact layer. Higher efficiency was noted with the addition of this compact layer when compared to bare cells due to the reduction in recombination at the FTO-electrolyte interface [17]. FESEM images of untreated and pretreated FTO plates are shown in Figure 8.3 [18].

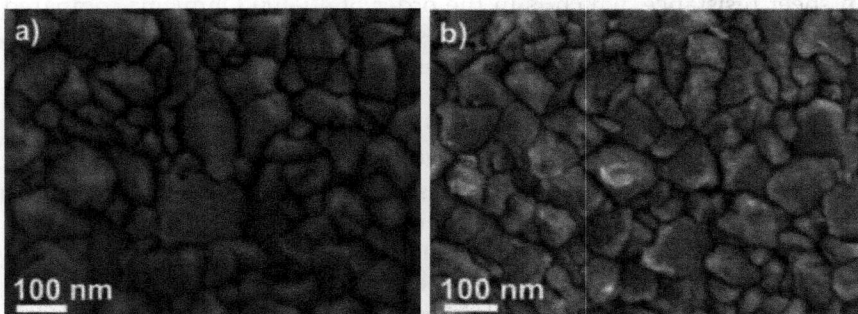

Figure 8.3   Field-emission scanning electron microscopy images at high magnification for (a) untreated and (b) treated with $TiCl_4$ FTO substrates [18].

Researchers were also interested in a SnO_2-based compact layer for more electron mobility than a conventional TiO_2 compact layer made from $TiCl_4$ pre-treatment. From the Hall effect measurements, the maximum value of carrier concentration of 9.166×10^{16} cm^{-3} was obtained for SnO_2 compact layer prepared from 40 mM precursor solution, and it exhibited the lowest resistivity value of 1.967×10^{-2} Ω-cm and the mobility of the sample was also high compared to $TiCl_4$ pretreated sample as shown in Figure 8.4.

It is well established that, in addition to the compact layer, a blocking layer formed between mesoporous and sensitiser are known as $TiCl_4$ post-treated photoanode, which further boosts the DSSC performance. For instance, a ZnO blocking layer with different thicknesses from 55–310 nm was investigated; out of this, 120 nm offered more efficiency with an increase in short-circuit current and open-circuit voltage [19]. TiO_2 blocking layers with a high charge carrier density of $1.25 \times 1{,}019$ cm^{-3} have also been prepared using spray pyrolysis instead of the conventional dip coating method [20]. Other than $TiCl_4$-based blocking layers, ZnO, SnO_2, MgO, Al_2O_3, and $CaCo_3$-based higher CB edge materials have also been investigated [21–24]. The compact and blocking layers reduce recombination and offer a high surface area for anchoring more dye molecules. Compact and blocking layers can also be made of a combination of nanomaterials (e.g. ZnO as a compact layer material and TiO_2 from $TiCl_4$ used as a blocking layer; similarly, TiO_2/ZnO, $SnO_2/TiCl_4$, $TiCl_4/SnO_2$ as a compact and blocking layer were used, respectively) [25].

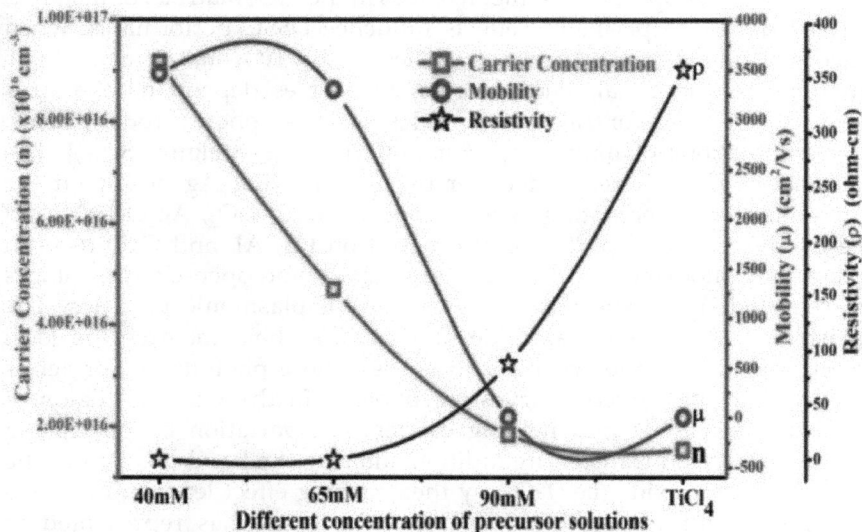

Figure 8.4 Electrical Parameters carrier concentration, mobility, and resistivity were measured using Hall effect SnO_2 pretreated and $TiCl_4$ pretreated FTO substrate.

8.5 MESOPOROUS ACTIVE LAYER

There are many metal oxides materials (i.e. ZnO, SnO_2, TiO_2, Fe_2O_3, V_2O_5, and Nb_2O_5) and perovskite materials (i.e. $ZnTiO_3$, $SrTiO_3$, $CaTiO_3$, and $BaTiO_3$), which have been utilised for the preparation of active layers in working electrodes [26, 27]. They were further used as doped materials with some elements, i.e. as N, S, or as hetero/core-shell structures such as CdS/TiO_2, $SrTiO_3/TiO_2$, ZnO/TiO_2, SnO_2/TiO_2, TiO_2/MgO, TiO_2/Fe_2O_3 [28–31]. While considering pure semiconductor oxide materials, TiO_2 is the most investigated due to its physiochemical stability against both temperature and environmental extremes and most importantly it is less toxic. The function of active layer materials is to collect the excited charge carriers from the sensitiser and transport them to the TCO. It can also act as a charge conversion agent by trapping photons directly from different regions of the electromagnetic spectrum (e.g. N-doping harvests visible photons, and Y- and Er-doping trap IR photons). Further, TiO_2 optical excitation energy is also reduced by various doping such as tin (Sn), iron (Fe), chromium (Cr), vanadium (V), zinc (Zn), silver (Ag), copper–nitrogen (CuN), magnesium (Mg), sulphur (S), barium (Ba), and cobalt (Co) and the resultant DSSC performances have been reported [32].

Plasmonic photoanodes have also been investigated by incorporating noble metal nanoparticles or metal oxides in the TiO_2 matrix. For instance, localised surface plasmonic effects influence DSSC performance when incorporating Ag and Au nanoparticles in the TiO_2 matrix as a single or bimetallic material. The reported efficiencies depend on the sizes and shapes of Ag or Au nanoparticles such as spheres, rods, prisms, or combinations of multi-shaped nanoparticles [see Figure 8.5(a)]. The plasmonic effect is realised either by immobilising Ag or Au on the TiO_2 matrix or preparating as a composite Au/Ag-TiO_2, Ag_2O-TiO_2, and Fe_2O_3-TiO_2. Other reports are also based on Cu, Al, and CuO used in plasmonic photoanodes. The increase in short- and open-circuit voltages is reported up to specific concentrations of plasmonic particles. The plasmonic metal or metal oxide nanoparticles help increase the local electric field near the sensitiser to harvest more photons and/or act as co-sensitiser by directly harvesting photons in the window region of the used sensitiser and helping carrier transportation by decreasing the back-electron transfer. In addition, altering the band position of the semiconductor oxide (i.e. TiO_2) by the charging effect leads to increases in open-circuit voltage [33–36]. The entire process is represented in Figure 8.5(b).

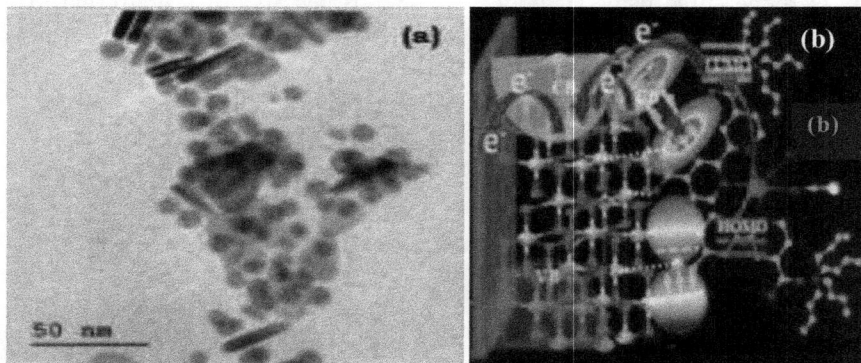

Figure 8.5 (a) HRTEM image of multi-shaped Ag nanoparticles. (b). Schematic representation of charge transfer and local field enhancement due to the presence of metal nanoparticles in the TiO_2 matrix [34].

In addition, composites of TiO_2 with carbon/graphene/rGO carbon-based materials have been investigated and resulted in an increase in conductivity compared with pure semiconductor oxide materials.

Recently reports show that graphitic carbon materials (e.g. C_3N_4 and C_3N_5) can make a cascading band alignment with TiO_2, which decreases the charge transport resistance. This heterojunction minimises the back-electron transfer [37, 38].

8.6   MORPHOLOGY

The morphology of the metal oxide material plays an essential role in the overall performance of DSSCs. The commonly reported morphological features in TiO_2 are nanoparticles, rods/wires/tubes/fibres, 2D sheets, belts, and 3D nano/microstructures. Each morphology has unique impacts on the DSSC performance. The TiO_2 particle [Figure 8.6(a)] morphology at the nanoscale offers a higher surface area, which enables it to adsorb a more significant number of sensitisers, but it also offers higher charge transport resistance as they have more grain boundaries with dead ends for charge transport. The one-dimensional structure [Figure 8.6(b)] offers a unidirectional path for charge transport, but the surface area is less when compared with particles. In the case of three-dimensional structures, several aspects need to be considered. Three-dimensional structures reported include solid spheres, hollow spheres, core-shell structures, and most importantly three-dimension structures made up of grains [Figure 8.6(c)], either small spherical particles, rod-like structures, or sheets [Figure 8.6(d)]. 3D spheres possess unique characteristics of offering optimum surface area, charge transport via its grains, and the ability to optimise size. They also serve as a scattering centre by increasing the optical path length of the incident photons inside the photoanode active layer. There were also investigations based on mixed morphologies (e.g. rod/particles, rod/sphere, sphere/particles, etc.). This mixed morphology [Figures 8.6(e) and 8.6(f)] can be obtained by mixing different weight ratios of different morphologies or by direct single-step synthesis. In addition to the high surface area and good charge transport, mesoporous photo anode material is equally essential for the diffusion of the electrolyte to rejuvenate the sensitiser molecules. The effect of the morphology of particles, tubes, and rods has been investigated, and it is expected that nanotubes offer advantages for light scattering, electron transportation, and fewer trap sites [39]. For instance, TiO_2 as nanoparticles, nanofibers, hierarchical spheres, and ellipsoid spheres are successfully obtained via titanium n-butoxide and acetic acid solvothermal reactions. DSSC consisting of nanoparticles, nanoparticles/nanowire, nanoparticles/nanotubes TiO_2 were sensitised with the sensitisers N719 and 3,7'-bis(2-cyano-1-acrylicacid)-10-ethyl-phenothiazine with chenodeoxycholic acid as co-adsorbent. Nanotube-nanoparticle used to produce photoanodes exhibited better performance among the investigations due to high active surface area confirmed by dye loading and facile charge transfer [40]. Nanoparticle (NP), nanofiber (NF), hierarchical TiO_2 sphere (HTS), and ellipsoid TiO_2 sphere (ETS) were investigated, and the HTS morphology achieved higher efficiency due to large surface area and superior light scattering ability [41].

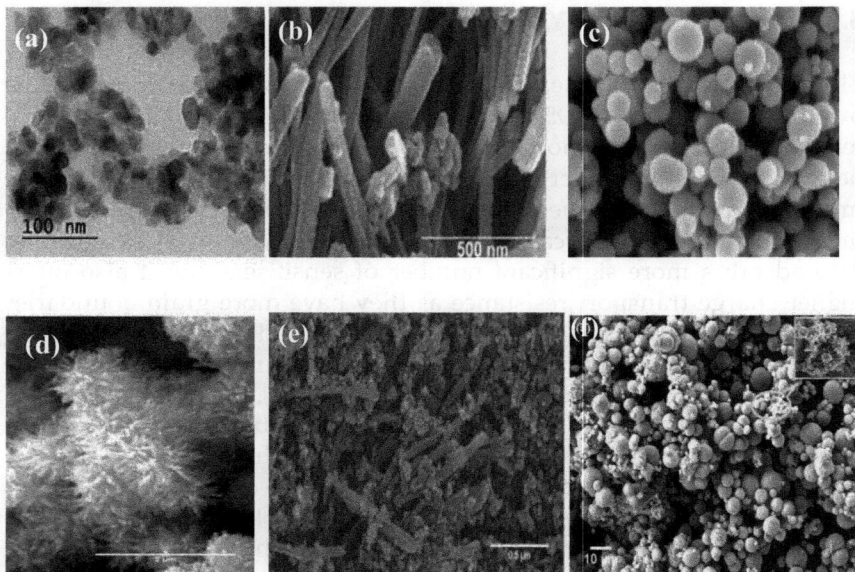

Figure 8.6   (a) particle morphology, (b) rod-like morphology, (c) sphere-like morphology, (d) 3D morphology with rod granules, (e) mixed morphology of rods and particles, and (f) mixed morphology of spheres and particles.

8.7   ACTIVE LAYER PREPARATION

There are two approaches in preparing active layers for DSSCs:

1. direct growth;
2. making thin film layers using appropriate coating techniques from semiconductor oxide paste.

8.7.1   Direct Growth

Direct mesoporous layers can be obtained using the hydrothermal method. Within a Teflon-coated vessel autoclave, FTO plates were directly immersed in the precursor solution and subjected to various temperatures ranging from 110°C to 240°C with different time duration starting from two to several hours. Vertically grown rods were usually obtained in the natural growth process. Dip coating, spray pyrolysis, liquid jet via electrospinning, and electrochemical anodisation are other possible approaches to obtaining TiO_2 layers directly on the conductive substrate.

8.7.2 TiO_2 Nanostructures Powder Preparation

For the thin film TiO_2 coating, TiO_2 is required in powder form. There are numerous well-reported methods, including sol-gel, hydrothermal, solvothermal, reflex, and microwave techniques to obtain TiO_2 powders from different precursors, mostly from titanium isopropoxide, titanium butoxide, titanium nitrate, and titanium oxalate. Depending on the initial concentration of precursor and solvents (water, ethanol, acetone, propanol, glycerol, ethylene glycol, ionic solvents, and acetic acid) with and without surfactant/morphology tuning agents, one can obtain different TiO_2 morphology and phases as individuals or mixtures. Sometimes commercially available TiO_2 powders (P25) were also used for the active layer paste preparation, and the performance of DSSCs made from user-prepared TiO_2 powders have been compared. Microwave-based synthesis techniques have emerged as one of the most efficient, simple, and fastest ways to create TiO_2 nanostructures, which provide high yields in short reaction times using domestic microwave ovens [Figure 8.7(a)], user-modified domestic microwave ovens and commercial microwave ovens [42–48]. In the case of the domestic microwave oven, synthesis can be done at different microwave powers, but temperature cannot be controlled as in a commercial microwave reactor. Using low-cost and readily available domestic microwave ovens is appealing, but there are restrictions on the choice of solvents.

8.7.2.1 *TiO_2 Paste Preparation*

Alpha terpinol, ethyl cellulose, acetic acid, water, ethanol, triton-X, tween-80, acetylacetone, polyethylene glycol, nitric acid and PVA, PVP, polystyrene, and natural polymers are primarily used to prepare paste either as individuals or mixtures by using a ball milling process, paint shaker or mortar and pistol [49–53]. Paste quality is essential to ensure that the final cell has a crack-free active layer. The solvent/dispersing agent choice depends on the nature of the substrate on which the TiO_2 paste will be coated as a film. Low-temperature curing is a critical parameter for flexible substrates like PET and PEN, while ITO-based DSSCs offer better performance if the curing temperature is less than 350 °C, unlike conventional FTO plates. It is because the sheet resistance increases for ITO at curing temperatures of 450 °C. The viscosity of the paste is also an important consideration in obtaining uniform layers without cracks, as it is one of the factors that decide the interparticle connection. Cracks in the TiO$_2$-coated thin film are shown in Figure 8.7(b).

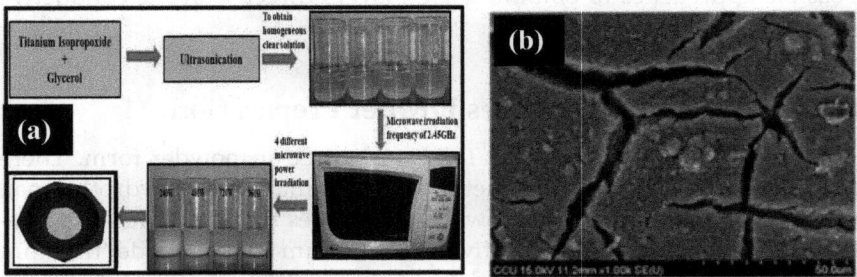

Figure 8.7   (a) TiO$_2$ nanopowders were obtained using the domestic microwave oven (b) Cracks in the TiO$_2$-coated film [55].

Moreover, different printing techniques require different viscosity (e.g. screen printing requires highly viscous pastes compared to spin coating) [54]. In addition, the choice of solvents decides the adhesion of TiO$_2$ film with the substrate. Further interconnection between TiO$_2$ particles can be enhanced by mechanical or cold pressing of the TiO$_2$ layer, which reduces the dead ends for the charge carrier transport [55].

8.7.2.2   Thin Film Active Layer Preparation

Thin film active layers deposited from the prepared pastes usually are obtained via spin coating, inkjet printing, digital printing, doctor blade method, screen printing, and roll-to-roll printing methods [56]. Uniform layers can be obtained with the spin coating technique with user-defined rotation per minute and a restricted active area. High viscous pastes cannot be used, however, for spin coating. Viscosity also plays a significant role in determining the thickness of the coated layer. The thickness can also be tuned for thicker films by repeating the coating process. Among the methods mentioned above, screen printing and roll-to-roll printing are utilised in the commercial sectors to obtain larger uniform and highly reproducible active areas from the prepared pastes.

Similarly, the thickness can also be varied by multiple coating and floating processes. Roll-to-roll printing is suited for flexible commercial substrates. G24 power, a leading DSSC module manufacturer, has adopted roll-to-roll printing to prepare flexible DSSC modules [57]. Small active areas (0.25 cm^2 to 1 cm^2) of working electrodes are prepared for laboratory investigation, but the output power from these active areas cannot be used for most practical/commercial applications. Hence, the active area has to be increased to power the most devices of interest. When increasing the active area, it has to be patterned for effective charge carriers collection, like conductive finger and bus bars in silicon-based solar cells. For the current collection, conductive fingers have to be patterned depending on the chosen area and design of the active layer. Usually, silver conductive fingers are deposited either by screen printing [58, 59]. Silver conductive fingers are efficient in charge transport, but stability is an issue when exposed to iodine-based liquid electrolytes. Hence, it needs a proper mask in such a way as to avoid contact between the electrolyte to silver conductive fingers. There are also other investigations based on carbon-based conductive fingers for charge collection. This direction of investigation may help to reduce the cost of DSSC production. The corrosion of metal conductive fingers can be reduced using quasi-solid electrolytes or solid electrolytes. Recently, most liquid electrolytes have been replaced by gel polymer electrolytes because of their higher ionic conductivity, and the possibility of tuning ionic conductivity by adding a couple of precursors compared to solid electrolytes [60]. The Japanese electronics company Ricoh, which has launched a commercial DSSC for indoor applications, found a solution to rectify the shortcoming of solid-state DSSC by the hole transport layer, consisting of an organic p-type semiconductor and a solid additive. Their DSSC module of 5.2 cm × 8.4 cm produced 230 microwatts of power, and they have extended the application of solid-state DSSC modules to charge IoT sensing devices and LEDs even under shadow conditions [61, 62].

8.7.3 Small Lab-Scale DSSC Towards Large Areas for Practical Applications

Protype DSSC is fabricated in an open and closed configuration so that working and counter electrodes are sandwiched using a binder clip with a separator. The liquid electrolyte is poured in between the layers of the sandwich. In the case of a closed configuration, the holes are drilled in the counter electrode to fill the electrolyte. Then polymer films (surlyn), available in different thicknesses, are utilised as separators to avoid shorts between the working and counter electrodes and blockers from the liquid electrolyte leakage in small active area DSSC devices. For large surface area DSSCs, the following steps must be followed both in the working electrode, counter electrode, and device assembly. The steps for the counter electrode include: (i) holes are drilled for electrolyte injection, (ii) cleaning the substrate, (iii) platinisation either using paste or drop-casting of hexachloroplatinic, (iv) acid solution, (v) paste curing/ reduction at 450 °C, (vi) conductive finer patterning, and (vii) masking.

Similarly for working electrode steps needed are: (i) substrate cleaning, (ii) making a compact layer, (iii) active layer coating in the predefined pattern, (iv) paste curing, (v) post-treatment and annealing,

(vi) coating of silver conductive fingers, and (vii) masking. The working electrode preparation steps are represented in Figure 8.8. Then both working and counter electrodes are sandwiched together using a hot melt press, and the electrolytes are injected through counter electrode holes. Finally, the holes are sealed. The schematic representation for the assembled DSSC device is given in Figure 8.8(g).

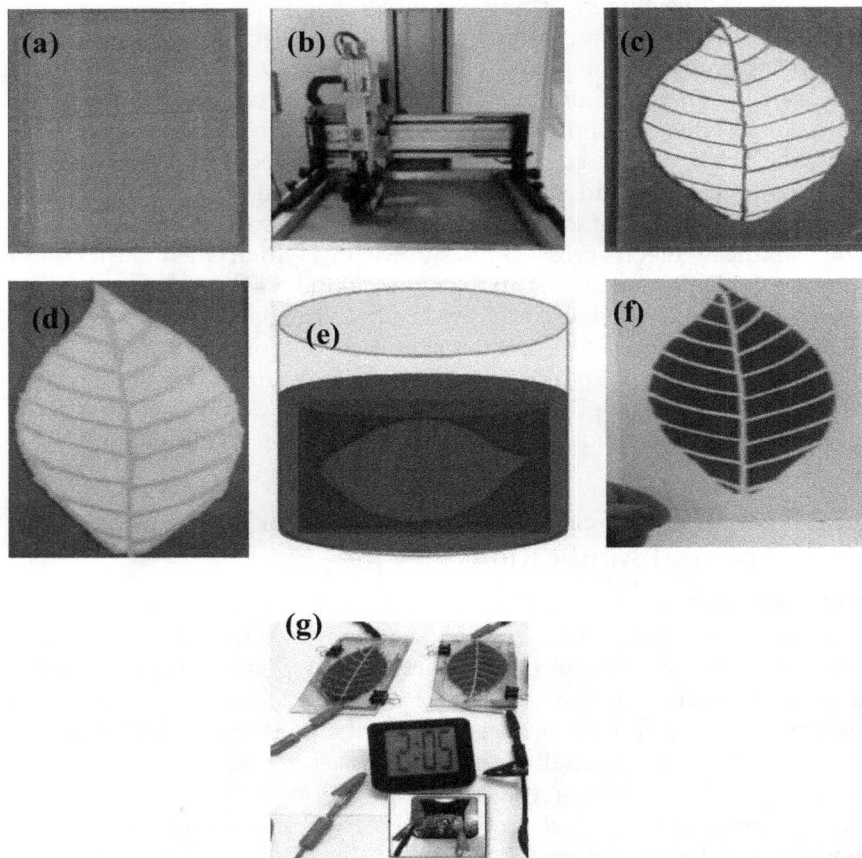

Figure 8.8 (a) FTO plate, (b) screen printer used for the preparation, (c) active layer obtained, (d) active layer with a silver conductive finger, (e) sensitisation process, (f) sensitised working electrode, and (g) fabricated DSSC device powering digital watch.

8.7.4 DSSC Cell to Module Towards Commercialisation

The assembled DSSC large-area devices (cells) can be integrated for the efficient utilisation of available light dependent on the practical requirements of the application (e.g. whether the device needs more current or more voltage to enable the working of the devices). Based on the required criteria to power the devices with DSSCs, the module is assembled using cells in parallel or series in combination. It can be done as monolithic, Z, and W configurations, as shown in Figure 8.9 [63].

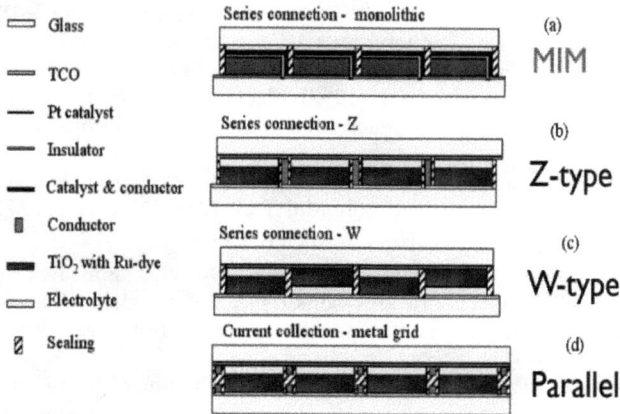

Figure 8.9   Schematic configuration to integrate DSSC cells into modules [63].

8.8   CONCLUSION

This chapter discusses how DSSCs can provide sustainable solutions for an uninterrupted power supply for low-power-consuming devices, even under diffused room ambient light conditions. The architecture of DSSCs and details of working electrodes, including conductive substrates, compact layers, and mesoporous active layers, were also discussed. The preparation of TiO_2 powders, doping, composite structures, junction formations, and unique and mixed morphological influences on the DSSC device performance was also reviewed. Paste preparation and coating methods were also discussed, as well as considerations for scaling up from small-area lab-scale DSSC preparation to large-area DSSC devices and DSSC devices integrated into modules for real-world applications.

REFERENCES

[1]   Grafman, L. and J.M. Pearce. 2021. To Catch The Sun. Arcata. CA: Humbodlt State University Press.

[2]   Dhruba, B.K., X. Chen, T.R. Rana and S. Kim. 2020. Chalcogenide semiconductors based thin-film solar cell. Int. J. Photoenergy. Special Issue.

[3] Heriche, H., I. Bouchama, N. Bouarissa, Z. Rouabah and A. Dilmi. 2017. Enhanced efficiency of Cu (In, Ga) Se_2 solar cells by adding Cu_2ZnSn (S, Se)$_4$ absorber layer. Optik 144: 378–386.

[4] Nakamura, M., K. Yamaguchi, Y. Kimoto, Y. Yasaki, T. Kato and H. Sugimoto. 2019. Cd-free Cu (In, Ga)(Se, S)$_2$ thin-film solar cell with record efficiency of 23.35%. IEEE J. Photovoltaics 9: 1863–1867.

[5] Wu, C., B. Chen, X. Zheng and S. Priya. 2016. Scaling of the flexible dye sensitised solar cell module. Sol. Energy Mater. Sol. Cells 157: 438–446.

[6] Tributsch, H. and M. Calvin. 1971. Electrochemistry of excited molecules: photo-electrochemical reactions of chlorophylls. Photochem. Photobiol. 14: 95–112.

[7] O'regan, B. and M. Gratzel. 1991. A low-cost, high-efficiency solar cell based on dye-sensitised colloidal TiO_2 films. Nature 353: 737–740.

[8] https://www.nrel.gov/pv/cell-efficiency.html

[9] Kakiage, K., Y. Aoyama, T. Yano, K. Oya, J.-I. Fujisawa, M. Hanaya. 2015. Highly-efficient dye-sensitised solar cells with collaborative sensitisation by silyl-anchor and carboxy-anchor dyes. Chem. Commun. 51: 15894–15897.

[10] Patni, N., P. Sharma, M. Parikh, P. Joshi and S.G. Pillai. 2018. Cost effective approach of using substrates for electrodes of enhanced efficient dye sensitised solar cell. Mater. Res. Express 5: 095509.

[11] Su, H., M. Zhang, Y.-H. Chang, P. Zhai, N.Y. Hau, Y.-T. Huang, et al. 2014. Highly conductive and low cost Ni-PET flexible substrate for efficient dye-sensitised solar cells. ACS Appl. Mater. Interfaces 6: 5577–5584.

[12] 천. 김. 안성훈. 2010. Flexible dye-sensitised solar cell and preparation method thereof. South Korea Patent KR100994902B1, 16.11.2010.

[13] Zhao, B., Z. He, X. Cheng, D. Qin, M. Yun, M. Wang, et al. 2014. Flexible polymer solar cells with power conversion efficiency of 8.7%. J. Mater. Chem. C 2: 5077–5082.

[14] Chowdhury, M.S., K.S. Rahman, V. Selvanathan, A.K.M. Hasan, M.S. Jamal, N.A. Samsudin, et al. 2021. Recovery of FTO coated glass substrate via environment-friendly facile recycling perovskite solar cells. RSC Adv. 11: 14534–14541.

[15] Hu, Z., J. Zhang, Z. Hao, Q. Hao, X. Geng and Y. Zhao. 2011. Highly efficient organic photovoltaic devices using F-doped SnO_2 anodes. Appl. Phys. Lett. 98: 66.

[16] Balasingam, S.K., M.G. Kang and Y. Jun. 2013. Metal substrate based electrodes for flexible dye-sensitised solar cells: fabrication methods, progress and challenges. Chem. Commun. 49: 11457–11475.

[17] Choi, H., C. Nahm, J. Kim, J. Moon, S. Nam, D.-R. Jung, et al. 2012. The effect of $TiCl_4$-treated TiO_2 compact layer on the performance of dye-sensitised solar cell. Curr. Appl Phys. 12: 737–741.

[18] Lokman, M.Q., S. Shaban, S. Shafie, F. Ahmad, H. Yahaya, R.M. Rosnan, et al. 2021. Improving Ag-TiO_2 nanocomposites' current density by $TiCl_4$ pretreated on FTO glass for dye-sensitised solar cells. Micro Nano Lett. 16: 381–386.

[19] Yeoh, M.-E. and K.-Y. Chan. 2019. Efficiency enhancement in dye-sensitised solar cells with ZnO and TiO_2 blocking layers. J. Electron. Mater. 48: 4342–4350.

[20] Musila, N., M. Munji, J. Simiyu, E. Masika, R. Nyenge and M. Kineene. 2018. Characteristics of TiO_2 compact layer prepared for DSSC application. Traektoria Nauki= Path of Science 4: 3006–3012.

[21] Law, M., L.E. Greene, A. Radenovic, T. Kuykendall, J. Liphardt and P. Yang. 2006. ZnO^- Al_2O_3 and ZnO^- TiO_2 core-shell nanowire dye-sensitised solar cells. J. Phys. Chem. B 110: 22652–22663.

[22] Kay, A. and M. Grätzel. 2002. Dye-sensitised core-shell nanocrystals: improved efficiency of mesoporous tin oxide electrodes coated with a thin layer of an insulating oxide. Chem. Mater. 14: 2930–2935.

[23] Kaur, M. and N.K. Verma. 2014. $CaCO_3/TiO_2$ nanoparticles based dye sensitised solar cell. J. Mater. Sci. Technol. 30: 328–334.

[24] Jayabal, P., S. Gayathri, V. Sasirekha, J. Mayandi and V. Ramakrishnan. 2015. Effect of electronic-insulating oxides overlayer on the performance of zinc oxide based dye sensitised solar cells. J. Photochem. Photobiol. A: Chemistry 305: 37–44.

[25] Bhuiyan, MMH., F. Kabir, M.S. Manir, M.S. Rahaman, M.R. Hossain, P. Barua, et al. 2021. Effect of combination of natural dyes and the blocking layer on the performance of DSSC. Solar Cells: Theory, Mater. Recent Adv. 313.

[26] Okamoto, Y. and Y. Suzuki. 2014. Perovskite-type $SrTiO_3$, $CaTiO_3$ and $BaTiO_3$ porous film electrodes for dye-sensitised solar cells. J. Ceram. Soc. Jpn. 122: 728–731.

[27] Sasikala, R., M. Kandasamy, V. Ragavendran, S. Suresh, V. Sasirekha, S. Murugesan, et al. 2022. Perovskite zinc titanate-reduced graphene oxide nanocomposite photoanode for improved photovoltaic performance in dye-sensitised solar cell. Physica B: Condensed Matter. 646: 414300.

[28] Guo, E. and L. Yin. 2015. Tailored $SrTiO_3/TiO_2$ heterostructures for dye-sensitised solar cells with enhanced photoelectric conversion performance. J. Mater. Chem. A 3: 13390–13401.

[29] Taguchi, T., X.-T. Zhang, I. Sutanto, K.-I. Tokuhiro, T.N. Rao, H. Watanabe, et al. 2003. Improving the performance of solid-state dye-sensitised solar cell using MgO-coated TiO_2 nanoporous film. Chem. Commun. 19: 2480–2481.

[30] Qureshi, A.A., S. Javed, H.M.A. Javed, A. Akram, MS. Mustafa, U. Ali, et al. 2021. Facile formation of $SnO_2–TiO_2$ based photoanode and Fe_3O_4@rGO based counter electrode for efficient dye-sensitised solar cells. Mater. Sci. Semicond. Process. 123: 105545.

[31] Hussein, A.M., A.V. Iefanova, R.T. Koodali, B.A. Logue and R.V. Shende. 2018. Interconnected ZrO_2 doped ZnO/TiO_2 network photoanode for dye-sensitised solar cells. Energy Rep. 4: 56–64.

[32] Dubey, R.S., S.R. Jadkar and A.B. Bhorde. 2021. Synthesis and characterisation of various doped TiO_2 nanocrystals for dye-sensitised solar cells. ACS Omega 6: 3470–3482.

[33] Song, D.H., H.-S. Kim, J.S. Suh, B.-H. Jun and W.-Y. Rho. 2017. Multi-shaped Ag nanoparticles in the plasmonic layer of dye-sensitised solar cells for increased power conversion efficiency. Nanomaterials 7: 136.

[34] Selvapriya, R., T. Abhijith, V. Ragavendran, V. Sasirekha, V.S. Reddy, J.M. Pearce, et al. 2022. Impact of coupled plasmonic effect with multishaped silver nanoparticles on efficiency of dye sensitised solar cells. J. Alloys Compd. 894: 162339.

[35] Kaur, N., V. Bhullar, D.P. Singh and A. Mahajan. 2020. Bimetallic implanted plasmonic photoanodes for TiO_2 sensitised third generation solar cells. Sci. Rep. 10: 7657.

[36] Zarick, H.F., W.R. Erwin, A. Boulesbaa, O.K. Hurd, J.A. Webb, A.A. Puretzky, et al. 2016. Improving light harvesting in dye-sensitised solar cells using hybrid bimetallic nanostructures. ACS Photonics 3: 385–394.

[37] Xu, J., G. Wang, J. Fan, B. Liu, S. Cao and J. Yu. 2015. g-C_3N_4 modified TiO_2 nanosheets with enhanced photoelectric conversion efficiency in dye-sensitised solar cells. J. Power Sources 274: 77–84.

[38] Nien, Y.-H., Z.-R. Yong, J.-C. Chou, C.-H. Lai, P.-Y. Kuo, Y.-C. Lin, et al. 2021. Improving photovoltaic performance of dye-sensitized solar cell by modification of photoanode with gC_3N_4/TiO_2 nanofibers. IEEE Trans. Electron Devices 68: 4982–4988.

[39] Arla, S.K., N.S.S. Konidena, S.S. Sana, S.K. Godlaveeti and V.K.N. Boya. 2022. Structural and morphological effect of multidimensional TiO_2 nanostructures on the dye-sensitised solar cells performance. Sādhanā 47: 129.

[40] Gnida, P., P. Jarka, P. Chulkin, A. Drygała, M. Libera, T. Tański, et al. 2021. Impact of TiO_2 nanostructures on dye-sensitised solar cells performance. Materials 14: 1633.

[41] Liao, J.-Y., J.-W. He, H. Xu, D.-B. Kuang and C.-Y. Su. 2012. Effect of TiO_2 morphology on photovoltaic performance of dye-sensitised solar cells: nanoparticles, nanofibers, hierarchical spheres and ellipsoid spheres. J. Mater. Chem. 22: 7910–7918.

[42] Vinodhini, J., J. Mayandi, R. Atchudan, P. Jayabal, V. Sasirekha and J.M. Pearce. 2019. Effect of microwave power irradiation on TiO_2 nanostructures and binder free paste screen printed dye sensitised solar cells. Ceram. Int. 45: 4667–4673.

[43] Ramakrishnan, V.M., S. Pitchaiya, N. Muthukumarasamy, K. Kvamme, G. Rajesh, S. Agilan, et al. 2020. Performance of TiO_2 nanoparticles synthesised by microwave and solvothermal methods as photoanode in dye-sensitised solar cells. Int. J. Hydrogen Energy 45: 27036–27046.

[44] Ramakrishnan, V.M., S. Sandberg, N. Muthukumarasamy, K. Kvamme, P. Balraju, S. Agilan, et al. 2019. Microwave-assisted solvothermal synthesis of worms-like TiO_2 nanostructures in submicron regime as light scattering layers for dye-sensitised solar cells. Mater. Lett. 236: 747–751.

[45] Huang, C.-H., Y.-T. Yang and R.-A. Doong. 2011. Microwave-assisted hydrothermal synthesis of mesoporous anatase TiO_2 via sol–gel process for dye-sensitised solar cells. Microporous Mesoporous Mater. 142: 473–480.

[46] Hart, J.N., R. Cervini, Y.-B. Cheng, G.P. Simon and L. Spiccia. 2004. Formation of anatase TiO_2 by microwave processing. Sol. Energy Mater. Sol. Cells 84: 135–143.

[47] Hart, J.N., D. Menzies, Y.-B. Cheng, G.P. Simon and L. Spiccia. 2017. A comparison of microwave and conventional heat treatments of nanocrystalline TiO_2. Sol. Energy Mater. Sol. Cells 91: 6–16.

[48] Sahu, K., M. Dhonde and V.V.S. Murty. 2021. Microwave-assisted hydrothermal synthesis of Cu-doped TiO_2 nanoparticles for efficient dye-sensitised solar cell with improved open-circuit voltage. Int. J. Energy Res. 45: 5423–5432.

[49] Gemeiner, P. and M. Mikula. 2013. Efficiency of dye sensitised solar cells with various compositions of TiO_2 based screen printed photoactive electrodes. Acta Chimica Slovaca 6: 29–34.

[50] Mori, R., T. Ueta, K. Sakai, Y. Niida, Y. Koshiba, L. Lei, et al. 2011. Organic solvent based TiO_2 dispersion paste for dye-sensitised solar cells prepared by industrial production level procedur. J. Mater. Sci. 46: 1341–135.

[51] Karthick, S.N., K.V. Hemalatha, C.J. Raj, A. Subramania and H.-J. Kim. 2012. Preparation of TiO_2 paste using poly (vinylpyrrolidone) for dye sensitised solar cells. Thin Solid Films 520: 7018–7021.

[52] Arla, S.K., S.S. Sana, V. Badineni and V.K.N. Boya. 2020. Effect of nature of polymer as a binder in making photoanode layer and its influence on the efficiency of dye-sensitised solar cells. Bull. Mater. Sci. 43: 1–7.

[53] Alwin, S., X.S. Shajan, K. Karuppasamy and K.G.K. Warrier. 2017. Microwave assisted synthesis of high surface area TiO_2 aerogels: a competent photoanode material for quasi-solid dye-sensitised solar cells. Mater. Chem. Phys. 196: 37–44.

[54] Weerasinghe, H.C., F. Huang and Y.-B. Cheng. 2014. Fabrication of flexible dye sensitised solar cells on plastic substrates. Nano Energy 2: 17–189.

[55] Meen, T.H., J.K. Tsai, Y.S. Tu, T.C. Wu, W.D. Hsu and S.-J. Chang, 2014. Optimization of the dye-sensitized solar cell performance by mechanical compression. Nanoscale Res. Lett. 9: 523.

[56] Mariani, P. and L. Vesce and A. Di Carlo. 2015. The role of printing techniques for large-area dye sensitised solar cells. Semicond. Sci. Technol. 30: 104003.

[57] GCell by G24 Power, https://gcell.com/about-g24-power/manufacturing -process.

[58] Chandrakala, K.R.M.V., K.R. Teja, N.S. Kumar, P.V.P. Raghavendra and L.N.V.S.B. Majji. 2017. Fabrication of high efficient dye sensitized solar cell using eosin blue sensitizer. International Journal on Electrical Engineering and Informatics 9: 185–183.

[59] Gonzalez-Flores, C.A., D. Pourjafari, R. Escalante, E.J. Canto-Aguilar, A.V. Poot, J.M. Andres Castán, et al. 2021. Influence of redox couple on the performance of ZnO dye solar cells and minimodules with benzothiadiazole-based photosensitizers. ACS Appl. Energy Mater. 5: 14092–14106.

[60] Raut, P., V. Kishnani, K. Mondal, A. Gupta and S.C. Jana. 2022. A review on gel polymer electrolytes for dye-sensitised solar cells. Micromachines 13: 680.

[61] Ricoh (The Japanese electronics company). 2020. https://www.pv-magazine.com/2020/02/05/ricoh-launches-solar-cell-for-indoor-applications/. pv magazine - Photovoltaics Mark. Technol. (online).

[62] Ricoh (online). 2014. https://www.ricoh.com/technology/rd/f_runner/fr15.

[63] Aslam, A., U. Mehmood, M.H. Arshad, A. Ishfaq, J. Zaheer, A.U.l.H. Khan, et al. 2020. Dye-sensitised solar cells (DSSCs) as a potential photovoltaic technology for the self-powered internet of things (IoTs) applications. Sol. Energy 207: 874–892.

Chapter **9**

Nanostructured Metal Oxides for Photocatalytic Water Splitting

Anu Kumari[#], Shaswati Jyoti[#] and Sonalika Vaidya[*]

Institute of Nano Science and Technology, Knowledge City, Sector-81, Sahibzada
Ajit Singh Nagar, Punjab –140306, India

9.1   INTRODUCTION

At a global level, the issues about energy and the environment can be
solved by constructing a clean energy system. Hydrogen is considered
as one of the clean energy source and is expected to become one of
the significant source of energy due to its high energy capacity and
environmentally friendly byproducts. Hydrogen is used in fuel cells
and chemical industries [1]. It is primarily produced from fossil fuels
(e.g. natural gas) by steam reforming. The main disadvantage of these
methods is that fossil fuels are a non-renewable source, and there is
an emission of CO_2 during the production of hydrogen from fossil
fuels. Thus, to make the production of hydrogen more environment
friendly, it is desired that hydrogen be produced from a renewable
and abundant source, i.e. water. For producing hydrogen from water,

*For Correspondence: Sonalika Vaidya (svaidya@inst.ac.in)
[#]The authors have an equal contribution

a catalyst is required. The term "catalyst" is a derivative of the Greek word καταλύειν, *kataluō*, which means "loosen" or "untie". Catalysts are chemical species that enhance the reaction rate by forming intermediates without participating in the reaction. Catalysis is the process of speeding up a reaction. There are mainly two ways for producing hydrogen/oxygen from water using catalysis, viz. (i) electrolysis of water and (ii) photocatalytic or photoelectrochemical water splitting. In this chapter, we focus on water-splitting using photocatalysis.

9.2 MECHANISM OF PHOTOCATALYTIC WATER SPLITTING

Photocatalytic water-splitting reaction (artificial photosynthesis) is an uphill reaction that requires Gibb's energy more significant than 237 kJ to split water into hydrogen and oxygen, which is overcome by a catalyst. The first demonstration of overall water splitting through UV-light-induced electrocatalysis (photoelectro-catalysis) was introduced by Fujishima and Honda on TiO_2 electrodes in 1972 [2]. After that, various other metal oxides, like titanates, niobates, etc., were reported, which have been used for photocatalytic hydrogen evolution. The most common feature of a heterogeneous catalyst is that the semiconducting metal oxide and other kinds of semiconductors should have a suitable band gap. Figure 9.1 shows schematic for the photocatalytic process. In the photocatalytic process, when the light of a suitable wavelength is made to fall on semiconductors having an appropriate band gap, electrons move from the valence band to the conduction band, creating holes in the valence band. These electrons and holes drift to the surface and thus participate in the reaction. The electrons cause a reduction of the water molecules leading to the formation of hydrogen, while the holes cause oxidation of the water molecules to form oxygen [3]. The entire process takes place in three main steps:

(i) Generation of photo-induced charge carriers upon light irradiation.

(ii) Separation of charge carriers and migration to their respective reaction site.

(iii) Photo-generated charge carriers participated in the reactions on the surface of the catalyst.

The catalyst's size, shape, crystal structure, and crystallinity influence the separation of photo-generated electrons and holes and their migration to the reaction sites. With higher crystallinity, the amount of defects present is less. The photo-generated electrons and holes are trapped at the defect site. These sites may also act as recombination centers. Both these processes affect photocatalytic activity. When the size of the photocatalyst is decreased, the migration of photo-generated electrons and holes to the surface is facilitated. The distance between their generation site and the surface, i.e. the reaction site decreases [3].

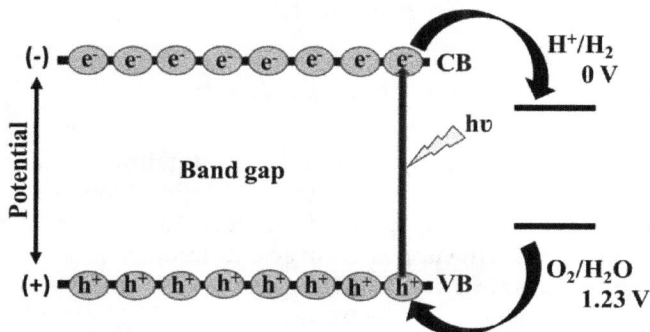

Figure 9.1 Schematic showcasing water splitting using a photocatalyst.

The presence of active sites at the surface is another important criterion for improving the photocatalytic performance of the catalyst. The electrons and holes generated by shining light are likely to recombine with each other, even if they possess suitable thermodynamic potentials for water splitting if no active sites are present over the surface for the reactions. Thus, the photocatalytic performance of the catalyst can also be promoted by co-catalysts, usually metal or metal oxide deposited on the surface of the semiconductor. The co-catalysts provide reaction sites and also help in the collection of charge carriers [3].

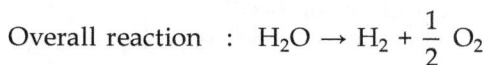

Oxidation   : $H_2O \rightarrow 2H^+ + \dfrac{1}{2} O_2 + 2e^-$

Reduction   : $2H^+ + 2e^- \rightarrow H_2$

Overall reaction   : $H_2O \rightarrow H_2 + \dfrac{1}{2} O_2$

9.3 ESSENTIAL CONDITIONS FOR THE MATERIAL FOR THE PHOTOCATALYST

The standard Gibbs free energy change for the water-splitting reaction is 237 kJ/mol or 1.23 eV. The essential criteria for a semiconductor photocatalyst for water splitting are a suitable band gap and the band edge positions (conduction and valence band edge levels) vs. NHE. Semiconductors having a band gap energy 3.0 eV > E_g > 1.23 eV show photocatalytic activity under UV and visible range of the solar spectrum. For hydrogen evolution, the position of the conduction band edges minimum should be more negative than 0 V vs NHE (pH = 0); whereas, for oxygen evolution, the position of the valence band maxima should be more positive than 1.23 V vs NHE (pH = 0). It suggests that a semiconductor's minimum band gap requirement or minimum photon energy required to act as a photocatalyst for water splitting is 1.23 eV.

9.4 HISTORY OF PHOTOCATALYSIS

In 1901, chemist Giacomo Ciamician while investigating whether chemical reactions could be carried out by "light and light alone", observed that certain chemical reactions responded to specific wavelengths only. The term "photocatalysis" first appeared in scientific journals as a keyword in 1911 [4]. In 1972, Fujishima and Honda reported photo-assisted production of H_2 from water using TiO_2 electrodes under UV light irradiation. The authors fabricated an electrochemical cell, wherein the TiO_2 electrode was connected to a black platinum electrode. Upon irradiation, the current flows from the Pt to the TiO_2 electrode, along with the evolution of O_2 at the TiO_2 electrode and hydrogen at the Pt electrode [2]. With this breakthrough in the 1970s, the application of metal oxides in photocatalysis became an important area of research. Several other photoanodes were also studied, like $SrTiO_3$ [5], SnO_2 [6], $KTaO_3$ [7], WO_3 [8], etc. The band gap of most of these semiconducting oxides is more significant than 3.2 eV, which means that the UV light was suitable for exciting electrons in these oxides. Thus, most semiconductors could not use sunlight due to their large band gaps. In 1977, Fujishima and Honda observed that CdS could be used as a visible light photocatalyst as it showed strong absorption at less than 520 nm. However, it was observed that the anodic reaction occurring using CdS caused its dissolution, which was prevented by using appropriate reducing agents, such as I^-, S^{2-} etc., into the electrolyte solution [9].

Thus, in general, certain elements have been used in heterogeneous photocatalysis [3], which were categorized into four classes:

(a) Metal sulfide, oxides, and nitrides-based catalysts with d^0 and d^{10} metal ions.

(b) Alkali, alkaline earth metals, and lanthanides-based catalysts.

(c) To introduce impurity levels as dopants using transition metal cations having partially filled d-orbitals such as (Rh^{3+}, Ni^{2+}, and Cr^{3+})

(d) To use co-catalysts, e.g., some transition metals and oxides (Pt, Rh, Au), NiO, and RuO_2.

We briefly discuss various kinds of oxides reported as photocatalysts for water-splitting reactions.

9.5 BINARY METAL OXIDES

9.5.1 Titanates

Researchers have worked extensively with titanates since TiO_2 was first reported to be used as a photocatalyst in 1972 [2]. Since the 1970s, much work has been done with TiO_2 by applying various modifications like different crystal structures, dopants, co-catalysts, etc. In the last seven years, many review articles have been published which are based on TiO_2 [10–12]. TiO_2 crystallizes in three structures: rutile, anatase, and brookite, containing TiO_6 octahedra. They are connected through distinct edges and shared corners. The typical edges are two in rutile, three in brookite, and four in anatase. The band gap of rutile is 3.0 eV, and anatase is 3.15 eV. Due to the large band gap in TiO_2, it is unsuitable as a visible light photocatalyst. In general, titanates have a band gap greater than 3 eV. These exhibit good photostability and corrosion resistance and are n-type semiconductors, thus serving as photoanodes. Doping is one of the strategies to improve the photocatalytic activity of TiO_2 and other titanates, as well as to shift their band gap in the visible light region. Single-element, double-element, and tri-element doping are reported. Doping results in the creation of donors or acceptor levels in the forbidden region. It also decreases the band gap. Transition metals, alkaline, and rare earth metals are commonly used as dopants. Chemical doping of TiO_2 with transition metal ions having partially filled d-orbitals induces a visible light response in TiO_2 [13]. These ions do not show any significant reactivity for water splitting. In 2002, Kato and Krudo [14] that simultaneous co-doping of Sb^{5+} and Cr^{3+} in TiO_2 resulted in its use as a photocatalyst for oxygen evolution in visible light. Pt or Au loading on conducting TiO_2 has been shown to improve the performance of the modified TiO_2 in visible light toward hydrogen production from a water-ethanol solution [15]. Doping with nitrogen in TiO_2 [16, 17] is found to be most effective for photocatalysis, which has been attributed to the comparable atomic sizes of nitrogen and oxygen, small ionization energy, and stability. Preethi, L.K. et al. [18] synthesized biphasic TiO_2 containing anatase and rutile and triphasic TiO_2 nanotubes containing all three phases, viz. anatase, rutile, and brookite. They evaluated the efficiency of the catalysts toward hydrogen evolution, wherein they observed that the triphasic system of TiO_2 was more efficient than the biphasic system. Zhang Yan et al. [19] worked with blue TiO_2(B) single-crystalline nanorods and showed that the photocatalytic activity is enhanced due to their rod shape and oxygen vacancies in TiO_2, which prevented the recombination of electrons and holes. Diaz L. et al. [20] reported using non-noble metals as co-catalysts instead of costly noble metals like Pt, Au, Ag, or Pd to prepare low-cost M/TiO_2 using the impregnation method. In the presence of both UV and visible light, M/TiO_2 was observed to give higher rates of hydrogen evolution than bare TiO_2, wherein it was observed that Cu/TiO_2 achieved almost 80% of the H_2 production rate of that of Pt/TiO_2. In 2020, Yang Libin et al. [21] reported the photo-assisted deposition of clusters of Ru or Au as a co-catalyst on anatase TiO_2 to enhance the catalyst's performance toward water splitting. This enhancement was attributed to the chemical bonds, i.e. Ru-O-Ti and Au-O-Ti bonds, which acted as traps for the electron and showed a surface plasmon resonance (SPR) effect.

There are numerous reports on titanates [22–24]. We take $SrTiO_3$, one of the most studied titanates, as an example to showcase their importance and strategies adopted in titanates in improving photocatalytic performance. Doping generates cation vacancies, which promote hydrogen evolution in the photocatalytic processes. In one of the studies, it has been reported [25] that doping by nitrogen, which replaces oxygen, in $SrTiO_3$ increases its visible light activity toward photocatalysis. Cr^{3+} and Ta^{5+} doped at Ti^{4+} sites in $SrTiO_3$ have been shown to increase the hydrogen evolution rate to almost twice that observed for $SrTiO_3$ doped with only Cr under visible light. Yu et al. [26] reported similar results showing that simultaneous doping of Cr and B in $SrTiO_3$ increased the hydrogen production rate to almost twice that of Cr-doped $SrTiO_3$. Co-doping of Cr/Ta ions in $SrTiO_3$ [27] resulted in the formation of a donor level due to the 3d orbital of Cr^{3+} above the O-2p valence band. Efforts to use co-catalysts by loading them with a catalyst and forming a heterojunction without changing the perovskite structure were also used to enhance the photocatalytic activity in visible light. It is shown by citing one example wherein a composite formed from N-doped $SrTiO_3$ and TiO_2 resulted in improving the photocatalytic performance due to the creation of the unique energy band structure [28]. In another example, a multijunction formed from $SrTiO_3/TiO_2$ nanotube/N-doped TiO_2 composite resulted in forming a particular band in the depletion region that was formed due to TiO_2 nanotubes [29]. It resulted in improved photocatalytic performance. Manchala et al. reported a $SrTiO_3/CdS/carbon$ nanospheres-based photocatalytic system that showed hydrogen evolution exhibiting a rate of 3085 $\mu mol\ h^{-1}\ gcat^{-1}$ in the sunlight [30]. Tuning the nature of exposed surfaces is another strategic approach to improve the photocatalytic performance of the oxide. We have showcased using $SrTiO_3$ wherein the nature of exposed surfaces were tuned by changing the solvent used during synthesis [5]. We discuss this example in detail in subsequent section.

9.5.2 Tantalates and Niobates

Kato and Kudo have reported [31, 32] that alkali metal-based tantalates having a general formula, $ATaO_3$ (A: Li, Na, and K), showed increased photocatalytic activity, having order Li> Na>>K, for water splitting under UV irradiation. They further doped various lanthanides and alkaline earth metals in $NaTaO_3$, wherein they observed enhanced activity which they attributed to the decreased particle size and ordered surface area. Among all the different doping studied by them, doping of La resulted in the highest activity having a quantum yield of about 56% at 270 nm [32]. Shimizu and co-workers first developed hydrated perovskite having a layered structure and general formula as $A_2A'Ta_2O_7$ wherein A was taken as H, K, Rb, and A' were taken as Sr and La. These showed increased photocatalytic activity without even using co-catalysts [33, 34]. Many transition metal oxides-based tantalates containing metals, such as Cd, Pb, Ag, and Bi, have been studied for photocatalytic water-splitting [35, 36].

Pure Nb_2O_5 with a band gap of 3.4 eV is inactive under the threshold of the UV region, whereas Ta-based oxides with a band gap under 4 eV, are active under UV irradiation. Use of co-catalysts and intercalation in Nb_2O_5 has increased the system's efficiency in participating in HER [37]. Domen and co-workers, in 1986, first developed $K_4Nb_6O_{17}$ that efficiently produced hydrogen from an aqueous methanol solution without any aid under a Xe light source [38, 39]. On using co-catalysts like NiO [40], Au [41], Pt [42], and Cs [43], this niobate was able to evolve hydrogen and oxygen from water. Similarly, $ANbO_3$ where A was taken as Na, K [44, 45], $ANbWO_6$ wherein A was taken as Rb, Cs [46], and nickel oxide [47] loaded catalyst showed photocatalytic activity toward overall splitting of water under UV irradiation. Under UV light, strontium niobates, viz. $SrNb_2O_6$ [48], $Sr_2Nb_2O_7$ [48–50], and $Sr_5Nb_4O_{15}$ [51] are observed to be efficient photocatalysts for the evolution of hydrogen and oxygen. $Ca_2Nb_2O_7$ [51] and $Ba_5Nb_4O_{15}$ [50, 51] are other examples of niobates that were shown to act as photocatalysts for water splitting. However, the quantum yields for water-splitting reactions were found to be less than their strontium counterparts. Layered perovskite-based niobates that were capable of exchanging and having a general formula as $A(M_{n-1}Nb_nO_{3n+1})$ wherein A was taken as Na, K, Rb, Cs, and M was taken as La, Ca, and Sr have been shown to give photocatalytic activity for H_2 evolution under Hg lamp from a mixture of water and methanol [52, 53]. Reports about changes in the photocatalytic activity due to varying crystal structures are also made. Chen and co-workers reported the activity order of a series of perovskites, synthesized using the solid-state method and having general formula as $ABi_2Nb_2O_9$ (A: Ca, Sr, Ba) to be varying as Sr > Ba > Ca by conventional solid-state method [54].

Though there are plethora of examples of titanates, niobates and tantalates being used as photocatalysts [55–61]; we, by giving a few examples in this chapter, have tried to highlight the importance of these oxides in the field of material science, especially in the application of photocatalysis.

9.5.3   Other Metal Oxides

Under UV irradiation, various d^{10} metal oxides like In^{3+} [62], Ga^{3+} [63], Sn^{4+} [64], Sb^{5+} [65], and a group of p-block metal oxides with co-catalysts like RuO_2 or Pt show effective photochemical water splitting [66]. Tungstates and molybdates-based heterogeneous photocatalysts have shown activity only in the UV region [67]. Under UV irradiation and in the presence of sacrificial reagents, $Na_2W_4O_{13}$ [68] and $Bi_2W_2O_9$ [69] were observed to evolve H_2 and O_2. f-block elements, in combination with other metals, behave as promising photocatalysts. For instance, CeO_2 was reported to produce O_2 from aqueous solutions which contained Fe^{3+} and Ce^{4+} as electron acceptors [70]. $BaCeO_3$ in the presence of CH_3OH and $AgNO_3$ as sacrificial agents resulted in H_2 and O_2 respectively [71]. Crystal structure and electronic states at the surface play an active role in modulating the catalytic activity of the material. Zhang Xiandi et al. reported the synthesis of WO_3 via a wet chemical approach, wherein introducing oxygen vacancy in the WO_3 band structure helped to tune the photocatalytic efficiency of the system [72]. Jun Wang et al. used monoclinic and hexagonal crystalline structures of WO_3 to check the variation in the catalytic properties and found that monoclinic WO_3 shows higher catalytic activity than hexagonal WO_3 [73]. They also used graphene oxide to increase the charge transfer efficiency. These systems show more excellent stability in the durability test, wherein the monoclinic system was more efficient than the hexagonal one. Wen-Cheng Ke et al. reported [74] that p-n heterojunction of $SiO_2/Ag_2O/Zn(O, S)$ formed from p-type Ag_2O and n-type Zn (O, S) loaded on mesoporous silica resulted in the evolution of hydrogen. There was a significant improvement in the catalytic activity in this case, which was attributed to the synergetic effects of various factors such as band structure, active surface oxygen, increased photo absorption, increased electron-hole separation, and low charge transfer resistance. Another p-n heterojunction system reported by Gang Zhou et al. [75], with multiple components (CuO, ZnO, and Au) wherein large active surface area and synergistic effects of CuO, ZnO and formed heterojunctions extended the light absorption range and photocatalytic activity of the material.

9.6 ROLE OF STRUCTURAL PARAMETERS IN ENHANCING PHOTOCATALYTIC EFFICIENCY

There has been development of many novel catalysts for photocatalysis. The size and morphology of the catalyst, surface area, band gap, crystal structure, exposed facets, preferred orientation, towards a particular crystal plane are some critical structural parameters responsible for affecting the catalytic efficiency of the catalyst. External factors such as electric polarization, magnetic polarization, metal co-catalyst (discussed earlier), and type of scavenger or electrolyte used also affect the catalytic reaction rate. We discuss here the effect of a few structural parameters on the photocatalytic efficiency of the catalyst.

9.6.1 Effect of Size

Decreasing the catalyst size in the nano-regime provides a higher surface-to-volume ratio than the bulk catalyst, thereby allowing more reaction sites to react simultaneously at the surface. By decreasing the catalyst size, the transport of charge carriers from bulk to the surface is increased. Photon absorption is increased in smaller particles, which is supposed to enhance the photocatalytic activity of the catalyst [76]. However, a further decrease in particle size to quantum dots leads to photon scattering, thus decreasing the photocatalytic activity.

9.6.2 Effect of Morphology

The photocatalytic activity of the catalyst can be increased by tuning the morphology, which increases the active site and specific surface area of the catalyst. We explain this by discussing a few examples. Bora Seo et al. reported that spherical-shaped Ni_2P nanoparticles exhibited higher hydrogen evolution activity when compared with rods shaped Ni_2P [77]. Takeshi Kimijima et al. studied the photocatalytic activity of $SrTiO_3$ with varying morphology, viz. cube, sphere, and flake-like. They observed that nanoflakes were bound by {110} surface, which resulted in their higher photocatalytic activity than other kinds of morphology for H_2 evolution [78].

9.6.3 Effect of Crystal Structure

The crystal structure of the semiconductor plays an essential role in altering the band gap or band position of the semiconductor. Also, different surface arrangements of the atoms lead to a difference in the adsorption of reactant molecules at the surface of the catalyst, influencing the charge separation and reduction abilities in the surface reactions. Peng Li et al. [79] synthesized two different crystal structures of perovskite $NaNbO_3$ orthorhombic and cubic. They observed higher photocatalytic activity of cubic $NaNbO_3$, which they attributed to a more dispersive conduction band in $NaNbO_3$ with cubic crystal structure than orthorhombic-$NaNbO_3$ resulting in higher migration of photo-induced electrons. Gian Luca Chiarello et al. [80] studied the effect of using three different kinds of crystalline phases of TiO_2, viz. anatase, rutile, and brookite, on photocatalytic activity for hydrogen evolution. The authors observed that brookite modified by Pt nanoparticles showed increased photocatalytic activity for hydrogen production from the vapors of a mixture containing methanol and water under UV-visible irradiation compared to other crystalline phases. The brookite phase was also showcased to have high selectivity toward CO_2 formation and low CO production. We have highlighted this influence in Sr-Ti-O system, viz. $SrTiO_3$ (cubic perovskite) and $SrO-(SrTiO_3)_n$ (n = 1 and 2) (Ruddlesden-Popper) [81]. We observed that the order for photocatalytic hydrogen evolution activity of these three nanostructures was $Sr_3Ti_2O_7 > Sr_2TiO_4 > SrTiO_3$.

9.6.4 Effect of Exposed Facets

The properties of a catalyst are highly dependent on the nature of exposed surfaces as it involves the arrangement of atoms on the surface and their coordination. Adsorption of water molecules and their reduction/oxidation depends on the surface energy of the particular facet, which could further improve the reaction rates. Moreover, oxidation and reduction reactions occurring at different facets improve charge separation efficiency and thus photocatalytic activity. Moussab Harb et al. [82] investigated the impact of the exposed facets [including (010), (110), (001), and (121)] of $BiVO_4$ on photocatalytic water-splitting reaction. From DFT-based calculations, they predicted that the (001) surface of $BiVO_4$ is highly active for both hydrogen and oxygen evolution reactions, whereas the (010) surface is only active for OER reactions. Jie Meng et al. [83] reported that $BaZrO_3$ with cubic morphology and having {001}/{011} as exposed facets showed higher activity for photocatalytic hydrogen evolution in pure water. The authors attributed the increased activity to modified band structure resulting due to exposed {001}/{011} facets and also on the larger surface area. The high reduction in the energy obtained from DFT studies was attributed to the existence of {001} facets. Jianan Li et al. [84] reported that the development of facet junction between {001} and {111} facet in $ZnFe_2O_4$ nanoparticles improved the charge separation by decreasing the recombination rate of the charge carriers generated by light, thereby enhancing the photocatalytic degradation efficiency over gaseous toluene. In one of our studies, the size, morphology, and nature of exposed surfaces of $SrTiO_3$ were tuned by varying the synthetic conditions. Herein, the dielectric constant of the solvent (water and polyols) was varied [5]. The shape obtained was a cube with water, a flower-like structure formed by an assembly of hexagonal-shaped particles with ethylene glycol, edge truncated cuboids with PEG-300 and cuboids with PEG (polyethylene glycol)-400 as solvent. [001] was observed to be the major exposed facet of the synthesized $SrTiO_3$, wherein [01$\bar{1}$] surface was also observed along with the oxide synthesized using ethylene glycol and PEG-300 as the solvent. It was observed that the presence of [01$\bar{1}$] surface decreased the hydrogen evolution performance of the catalyst.

9.6.5 Effect of Electrical Polarization

The presence of the internal field within the crystal changes the bending of the bands at the interfaces. It also provides a driving force for separating the photo-generated electrons and holes, which helps to improve the catalytic efficiency of the catalyst. Applying an external electric field or electrical polarization helps in further transferring of charge carriers to opposite reaction sites and enhances the adsorption capacity of the charged ions on the surface. Here we cite examples from both the photocatalytic and electrocatalytic splitting of water. Yongfei Cui et al. [85] reported that ferroelectric $BaTiO_3$ showcased better catalytic activity than non-ferroelectric $BaTiO_3$ for the photocatalytic dye decolourization, which is due to the separation of photo-induced charge carriers as a result of the creation of space charge layer. Sangbaek Park et al. [86] studied the role of ferroelectric $K_{0.5}Na_{0.5}NbO_3$ catalyst on photocatalytic hydrogen evolution activity. The authors observed an enhancement in the hydrogen evolution activity by 7-fold when the catalyst was polarized. Xiaoning Li et al. [87] reported the effect of ferroelectric polarization in enhancing the electrochemical oxygen performance of $Bi_4Ti_3O_{12}\cdot(BiCoO_3)_2$. The authors observed a significant decrease in the overpotential of polarization. H.S. Kushwaha et al. [88] reported enhanced electrochemical oxygen evolution performance of $Bi_{0.5}Na_{0.5}TiO_3$ due to polarization.

9.7 CONCLUSION

Various metal oxides have the potential for being used as photocatalysts for water splitting. Studies on increasing their efficiency towards photocatalytic water splitting utilizing doping, forming heterojunctions, and co-catalysts have been reported widely in the literature. In this book chapter, we have tried to give a glimpse of the use of metal oxides for photocatalytic water splitting. We have also tried to highlight the influence of various structural parameters that affect the photocatalytic behavior of the oxides.

ACKNOWLEDGMENTS

AK and SJ thank INST for the fellowship. SV thanks CSIR (01(2943)/18-EMR-II), Govt. of India, for funding.

REFERENCES

[1] Dutta, S. 2021. Review on solar hydrogen: its prospects and limitations. Energy Fuels 35(15): 11613–11639.

[2] Fujishima, A. and K. Honda. 1972. Electrochemical photolysis of water at a semiconductor electrode. Nature 238: 37. online.

[3] Kudo, A. and Y. Miseki. 2009. Heterogeneous photocatalyst materials for water splitting. Chem. Soc. Rev. 38(1): 253–278.]

[4] Zhu, S. and D. Wang. 2017. Photocatalysis: basic principles, diverse forms of implementations and emerging scientific opportunities. Adv. Energy Mater. 7(23): 1700841.

[5] Vijay, A. and S. Vaidya. 2021. Tuning the morphology and exposed facets of SrTiO$_3$ nanostructures for photocatalytic dye degradation and hydrogen evolution. ACS Appl. Nano. Mater. 4(4): 3406–3415.

[6] Sun, C., J. Yang, Y. Cui, W. Ren and J. Zhang. 2022. Recent intensification strategies of SnO$_2$-based photocatalysts: a review. Chem. Eng. J. 427: 131564.

[7] Ishihara, T., H. Nishiguchi, K. Fukamachi and Y. Takita. 1999. Effects of acceptor doping to KTaO$_3$ on photocatalytic decomposition of pure H$_2$O. J. Phys. Chem. B. 103: 1–3.

[8] Zhang, X., G. Jin, D. Wang, Z. Chen, M. Zhao and G. Xi. 2021. Crystallographic phase and morphology dependent hydrothermal synthesis of tungsten oxide for robust hydrogen evolution reaction. J. Alloys. Compd. 875: 160054.

[9] Zhang, F., X. Wang, H. Liu, C. Liu, Y. Wan, Y. Long, et al. 2019. Recent advances and applications of semiconductor photocatalytic technology. Appl. Sci. 9(12): 2489.

[10] Ni, M., M.K.H. Leung, D.Y.C. Leung and K. Sumathy. 2007. A review and recent developments in photocatalytic water-splitting using TiO$_2$ for hydrogen production. Renewable Sustainable Energy Rev. 11(3): 401–425.

[11] Yamakata, A., T.-a. Ishibashi and H. Onishi. 2003. Kinetics of the photocatalytic water-splitting reaction on TiO_2 and Pt/TiO_2 studied by time-resolved infrared absorption spectroscopy. J. Mol. Catal. A: Chem. 199(1): 85–94.

[12] Gupta, N.M. 2017. Factors affecting the efficiency of a water splitting photocatalyst: A perspective. Renewable Sustainable Energy Rev. 71: 585–601.

[13] Belfaa, K., M.S. Lassoued, S. Ammar and A. Gadri. 2018. Synthesis and characterization of V-doped TiO_2 nanoparticles through polyol method with enhanced photocatalytic activities. J. Mater. Sci.: Mater. Electron. 29(12): 10269–10276.

[14] Kato, H. and A. Kudo. 2002. Visible-light-response and photocatalytic activities of TiO_2 and $SrTiO_3$ photocatalysts codoped with antimony and chromium. J. Phys. Chem. B 106(19): 5029–5034.

[15] Bamwenda, G.R., S. Tsubota, T. Nakamura and M. Haruta. 1995. Photo-assisted hydrogen production from a water-ethanol solution: a comparison of activities of $Au-TiO_2$ and $Pt-TiO_2$. J. Photochem. Photobiol., A: Chemistry 89(2): 177–189.

[16] Yang, G., Z. Jiang, H. Shi, T. Xiao and Z. Yan. 2010. Preparation of highly visible-light active N-doped TiO_2 photocatalyst. J. Mater. Chem. 20(25): 5301–5309.

[17] Lu, D., M. Zhang, Z. Zhang, Q. LI, X. Wang and J. Yang. 2014. Self-organized vanadium and nitrogen co-doped titania nanotube arrays with enhanced photocatalytic reduction of CO_2 into CH_4. Nanoscale Res. Lett. 9(1): 272.

[18] Preethi, L.K., R.P. Antony, T. Mathews, L. Walczak and C.S. Gopinath. 2017. A study on doped heterojunctions in TiO_2 nanotubes: an efficient photocatalyst for solar water splitting. Sci. Rep. 7(1): 14314.

[19] Zhang, Y., Z. Xing, X. Liu, X. Li, X. Wu, J. Jiang et al. 2016. Ti^{3+} Self-doped blue TiO_2(B) single-crystalline nanorods for efficient solar-driven photocatalytic performance. ACS Appl. Mater. Interfaces 8(40): 26851–26859.

[20] Díaz, L., V.D. Rodríguez, M. González-Rodríguez, E. Rodríguez-Castellón, M. Algarra, P. Núñez, et al.. 2021. M/TiO_2 (M = Fe, Co, Ni, Cu, Zn) catalysts for photocatalytic hydrogen production under UV and visible light irradiation. Inorg. Chem. Front. 8(14): 3491–3500.

[21] Yang, L., P. Gao, J. Lu, W. Guo, Z. Zhuang, Q. Wang, et al. 2020. Mechanism analysis of Au, Ru noble metal clusters modified on TiO_2 (101) to intensify overall photocatalytic water splitting. RSC Adv. 10(35): 20654–20664.

[22] Kudo, A. 2006. Development of photocatalyst materials for water splitting. Int. J. Hydrogen Energy 31(2): 197–202.

[23] Kudo, A. 2003. Photocatalyst materials for water splitting. Catal. Surv. Asia 7(1): 31–38.

[24] Bavykin, D.V., J.M. Friedrich and F.C. Walsh. 2006. Protonated titanates and TiO_2 nanostructured materials: synthesis, properties, and applications. Adv. Mater. 18(21): 2807–2824.

[25] Ishii, T., H. Kato and A. Kudo. 2004. H_2 evolution from an aqueous methanol solution on $SrTiO_3$ photocatalysts codoped with chromium and tantalum ions under visible light irradiation. J. Photochem. Photobiol. A: Chemistry 163(1): 181–186.

[26] Yu, H., J. Wang, S. Yan, T. Yu and Z. Zou. 2014. Elements doping to expand the light response of $SrTiO_3$. J. Photochem. Photobiol. A: Chemistry 275: 65–71.

[27] Wang, D., J. Ye, T. Kako and T. Kimura. 2006. Photophysical and photocatalytic properties of $SrTiO_3$ doped with Cr cations on different sites. J. Phys. Chem. B 110(32): 15824–15830.

[28] Yan, J.-H., Y.-R. Zhu, Y.-G. Tang and S.-Q. Zheng. 2009. Nitrogen-doped $SrTiO_3/TiO_2$ composite photocatalysts for hydrogen production under visible light irradiation. J. Alloys. Compd. 472(1): 429–433.

[29] Su, E.-C., B.-S. Huang and M.-Y. Wey. 2016. Enhanced optical and electronic properties of a solar light-responsive photocatalyst for efficient hydrogen evolution by $SrTiO_3/TiO_2$ nanotube combination. Sol. Energy 134: 52–63.

[30] Manchala, S., A. Gandamalla, V.N. Rao, S.M. Venkatakrishnan and V. Shanker. 2021. Solar-light responsive efficient H_2 evolution using a novel ternary hierarchical $SrTiO_3/CdS/carbon$ nanospheres photocatalytic system. J. Nanostruct. Chem. 12: 179–191.

[31] Kato, H. and A. Kudo. 1998. New tantalate photocatalysts for water decomposition into H_2 and O_2. Chem. Phys. Lett. 295(5): 487–492.

[32] Kato, H. and A. Kudo. 2001. Water splitting into H_2 and O_2 on alkali tantalate photocatalysts $ATaO_3$ (A = Li, Na, and K). J. Phys. Chem. B 105(19): 4285–4292.

[33] Shimizu, K.-i., Y. Tsuji, T. Hatamachi, K. Toda, T. Kodama, M. Sato, et al. 2004. Photocatalytic water splitting on hydrated layered perovskite tantalate $A_2SrTa_2O_7 \cdot nH_2O$ (A = H, K, and Rb). Phys. Chem. Chem. Phys. 6(5): 1064–1069.

[34] Shimizu, K.-i., S. Itoh, T. Hatamachi, T. Kodama, M. Sato and K. Toda. 2005. Photocatalytic water splitting on Ni-intercalated Ruddlesden–Popper tantalate $H_2La_{2/3}Ta_2O_7$. Chem. Mater. 17(20): 5161–5166.

[35] Yang, H., X. Liu, Z. Zhou and L. Guo. 2013. Preparation of a novel $Cd_2Ta_2O_7$ photocatalyst and its photocatalytic activity in water splitting. Catal. Commun. 31: 71–75.

[36] Boltersdorf, J., T. Wong and P.A. Maggard. 2013. Synthesis and optical properties of Ag(I), Pb(II), and Bi(III) tantalate-based photocatalysts. ACS Catal. 3(12): 2943–2953.

[37] Lin, H.-Y., H.-C. Huang and W.-L. Wang. 2008. Preparation of mesoporous In–Nb mixed oxides and its application in photocatalytic water splitting for hydrogen production. Microporous Mesoporous Mater. 115(3): 568–575.

[38] Domen, K., A. Kudo, M. Shibata, A. Tanaka, K.-I. Maruya and T. Onishi. 1986. Novel photocatalysts, ion-exchanged $K_4Nb_6O_{17}$, with a layer structure. J Chem. Soc. Chem. Commun. 23: 1706–1707.

[39] Domen, K., A. Kudo and A. Shinozaki, A. Tanaka, K. -I. Maruya and T. Onishi. 1986. Photodecomposition of water and hydrogen evolution from aqueous methanol solution over novel niobate photocatalysts. J. Chem. Soc. Chem. Commun. 4: 356–357.

[40] Kudo, A. 2007. Photocatalysis and solar hydrogen production. Pure Appl. Chem. 79(11): 1917–1927.

[41] Iwase, A., H. Kato and A. Kudo. 2006. Nanosized Au particles as an efficient co-catalyst for photocatalytic overall water splitting. Catal. Lett. 108(1): 7–10.

[42] Sayama, K., A. Tanaka, K. Domen, K. Maruya and T. Onishi. 1991. Photocatalytic decomposition of water over platinum-intercalated potassium niobate ($K_4Nb_6O_{17}$). J. Phys. Chem. 95(3): 1345–1348.

[43] Chung, K.-H. and D.-C. Park. 1998. Photocatalytic decomposition of water over cesium-loaded potassium niobate photocatalysts. J. Mol. Catal. A: Chem. 129(1): 53–59.

[44] Li, G., T. Kako, D. Wang, Z. Zou and J. Ye. 2008. Synthesis and enhanced photocatalytic activity of $NaNbO_3$ prepared by hydrothermal and polymerized complex methods. J. Phys. Chem. Solids 69(10): 2487–2491.

[45] Kim, S., J.-H. Lee, J. Lee, S.-W. Kim, M.H. Kim, S. Park, et al. 2013. Synthesis of monoclinic potassium niobate nanowires that are stable at room temperature. J. Am. Chem. Soc. 135(1): 6–9.

[46] Ikeda, S., T. Itani, K. Nango and M. Matsumura. 2004. Overall water splitting on tungsten-based photocatalysts with defect pyrochlore structure. Catal. Lett. 98(4): 229–233.

[47] Sayama, K., A. Tanaka, K. Domen, K. Maruya and T. Onishi. 1990. Improvement of nickel-loaded $K_4Nb_6O_{17}$ photocatalyst for the decomposition of H_2O. Catal. Lett. 4(3): 217–222.

[48] Chen, D. and J. Ye. 2009. Selective-synthesis of high-performance single-crystalline $Sr_2Nb_2O_7$ nanoribbon and $SrNb_2O_6$ nanorod photocatalysts. Chem. Mater. 21(11): 2327–2333.

[49] Kudo, A., H. Kato and S. Nakagawa. 2000. Water splitting into H_2 and O_2 on new $Sr_2M_2O_7$ (M = Nb and Ta) photocatalysts with layered perovskite structures: factors affecting the photocatalytic activity. J. Phys. Chem. B 104(3): 571–575.

[50] Kim G.H., D.W. Hwang, J. Kim, Y.G. Kim and J.S. Lee. 1999. Highly donor-doped (110) layered perovskite materials as novel photocatalysts for overall water splitting. Chem. Commun. (12): 1077–1078.

[51] Miseki, Y., H. Kato and A. Kudo. 2009. Water splitting into H_2 and O_2 over niobate and titanate photocatalysts with (111) plane-type layered perovskite structure. Energy Environ. Sci. 2(3): 306–314.

[52] Domen, K., J. Yoshimura, T. Sekine, A. Tanaka and T. Onishi. 1990. A novel series of photocatalysts with an ion-exchangeable layered structure of niobate. Catal. Lett. 4(4): 339–343.

[53] Kazunari, D., N.J. Kondo, H. Michikazu and T. Tsuyoshi. 2000. Photo- and mechano-catalytic overall water splitting reactions to form hydrogen and oxygen on heterogeneous catalysts. Bull. Chem. Soc. Jpn. 73(6): 1307–1331.

[54] Li, Y., G. Chen, H. Zhang and Z. Lv. 2010. Band structure and photocatalytic activities for H_2 production of $ABi_2Nb_2O_9$ (A=Ca, Sr, Ba). Int. J. Hydrogen Energy 35(7): 2652–2656.

[55] Phoon, B.L., C.W. Lai, J.C. Juan, P.-L. Show and G.-T. Pan. 2019. Recent developments of strontium titanate for photocatalytic water splitting application. Int. J. Hydrogen Energy 44(28): 14316–14340.

[56] Bin Adnan, M.A., K. Arifin, L.J. Minggu and M.B. Kassim. 2018. Titanate-based perovskites for photochemical and photoelectrochemical water splitting applications: a review. Int. J. Hydrogen Energy 43(52): 23209–23220.

[57] Inoue, Y. 2009. Photocatalytic water splitting by RuO_2-loaded metal oxides and nitrides with d^0- and d^{10} -related electronic configurations. Energy Environ. Sci. 2(4): 364–386.

[58] Tee, S.Y., K.Y. Win, W.S. Teo, L.-D. Koh, S. Liu, C.D. Teng, et al. 2007. Recent progress in energy-driven water splitting. Adv. Sci. 4(5): 1600337.

[59] Navarro Yerga, R.M., M.C. Álvarez Galván, F. del Valle, A.J. Villoria de la Mano and J.L.G. Fierro. 2009. Water splitting on semiconductor catalysts under visible-light irradiation. ChemSusChem. 2(6): 471–485.

[60] Rajeshwar, K. 2007. Hydrogen generation at irradiated oxide semiconductor–solution interfaces. J. Appl. Electrochem. 37(7): 765–787.

[61] Zhang, P., J. Zhang and J. Gong. 2014. Tantalum-based semiconductors for solar water splitting. Chem. Soc. Rev. 43(13): 4395–4422.

[62] Yang, X., Z. Guo, X. Zhang, Y. Han, Z. Xue, T. Xie, et al. 2021. The effect of indium doping on the hydrogen evolution performance of g-C_3N_4 based photocatalysts. New J. Chem. 45(2): 544–550.

[63] Takashi, Y., S. Yoshihisa and I. Hayao. 2004. Photocatalytic decomposition of H_2O into H_2 and O_2 over Ga_2O_3 loaded with NiO. Chem. Lett. 33(6): 726–727.

[64] Manikandan, M., T. Tanabe, P. Li, S. Ueda, G.V. Ramesh, R. Kodiyath, et al. 2014 Photocatalytic water splitting under visible light by mixed-valence Sn_3O_4. ACS Appl. Mater. Interfaces 6(6): 3790–3793.

[65] Annamalai, A., R. Sandström, E. Gracia-Espino, N. Boulanger, J.-F. Boily, I. Mühlbacher, et al. 2018. Influence of Sb^{3+} as a double donor on hematite (Fe^{3+}) photoanodes for surface-enhanced photoelectrochemical water oxidation. ACS Appl. Mater. Interfaces 10(19): 16467–16473.

[66] Kadowaki, H., N. Saito, H. Nishiyama, H. Kobayashi, Y. Shimodaira and Y. Inoue. 2007. Overall splitting of water by RuO_2-loaded $PbWO_4$ photocatalyst with $d^{10}s^2$-d^0 configuration. J. Phys. Chem. C 111(1): 439–444.

[67] Hideki, K., M. Naoko and K. Akihiko. 2004. Photophysical and photocatalytic properties of molybdates and tungstates with a scheelite structure. Chem. Lett. 33(9): 1216–1217.

[68] Akihiko, K. and K. Hideki. 1997. Photocatalytic activities of $Na_2W_4O_{13}$ with layered structure. Chem. Lett. 26(5): 421–422.

[69] Akihiko, K. and H. Satoshi. 1999. H_2 or O_2 evolution from aqueous solutions on layered oxide photocatalysts consisting of Bi^{3+} with

$6s^2$ configuration and d^0 transition metal. Ions. Chem. Lett. 28(10): 1103–1104.

[70] Bamwenda, G.R., T. Uesigi, Y. Abe, K. Sayama and H. Arakawa. 2001. The photocatalytic oxidation of water to O_2 over pure CeO_2, WO_3, and TiO_2 using Fe^{3+} and Ce^{4+} as electron acceptors. Appl. Catal., A: General 205(1): 117–128.

[71] Yuan, Y., J. Zheng, X. Zhang, Z. Li, T. Yu, J. Ye, et al. 2008. $BaCeO_3$ as a novel photocatalyst with 4f electronic configuration for water splitting. Solid State Ionics 178(33): 1711–1713.

[72] Zhang, X., W. Hao, C.-S. Tsang, M. Liu, G.S. Hwang and L.Y.S. Lee. 2019. Psesudocubic phase tungsten oxide as a photocatalyst for hydrogen evolution reaction. ACS Appl. Energy Mater. 2(12): 8792–8800.

[73] Yang, J., X. Chen, X. Liu, Y. Cao, J. Huang, Y. Li, et al. 2021. From hexagonal to monoclinic: engineering crystalline phase to boost the intrinsic catalytic activity of tungsten oxides for the hydrogen evolution reaction. ACS Sustainable Chem. Eng. 9(16): 5642–5650.

[74] Gultom, N.S., H. Abdullah, D.-H. Kuo and W.-C. Ke. 2019. Oriented p-n heterojunction $Ag_2O/Zn(O,S)$ nanodiodes on mesoporous SiO_2 for photocatalytic hydrogen production. ACS Appl. Energy Mater. 2(5): 3228–3236.

[75] Zhou, G., X. Xu, T. Ding, B. Feng, Z. Bao and J. Hu. 2015. Well-steered charge–carrier transfer in 3D branched $Cu_xO/ZnO@Au$ Heterostructures for efficient photocatalytic hydrogen evolution. ACS Appl. Mater. Interfaces 7(48): 26819–26827.

[76] Amaechi, I.C., R. Katoch, G. Kolhatkar, S. Sun and A. Ruedieger. 2020. Particle size effect on the photocatalytic kinetics of barium titanate powders [10.1039/D0CY01358G]. Catal. Sci. Technol. 10(18): 6274–6284.

[77] Seo, B., D.S. Baek, Y.J. Sa and S.H. Joo. 2016. Shape effects of nickel phosphide nanocrystals on hydrogen evolution reaction. Cryst. Eng. Comm. 18(32): 6083–6089.

[78] Kimijima, T., K. Kanie, M. Nakaya and A. Muramatsu. 2014. Solvothermal synthesis of $SrTiO_3$ nanoparticles precisely controlled in surface crystal planes and their photocatalytic activity. Appl. Catal. B: Environmental 144: 462–467.

[79] Li, P., S. Ouyang, G. Xi, T. Kako and J. Ye. 2012. The effects of crystal structure and electronic structure on photocatalytic H_2 evolution and CO_2 reduction over two phases of perovskite-structured $NaNbO_3$. J. Phys. Chem. C. 116(14): 7621–7628.

[80] Chiarello, G.L., A.D. Paola, L. Palmisano and E. Selli. 2011. Effect of titanium dioxide crystalline structure on the photocatalytic production of hydrogen. Photochem. Photobiol. Sci. 10(3): 355–360.

[81] Vijay, A., K. Bairagi and S. Vaidya. 2022. Relating the structure, properties, and activities of nanostructured $SrTiO_3$ and $SrO–(SrTiO_3)_n$ (n = 1 and 2) for photocatalytic hydrogen evolution. Mater. Adv. 3: 5055–5063.

[82] Lardhi, S., L. Cavallo, M. Harb. 2020. Significant impact of exposed facets on the $BiVO_4$ material performance for photocatalytic water splitting reactions. J. Phys. Chem. Lett. 11(14): 5497–5503.

[83] Meng, J., Z. Lan, Q. Lin, T. Chen, X. Chen, X. Wei, et al. 2019. Cubic-like $BaZrO_3$ nanocrystals with exposed {001}/{011} facets and tuned electronic band structure for enhanced photocatalytic hydrogen production. J. Mater. Sci. 54(3): 1967–1976.

[84] Li, J., X. Li, Z. Yin, X. Wang, H. Ma and L. Wang. 2019. Synergetic effect of facet junction and specific facet activation of $ZnFe_2O_4$ nanoparticles on photocatalytic activity improvement. ACS Appl. Mater. Interfaces 11(32): 29004–29013.

[85] Cui, Y., J. Briscoe and S. Dunn. 2013. Effect of ferroelectricity on solar-light driven photocatalytic activity of $BaTiO_3$ - influence on the carrier separation and stern layer formation. Chem. Mater. 25(21): 4215–4223.

[86] Park, S., C.W. Lee, M.-G. Kang, S. Kim, H.J. Kim, J.E. Kwon, et al. 2014. A ferroelectric photocatalyst for enhancing hydrogen evolution: polarized particulate suspension. Phys. Chem. Chem. Phys. 16(22): 10408–10413.

[87] Li, X., H. Liu, Z. Chen, Q. Wu, Z. Yu, M. Yang, et al. 2019. Enhancing oxygen evolution efficiency of multiferroic oxides by spintronic and ferroelectric polarization regulation. Nat. Commun. 10(1): Article 1409.

[88] Kushwaha, H.S., A. Halder and R. Vaish. 2018. Ferroelectric electro-catalysts: a new class of materials for oxygen evolution reaction with synergistic effect of ferroelectric polarization. J. Mater. Sci. 53(2): 1414–1423.

Chapter 10

Nanofluidics for Heat Transfer System and Energy Applications

Nagendra S. Chauhan

Department of Applied Physics, Graduate School of Engineering,
Tohoku University, Sendai, Miyagi, Japan– 980-8579
Email: nagendra599@gmail.com; chauhan.nagendra.singh.b1@tohoku.ac.jp

10.1 INTRODUCTION

Heat transfer research has a long history of about 30 decades, owing to its persistent relevance in industrial processing that encompasses numerous exciting opportunities across a wide spectrum for developing critical technologies in energy, automobiles, electronics, and the medical sector. As a widely studied discipline of thermal engineering, any heat transfer process is a path function governed by three distinctive modes based on classical heat transfer laws, i.e., Fourier's law (conduction), Newton's law of cooling (convection), and Planck's law (radiation). In the context of temperature regimes, the subject of heat transfer is broadly classified into two sub-categories, i.e., (i) for temperatures well below room temperature (termed refrigeration or cryogenic technology); (ii) for temperatures above room temperature. Heat transfer fluids (HTFs), as a term for heat transfer media, is commonly reserved for heat transfer occurring above room temperature. While for refrigeration, the heat transfer media is commonly referred to as refrigerant.

Approaches for improving heat transmission in heat exchangers commonly include passive and active techniques, such as creating turbulence, extending the exchange surface (such as fins), or using a fluid with higher thermophysical properties. Over the years, heat transfer systems have evolved from geometrical modifications to compact liquid cooling systems, thus shifting the focus from the geometry of heat exchangers to HTFs. The engineering data for the process design targeting heat transfer enhancement and performances indicates heat transfer media to be the primary determinant. In the wide range of thermal applications, commonly used HTFs include liquids such as ethylene glycol, water, silicone oil, propylene glycol, etc., which offer an advantage over the air in terms of their higher specific heat and thermal conductivity. Thus, advanced thermal designs are focused on improving the characteristics of these HTFs for achieving key desirables, such as low viscosity, higher thermal conductivity and diffusivity, high boiling and low freezing points, thermal stability, and anti-corrosive behaviour.

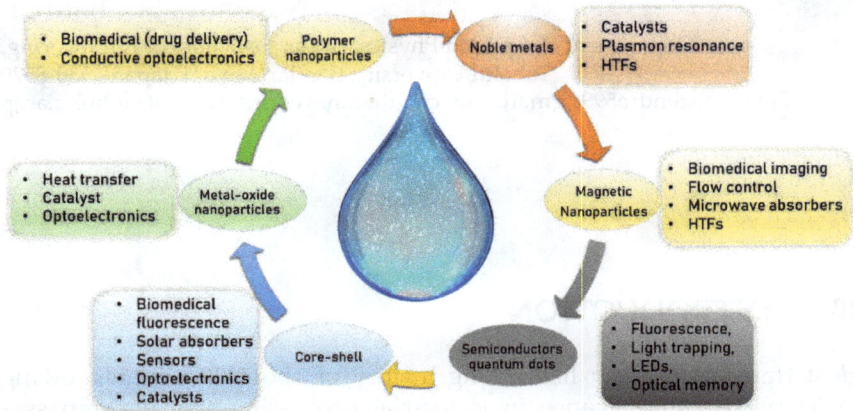

Figure 10.1 Nanomaterials for widespread applications of nanofluidics.

In this context, nanofluidics as an emerging thermal science and engineering domain encompasses the study and manipulation of fluids confined within nanostructures. Over the years, nanofluidics has integrated into a wide range of technological areas, where diverse formulation of nanomaterials is finding growing applicability in emerging applications, as displayed in Figure 10.1. Remarkably, thermophysical properties of nanofluids are being widely exploited with a growing understanding of the micro-nano interface of nanoparticles(NPs)-fluid and motion of NPs for use in applications ranging from lab-on-a-chip systems for chemical and biochemical analysis to ultra-efficient reactive processing and catalysis [1]. As a carrier fluid for drug delivery, nanofluids allow more efficient drug absorption by living cells. Moreover, the superior photo-thermal properties of nanofluids make them attractive options for use in direct solar absorption collectors [2, 3]. In tribological applications, lubrication properties are remarkably enhanced by using nanofluids as they enable the formation of a protective film with low hardness and elastic modulus of the worn surface [4]. These recent reviews can be referred to for a more detailed account and references to these emerging applications of nanofluids [1–6]. Here, the discussion is restricted only to the heat transfer capabilities of nanofluids.

Improving the heat transfer behaviour of traditional HTFs beyond the fundamental limits using NPs has attracted significant interest after the seminal work of Choi et al. [7, 8], who designated the NP's suspension in a fluid as "nanofluids". Tuning the thermophysical properties of traditional HTFs through control over particle size, shape, composition, and other parameters, thus motivated several studies in the recent past involving the formulation, characterisation, evaluation, and application of promising next-generation HTFs [9–13]. Remarkably, colloidal and classical thermal science has been established as the basis for these studies, wherein a comprehensive understanding of the underlying thermal transport phenomenon is widely sought for optimising the thermophysical design of nanofluids [14–16]. However, the insufficient description of the state of the NPs and their aggregates, besides diverse forces of interaction and fluid dynamics in the base fluid, has often skewed the overall understanding of nanofluids, impeding a rational design of such systems.

Although significant efforts were devoted to evaluating the heat transfer performance and its dependence on varied factors by investigating the innumerable possible combination of nanofluids, there remains a significant gap between fundamental research and the practical applicability of nanofluids for thermal management [11–13, 17]. It is noteworthy that the inherent complexity in the formulation of a consistent or standardised nanofluid sample necessitates the need for more benchmark studies in this domain [18–20]. Thus, several factors, including the lack of agreement between results, poor characterisation of suspensions, and the vague understanding of the underlying mechanisms, have remained the major bottlenecks for realising the potential of nanofluids as advanced, nano-engineered HTFs. In the realm of thermal science, large and anomalous thermophysical behaviour measured for several nanofluids has driven the idea of exploiting nanofluids as HTFs, which is the focus of this chapter with a threefold purpose. Primarily, the physical mechanisms, theoretical models, and parametric analysis of the thermal characteristics of nanofluids are comprehended to understand anomalous thermal transport observed in nanofluidics. Secondly, the extensive ongoing efforts for synthesis, stability, and accurate measurements are highlighted to aid researchers in developing nanofluids with superior thermal properties and enhanced performance. Finally, the prospects of nanofluids in varied applications are presented.

10.2 NANOFLUIDICS VS. MICRO-FLUIDICS

Maxwell pioneered the proposition of suspending particles in fluids to improve their thermal transport in the 19th century as a scheme [21]. In fluidics, at the very beginning, micron or larger scale particles were blended into the base fluids to make suspensions or slurries that were mostly unstable with inherent problems, thereby limiting their practical applicability. Thus, the route of suspending particles in liquid was a well-known but rejected option for heat transfer applications until the emergence of the idea of nanofluids in the mid-1990s proposed by Choi et al. [7, 8], which helped stimulate the re-examination of this option. Since then, nanofluidics has remained an active field of investigation in thermal engineering. Compared to microfluidics, nanofluidics offers advantages:

- The high specific surface area of NPs promotes interactions between particles and fluids.
- High dispersibility induced Brownian motion of NPs.
- Reduced pumping power to avoid coagulation.
- Higher adaptability to miniaturised heat transfer systems.

Nanofluids formulations aim to embody the Brownian agitation that overcomes any settling motion due to gravity. It is empirically found that a stable nanofluid is possible when particles are small enough in size (usually <100 nm). As the transport phenomena at the nanoscale bring high particle mobility and a large surface-to-volume ratio in a fluid flow. However, particle interactions at the nanoscale may lead to crowding by forming large particle agglomerates that settle out of the suspension. Thus, to overcome these inherent challenges and to develop nanofluids, over the years, numerous experimental and theoretical studies on heat transfer have been undertaken to gain insight into nanoscale thermal transport and explore the advantages of nanofluids under various conditions.

10.3 HEAT TRANSFER IN NANOFLUIDS

Heat is energy in transit whose flow is determined by temperature gradients and occurs in three modes, i.e. conduction, convection, and radiation. All heat transfer processes, including boiling and condensation, combine these modes. The dependence of heat transfer on temperature gradients was mathematically formulated using Fourier's law in both solids and fluids, while the efficiency is quantified using thermal conductivity (κ), which is regarded as a material property for pure materials. Heat transfer via conduction mode involves three types of energy carriers, i.e. phonons, electrons, and molecules. In the case of nanofluids, which constitutes nanoscale solid particles confined in a fluid, both NPs and based fluid tend to lose their identities, owing to comparable dimensionality of the nano-sized particles with the mean free path of the energy carriers, thereby leading to a quantum effect. Thus, effective thermal conductivity (κ_{NF}) in nanofluids depends on several parameters, and is not a material property, that is assumed only for quantification. The high κ_{NF} exhibited by nanofluids encompasses related and synergistic influences on specific heat (C_p), viscosity (μ), convective heat transfer coefficient (h), and agglomeration of NPs, most of which have been highlighted in several review articles [22–24]. Most studies from the onset has focused on understanding the anomalous heat transfer enhancement in nanofluids, which is paramount for its applicability as HTFs. Here, we present only a brief description of these mechanisms of heat transfer, proposed models, and governing parameters in nanofluidics to understand heat transfer.

10.3.1 Mechanism of Heat Transfer

Till date, several mechanisms have been proposed to explain the anomalous thermal transport behaviour observed in nanofluids. The practical mean theory (EMT) [25, 26] based continuum formulations were the pioneering description, which was modified with newer findings including Brownian motion of the NPs, molecular-level layering of the liquid at the liquid/particle interface, the nature of heat transport in the NPs, and the effects of NPs clustering [27, 28]. These mechanisms explain the microscopic and macroscopic transport of NPs in the static and dynamic state of nanofluids, as shown in Figure 10.2. The energy transport by dynamic motion of NPs is proposed to occur in four modes: (i) the collision between base fluid molecules; (ii) the thermal interactions of dynamic NPs with base fluid molecules; (iii) the collision between NPs due to Brownian motion; and (iv) the thermal diffusion of NPs in the fluid. These modes result in two kinds of Brownian motion in nanofluids:

collision between Brownian NPs and convection induced by Brownian NPs at the molecular and nanoscale levels [29–32]. Moreover, as clustering or aggregation is usual in nanofluids due to van der Waals forces, clustered NPs may also provide local percolation-like paths for rapid heat transport, which increases with the increasing volume fraction of NPs. In such a case, well-dispersed NPs in a fluid matrix result in the lowest κ_{NF} whereas interconnected NPs in the liquid enhance the κ_{NF} [33].

Studies to investigate the effects of aggregation on the nanofluids have shown that the aggregation time constant decreases rapidly with decreasing NPs size and that the κ enhancement increases with increasing levels of aggregation, levelling off after the optimum level of aggregation is reached [33, 34]. However, when Brownian motion is considered, such aggregation of NPs reduces the κ_{NF} of nanofluids as the random motion of aggregates is slower than that of a single nanoparticle. Moreover, clustering and formation of aggregates often reduces the efficiency of the energy transport enhancement of the suspended NPs [31, 32]. It is also observed that liquid molecules near a solid surface tend to form a layered structure with an intermediate physical state [35]. These layer structures act as a thermal bridge between solid NPs and bulk liquid in nanofluids and impact its thermophysical transport.

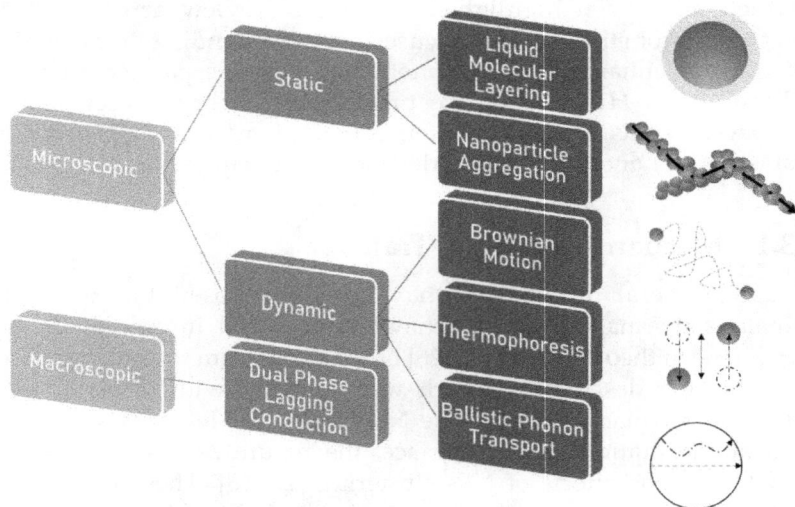

Figure 10.2 Mechanism of heat transfer in nanofluids with classification and their schematic illustration.

During the natural convection process, where the flow is driven by buoyancy and temperature, it was observed that when a mixture of two or more types of motile particles (i.e. particles with moving capability) is subjected to the force of a temperature gradient [36, 37] particles commonly travel in the direction of decreasing temperature. Also in such a case, the process of heat transfer increases with a decrease in bulk density. This phenomenon called "thermophoresis" explains the enhancement in κ_{NF} at low volume fraction of NPs, with a subsequent decrease with increasing volume fraction of particles. Dual-phase-lagging heat conduction is another macroscale phenomenon developed from first principles that differs from Fourier heat conduction, as it involves thermal oscillation and resonance originating from a coupled conduction between the NPs and the base fluids [38, 39]. The enhancement of the convective heat transfer performances revealed in several experimental investigations was described by the non-uniform distribution of thermal conductivity and viscosity field due to particle re-arrangement, shear-induced thermal conduction enhancement, reduction of thermal boundary layer thickness due to the chaotic movements, fluctuations, and interactions of NPs, thereby enhancing the convective heat transfer coefficient of nanofluids (h_{NF}).

10.3.2   Models of Heat Transfer

Several models were developed to predict anomalous thermal conduction behaviour based on proposed microscopic mechanisms, which can be classified into two broad categories i.e., static and dynamic. The static model accounts for the different geometrical, static structures of an NPs-fluid heterogeneous mixture. While the dynamic models, in addition, consider nano-convections phenomena. A summary of existing theoretical models and their significance are presented in Table 10.1 where κ_{NF} is the effective thermal conductivity of nanofluids, κ is the thermal conductivity of base fluids, and ϕ is the phase fraction of nanoparticles.

10.3.2.1   Classical Models

Classical nanofluids' conduction models were based on diffusive heat transfer through continuous matrix phase and dispersed phase, where NPs were considered motionless. Such static models take into consideration κ of the constituents, physical properties of NPs such as volume fraction, shape, and distribution, and the particle-particle interaction in dense suspensions. These models were based on (EMT) [25, 26] and provided a good description of fluidic phenomena occurring with microns and larger-size particle dispersions but failed to predict the anomalous thermal transport for nanofluids. Although firmly established and thoroughly investigated, these correlations considered no effect of aggregation, interface layer, and Brownian motion, which supposedly prevailed in nanofluids, thus resulting in conflicts between the experimental outcomes and theoretical predictions.

Table 10.1 A summary—Analytical models on thermal conductivity of nanofluids

Classical Models

Investigator	Formula	Remarks
Maxwell [21]	$$\frac{\kappa_{NF}}{\kappa} = \frac{\kappa_{NP} + 2\kappa + 2\varphi[\kappa_{NP} - \kappa]}{\kappa_{NP} + 2\kappa - \varphi[\kappa_{NP} - \kappa]}$$	• Based on the conduction solution through a stationary random suspension of spheres. • Spherical Particles.
Bruggeman [25]	$$\frac{\kappa_{NF}}{\kappa} = (3\varphi - 1)\frac{\kappa_{NP}}{\kappa} + [3(1-\varphi) - 1] + \sqrt{\Delta}$$ $$\Delta = \left[\left\{(3\varphi-1)\frac{\kappa_{NP}}{\kappa} + \{3(1-\varphi)-1\}\right\}^2 + 8\frac{\kappa_{NP}}{\kappa}\right]$$	• Based on the differential effective medium [DEM] theory to estimate the effective thermal conductivity of composites at high particle concentrations. • It consists in building up the composite medium through a process of incremental homogenisation. • Applicable to the high volume fraction of spherical particles.
Hamilton and Crosser [26]	$$\frac{\kappa_{NF}}{\kappa} = \frac{\kappa_{NP} + [n-1]\kappa + [n-1]\varphi[\kappa_{NP} - \kappa]}{\kappa_{NP} + [n-1]\kappa - \varphi[\kappa_{NP} - \kappa]}$$ $$= 4.97\varphi^2 + 2.72\varphi + 1$$	• Based on the effective thermal conductivity of a two-component mixture. • Applicable to spherical as well as non-spherical particles. • $n = 3$ [Spheres]; $n = 6$ [Cylinders].
Wasp et al. [40]	$$\frac{\kappa_{NF}}{\kappa} = \frac{\kappa_{NP} + 2\kappa + 2\varphi[\kappa_{NP} - \kappa]}{\kappa_{NP} + 2\kappa - \varphi[\kappa_{NP} - \kappa]}$$	• Based on the effective thermal conductivity of a two-component mixture. • Special Case of Hamilton and Crosser Model with $n = 3$.
Davis [41]	$$\frac{\kappa_{NF}}{\kappa} = 1 + \frac{3[\kappa-1]}{(\kappa-2) - \varphi[\kappa-1]}[\varphi + f(\kappa)\varphi^2 + O(\varphi^3)]$$	• Green's theorem was applied to the space occupied by the matrix material [spherical inclusions]. • A decaying temperature field was used.
Lu et al. [42]	$$\frac{\kappa_{NF}}{\kappa} = 1 + a\varphi + b\varphi^2$$	• The effective conductivity of composites containing aligned spheroids of finite conductivity was modelled with the pair interaction.

Investigator	Formula	Remarks
Jeffery[43]	$$\frac{\kappa_{NF}}{\kappa} = 1 + 3\eta\varphi + \varphi^2\left[3\eta^2 + \frac{3\eta^2}{4} + \frac{9\eta^3}{16}\frac{\kappa_1+2}{2\kappa_1+3} + \cdots\right]$$ $$\eta = \frac{\kappa_1-1}{\kappa_1+2}, \quad \kappa_1 = \frac{\kappa_{NP}}{\kappa}$$	• Based on a conduction solution through a stationary random suspension of spheres. • High order terms represent pair interactions of randomly dispersed spherical particles.
Xue et al. [35]	$$\frac{\kappa_{NF}}{\kappa} = \frac{1 - \varphi + 2\varphi\left[\dfrac{\kappa_{NP}}{\kappa_{NP}-\kappa}\ln\dfrac{\kappa_{NP}+\kappa}{2\kappa}\right]}{1 - \varphi + 2\varphi\left[\dfrac{\kappa_{NP}}{\kappa_{NP}-\kappa}\ln\dfrac{\kappa_{NP}+\kappa}{2\kappa}\right]}$$	• For CNTs-based nanofluids and including the axial ratio and the space distribution.
Bhattacharya et al. [29]	$$\frac{\kappa_{NF}}{\kappa} = \frac{\kappa_{NP}}{\kappa}\varphi + [1 - \varphi]$$	$$\kappa_{NP} = \frac{1}{k_B T^2 V}\sum_{j=0}^{n}[Q(0)Q(j\Delta T)]\Delta T$$ •

Theoretical Models for Nanofluids Effective Thermal Conductivity–Based on Nanolayer Effect

t = nanolayer thickness; – radius of nanoparticles

Investigator	Formula	Remarks
Xie et al.[44]	$$\frac{\kappa_{NP}-\kappa}{\kappa} = 3\theta\varphi_T + \left(\frac{3\varphi_T^2\theta^2}{1-\theta\varphi_T}\right)$$ $$\varphi_T = \frac{4}{3}\pi[r_p + t]^3 N_p = \varphi[1+\beta]^3, \ \beta = \frac{t}{r_p}$$	• Based on Fourier's Law of Heat Conduction • Low volume fraction • Nanolayer

(Contd.)

Table 10.1 A summary—Analytical models on thermal conductivity of nanofluids (*Contd.*)

Investigator	Formula	Remarks
Yu and Choi [45]	$$\kappa_{NF} = \frac{k_{pe} + 2\kappa + 2\varphi(k_{pe} - \kappa)(1+\beta)^3}{k_{pe} + 2\kappa - \varphi(k_{pe} - \kappa)(1+\beta)^3}\kappa$$ $$k_{pe} = \frac{2(1-\gamma) + (1+\beta)^3[1+2\gamma]\gamma}{-(1-\gamma) + (1+\beta)^3[1+2\gamma]}\kappa_{NP}$$ $$\beta = \frac{t}{r_p} \text{ and } \gamma = \frac{\kappa_{layer}}{\kappa_{NP}}$$	• Modified Maxwell model • Spherical particles • Nanolayer

Theoretical Models for Nanofluid's Effective Thermal Conductivity—Based on Brownian Effect

Investigator	Formula	Remarks
Xuan et al. [34]	$$\frac{\kappa_{NF}}{\kappa} = \frac{\kappa_{NP} + 2\kappa + 2\varphi[\kappa_{NP} - \kappa]}{\kappa_{NP} + 2\kappa - \varphi[\kappa_{NP} - \kappa]} + \frac{\rho_p \varphi C_p}{2k_f}\sqrt{\frac{k_B T}{3\pi\mu r_c}}$$	• Based on the Maxwell model with the theory of Brownian motion and diffusion-limited aggregation applied to simulate random motion and the aggregation process of the nanoparticles. • Includes the effect of particle size, random motion, concentration, and Temperature.
Koo and Kleinstreuer [46, 47]	$$\frac{\kappa_{NF}}{\kappa} = \frac{\kappa_{NP} + 2\kappa + 2\varphi(\kappa_{NP} - \kappa)}{\kappa_{NP} + 2\kappa - \varphi(\kappa_{NP} - \kappa)} + 5\times10^4 \beta\rho_p C_p \sqrt{\frac{k_B T}{\rho_p C_p}} f(T, \varphi)$$ $$f(T, \varphi) = (-134.63 + 1722.3\varphi) + [0.4705 - 6.04\varphi]\frac{T}{T_o}$$ $$\beta = \begin{cases} 0.0137(100\varphi)^{-0.8229} & \varphi < 0.01 \\ 0.0011(100\varphi)^{-0.7272} & \varphi > 0.01 \end{cases}$$	• Based on Maxwell model. • Curve fitting of the available experimental data to determine the effective conductivity due to Brownian motion. • Considered surrounding liquid travelling with randomly moving nanoparticles.

Investigator	Formula	Remarks
Prasher et al. [31]	$$\frac{\kappa_{NF}}{\kappa} = [1 + A.Re^m\, Pr^{0.333}\, \varphi]\,\frac{\kappa_{NP} + 2\kappa + 2\varphi[\kappa_{NP} - \kappa]}{\kappa_{NP} + 2\kappa - \varphi[\kappa_{NP} - \kappa]}$$	• Based on Maxwell model and heat transfer in fluidised beds. • Accounts for convection caused by the Brownian motion from multiple particles.
Chon et al. [48]	$$\frac{\kappa_{NF}}{\kappa} = 1 + 64.7\varphi^{0.74}\left(\frac{d_f}{d_p}\right)^{0.369}\left(\frac{\kappa_{NP}}{\kappa}\right)^{0.747} \times Pr^{0.9955}Re^{1.2321}$$	• Based on curve fitting of experimental data. • Reynolds number is based on the Brownian motion velocity. • Role of temperature and particle size.
Jang and Choi. [30]	$$\frac{\kappa_{NF}}{\kappa} = [1 - \varphi] + \varphi + 3C\frac{d_f}{d_p}Pr\varphi Re^2$$	• Based on kinetics, Kapitza resistance, and convection. • General expression for thermal conductivity involving four modes of energy transport in nanofluids: collision b/w fluid molecules, thermal diffusion of nanoparticles due to Brownian motion, and the thermal interactions of dynamic nanoparticles with fluid molecules.

10.3.2.2 Dynamic Models

Dynamic models were based on particle dynamics, including particle geometrical and directional clustering/percolation, matrix-particle layering effects, and the role of aggregation and interfacial thermal resistance, all of which were taken into account in addition to the particle's conventional static part. Convection phenomena induced by contribution from thermal Brownian motion, thermophoresis, diffusiophoresis, and other electromagnetic phenomena, including near field radiation, thermal waves, dual-phase lagging, and other unique phenomena, like ballistic phonon transport in NPs, occurring at mesoscale were thus given due considerations. Empirical correlations were routinely proposed for Nusselt number (Nu: the ratio of convective to conductive heat transfer across a boundary), Prandtl number (Pr: the ratio of momentum diffusivity (kinematic viscosity) to thermal diffusivity), Reynolds number (Re: a dimensionless quantity that is used to determine the type of flow pattern), and Peclet number (Pe: the ratio of the convection rate over the diffusion rate in the convection-diffusion transport system) in many of these studies based on the measured flow and convective heat transfer characteristics by regression analysis. These models and correlations stem from the mechanism, experimental investigations, and assumptions. However, due to the complexity of diverse NPs, additives, fluids, and their dynamic interfacial interactions, including the inter-coupling of many phenomena, these studies on heat transport in nanofluids requires a more systematic experiments. Well-dispersed and accurately characterised nanofluids are thus prerequisities for a better understanding of the physics of fluid flow and heat transfer at the nanoscale by which more accurate and precise models to predict the thermal transports in nanofluids can be established.

10.4 NANOMATERIALS FOR NANOFLUIDICS IN HEAT TRANSFER

Nanofluids can be formulated from any NPs by blending them into a fluid (liquid or gas), thus offering innumerable combinations. It is now well recognised that the thermophysical properties of conventional base fluids can be improved by adding a low concentration of NPs. However, formulating stable nanofluids with controlled thermophysical properties for heat transfer applications is a formidable challenge [49, 50]. For this purpose, several types of NPs based on pure metals (Au, Ag, Cu, Al, and Fe), metal oxides (Al_2O_3, CuO, Fe_3O_4, SiO_2, ZrO_2, TiO_2, and ZnO), carbides (SiC and TiC), and a variety of carbon materials (diamond, graphite, and single/multi-wall carbon nanotubes) were studied [4, 51–55].

Interestingly, high κ nanomaterials may not continuously improve the thermal performance of the base fluids, as κ_{NF} enhancement depends on the stability condition and dispersibility of NPs in base fluid [27]. Among studied nanofluids, metals and allotropes of carbon (such as diamond) were found to be most effective in κ_{NF} enhancement. Also, higher κ_{NF} for metallic nanofluids were observed compared to oxide nanofluids.

Based on the temperature regimes, the HTFs are commonly classified as low-temperature fluids (i.e. hydrocarbons, silicone oils, and salt brine), medium-temperature fluids (i.e. glycols like ethylene glycol, Bio-glycol (1,3 propanediol) and propylene glycol), and high-temperature fluids (i.e. paraffin oils, synthetic organic oils, molten nitrate salts). Traditional HTFs, such as water (0.609 $Wm^{-1}K^{-1}$), ethylene glycol (0.258 $Wm^{-1}K^{-1}$), and engine oil were investigated most commonly as base fluids for their heat transfer aspect in nanofluidics. Also, non-Newtonian fluids such as carboxy methyl cellulose (CMC) were examined [3, 56, 57]. The inverse relationship between the κ and particle size is experimentally validated and theoretically supported by mechanisms of Brownian motion of NPs and liquid layering around NPs for κ_{NF} enhancement [30, 48]. The concentration range of NPs lies typically between 0.01–5 wt.%, while the mean particle size is usually 10–100 nm. The success of effectively developing nanofluids depends much on our understanding of the effects of these interrelated aspects and parameters affecting thermophysical transport, as shown schematically in Figure 10.3.

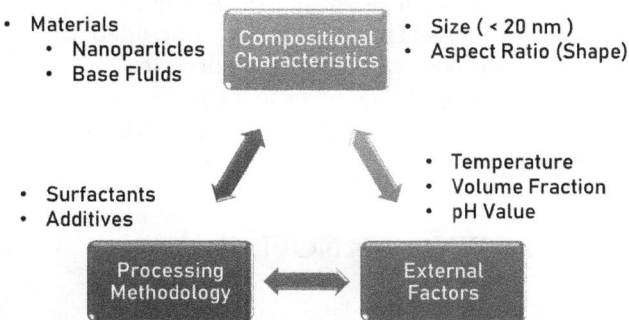

- Materials
  - Nanoparticles
  - Base Fluids

Compositional Characteristics

- Size (< 20 nm)
- Aspect Ratio (Shape)

- Surfactants
- Additives

- Temperature
- Volume Fraction
- pH Value

Processing Methodology

External Factors

Figure 10.3 Parameters affecting heat transfer behaviour in nanofluids.

The concentration of NPs determines the functionality of nanofluids. In most studies, a fascinating increase in the κ with low NPs volume fraction (<1%) is observed, which is in contrast with the traditional particle-liquid suspensions that require high concentrations (>10%) of particles to achieve such dramatic κ_{NF} enhancement. As a parameter investigated in almost all the experimental studies, it was found that at higher concentrations, there exists a strong tendency among NPs to form agglomerations in liquid media that affect the stability and NPs transport in nanofluids. Thus, the dispersibility of NPs in nanofluids is favoured by their small size, higher aspect ratio, and low volume fraction. Moreover, dispersants, pH adjustments, or various mixing methods can be employed to prevent agglomeration at higher volume fractions. A dynamic behavioural explanation for nanofluids is complex as the viscosity of the base fluid affects the Brownian motion of NPs which in turn affects the κ_{NF} of the nanofluids. The size and aspect ratio of the NPs are critical compositional characteristic that affects thermophysical transport, wherein κ_{NF} and h_{NF} increases with decreasing particle size [48, 58] and increasing aspect ratio of dispersed NPs [56, 59–61].

In nanofluids, change in temperature affects the Brownian motion of NPs and the clustering of NPs, which results in a dramatic change in κ_{NF} of nanofluids, thereby indicating the temperature dependence of thermal transport in nanofluids [62–64]. It is also found that the increase in the difference between the pH value of the nanofluids solution and the isoelectric point (pH at which a particular molecule or surface carries no net electrical charge) of NPs resulted in the enhancement of κ_{NF}, as with increasing differential, the mobility of NPs increases, which improves the micro-convection effect. At the optimum pH value, the surface charge of NPs increases, creating repulsive forces between NPs, thus preventing severe clustering of NPs (excessive clustering may result in sedimentation, which decreases κ enhancement). Thus, to obtain stable nanofluids, the pH value of the suspension must be far from the isoelectric point of the particles, where the overall charge on the NPs becomes zero. Thus, it can be concluded that nanofluids' thermal transport behaviour is strongly correlated to the processing methodology, compositional characteristics, and external factors, which can be utilised to interpret conflicting results and inconsistent findings via parametric analysis.

10.5   FORMULATING NANOFLUIDS

The synthesis methodology regulates the stability and size distribution of NPs, whereas functionality determines the combinations of nanomaterials and their concentration in base fluids. In nanofluids, agglomeration, chemical stability, and homogeneity have remained the major bottlenecks that advancements have consistently addressed in the pre-existing methodology for nanofluids synthesis [50, 65, 66]. Extensive research efforts on nanofluid synthesis were focused on improving thermal stability, dispensability in diverse media, chemical compatibility, and ease of chemical manipulation. These methods are broadly classified into two types: the single-step and the two-step method [27, 50, 67].

10.5.1   Two-Step Method

The two-step method is the most economical way to formulate the nanofluids on a large scale, as nano-powder synthesis techniques have already been scaled up to industrial production levels. Most studies used a two-step method, including the earliest investigations of nanofluids. In the first step, nanoparticles, nanofibers, nanotubes, or other nanomaterials are produced as dry powders employing both top-down and bottom-up approaches, as shown in Figure 10.4. General synthetic methods for NPs production includes transition metal salt reduction, thermal decomposition and photochemical methods, ligand reduction and displacement from organometallics, metal vapour synthesis, and electrochemical synthesis. The produced nano-sized powder is then dispersed in a fluid in the second processing step with the help of intensive magnetic force agitation, ultrasonic agitation, high-shear mixing, or ball milling.

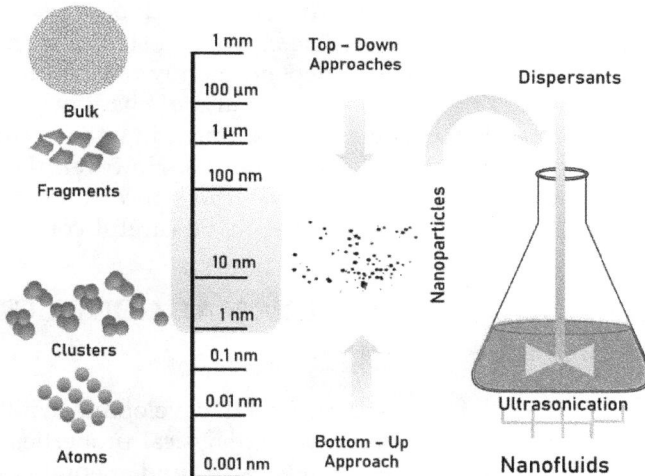

Figure 10.4   Length scale indicates bottom-up and top-down approaches for preparing nanomaterials, which are subsequently dispersed in fluids to formulate nanofluids.

10.5.2   Single-Step Method

To minimise the agglomeration of NPs, single-step nanofluids processing methods have also been developed. Few methods exist today for the preparation of nanofluids by single-step synthesis. In these methods, NPs are simultaneously produced and dispersed into the base fluid. Thus, the drying, storage, transportation, and dispersion of NPs are avoided, minimising the agglomeration of NPs, and increasing nanofluids' stability. The one-step processes can prepare uniformly dispersed NPs, and the particles can be stably suspended in the base fluid. The one-step physical method cannot synthesise nanofluids on a large scale, and the cost is also high so the one-step chemical method is developing rapidly. However, there are some disadvantages to the one-step method. The most important one is that the residual reactants are left in the nanofluids due to incomplete reaction or stabilisation. It is challenging to elucidate the NPs effect without eliminating this impurity effect. Compared to the single-step method, the two-step technique works well for oxide NPs, while it is less successful with metallic particles.

Thus, the nanofluids preparation method is crucial in achieving stable suspension. In addition, energetically powerful instruments such as ultrasonic baths, homogenisers, and the processor can be utilised to achieve well-dispersed suspensions. Other techniques, such as control of pH or addition of surface-active agents, are also used to attain stability of the suspension of the nanofluids against sedimentation. These methods change the surface properties of the suspended particles and thus suppress the tendency to form particle clusters. It should be noted that the selection of surfactants should depend mainly on the properties of the solutions and particles. In general, the aim of changing the surface properties of suspended particles and suppressing the formation of particles cluster is to obtain stable suspensions. However, the addition of dispersants can affect the heat transfer performance of the nanofluids, especially at high temperatures, thus requiring careful consideration.

10.6 MEASUREMENT OF THERMAL CONDUCTIVITY OF FLUIDS

Accurate measurement is a prerequisite for developing nanofluids, as inconsistencies in the evaluation of thermophysical properties made by separate groups for identical nanofluids raise fundamental doubts and pose a hindrance to the potential applicability of nanofluids. In general, there are two methods of measuring the thermal conductivity of liquids: steady-state methods and transient methods. Steady-state methods are simple and direct but lack accuracy and require extensive and highly complex experimental setups and procedures.

The convective heat transfer characteristics in different flows (turbulent flow; laminar flow in both the developing and fully developed regions) were investigated for various nanofluids. The experimental setup in these studies typically consists of a flow loop consisting of three sections: cooling unit, test section, and measuring units. The test section comprises a circular tube with specification length, inner diameter, and outer diameter. A constant wall temperature of pipes is maintained through constant heat flux conditions as a boundary condition while measuring the convective heat transfer coefficients. Besides the characteristics of nanofluids, other determinants also play a role in the indirect heat transfer process. For instance, average fluid velocity and geometry of the system, such as tube dimensions and their arrangement, affect the h_{NF} for forced convection. For this consideration, the heat transfer performance of nanofluids is measured directly under flow conditions.

Reviews on techniques for measuring the κ_{NF} of nanofluids indicates the shortcomings of steady-state methods, as the heat lost cannot be quantified, giving considerable inaccuracy and the possibility of natural convection, which gives higher conductivity values [68]. Among the steady-state thermal conductivity measurement methods, i.e. guarded hot plate, heat flow metre, temperature oscillation method, and the temperature oscillation method is mainly used. Different techniques used to measure the thermal conductivity of nanofluids are shown in Figure 10.5. The critical issues related to the realisation of an appropriate and comparative experimental protocol among several techniques to

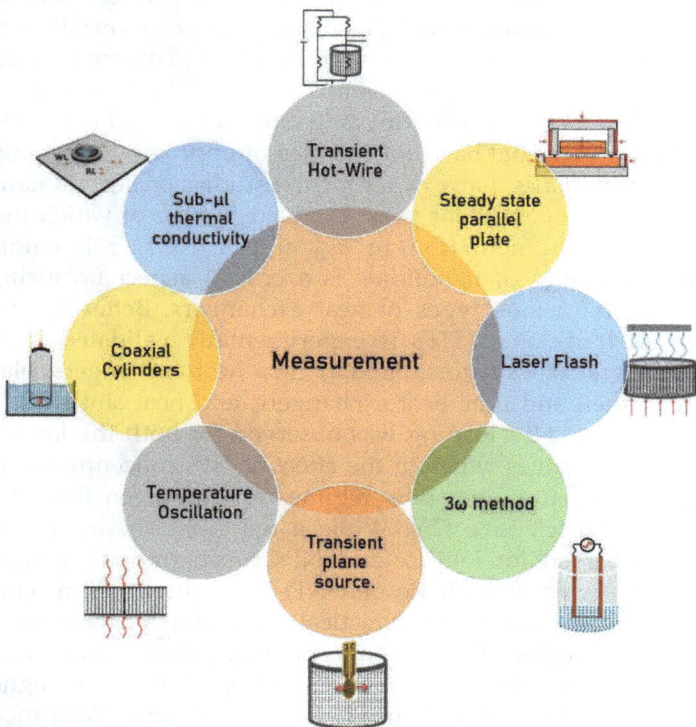

Figure 10.5 Different measurement techniques for thermal properties of nanofluids.

measure thermal transport in nanofluids and their adaptations, as discussed in the recent reviews [69, 70] includes following aspects:

- Calibration of the sensors and/or thermocouples.
- Application of uniform heat flux during measurement.
- Accounting dispersion-related changes in nanofluids, including pH variations, use of surfactants, etc.
- Consideration for the direction of the heat flow (upstream/ downstream)

10.7 NANOFLUIDICS FOR HEAT TRANSFER SYSTEMS

A stable and easily synthesised, inexpensive, and nontoxic fluid with excellent thermophysical properties (i.e. high thermal conductivity and low viscosity) and long service life is highly desirable for employability in heat transfer applications. Most of the experimental and numerical studies showed that nanofluids exhibit an enhanced h_{NF} compared to its base fluid that increases with increasing NPs concentration and Re. Remarkably, enhanced Nu and Re with increasing volume concentration of ultrafine metallic oxide (Al_2O_3 and TiO_2) particles in water, revealed in the pioneering experimental investigation by Pak and Cho (1998) [71], provided a fundamental basis for the idea of improving convective heat transfer characteristics. Later experimental studies on several nanofluids with varying flow conditions were conducted, most of which indicated a considerable augmentation in h_{NF} and κ_{NF} [51, 53]. Anomalous enhancement in κ_{NF} of nanofluids is a crucial aspect favouring their applicability in various types of heat exchangers. Better h_{NF} and κ_{NF} than that of traditional HTFs is experimentally validated in various heat exchangers, i.e. radiators, circular tube heat exchangers, plate heat exchangers, shell and tube heat exchangers, and heat sinks [72–74].

Increasing h_{NF} NPs loading was observed for both the laminar and turbulent flow regimes, wherein the effect of NPs concentration is more significant in the turbulent flow regimes for the given flow Reynolds number and particle size [75]. While a marginal change in pressure drop was observed for low Re and NPs concentrations, a significant deviation occurring at high Re and NPs concentrations in nanofluids over base fluids highlights the practical limitations of their use [74]. In the developing region, the h_{NF} enhancement reduces with increasing axial distance from the test section entrance, wherein for higher NPs loading, a longer thermal entrance length is evaluated. With increasing particle loading, an induced wall shear stress was also evaluated, while some studies showed a very marginal change in the heat performance of nanofluids, much lower than that predicted from a conventional correlation [76, 77].

An optimal particle loading for maximum heat transfer at a minimum cost of operation is desirable. Interestingly, in several nanofluids, the optimal concentration of suspended NPs was found to be higher for increased nanofluid bulk temperature, increased Re of the base fluid, and the increased diameter-to-length ratio of the pipe. At the same time, it is practically independent of the nanoparticle diameter [78]. The h_{NF} and κ_{NF} in most studies is found to hold strong relation with the flow conditions and typically increases with volume flow rate, showing a nonlinear enhancement with NPs volume fraction. Such enhancements of nanofluids were directly proportional to the NPs concentration and Pe. Remarkably, h_{NF} shows a more significant enhancement than the κ_{NF} for nanofluids and has been ascribed to NPs' chaotic movements, fluctuations, and interactions that reduce the thermal boundary layer thickness. Optimisation assessments recommend an optimal NPs loading of about ≤1% for different NPs and applications. Stable nanofluids with the optimum NPs selection at an optimal volume concentration can thus be employed in thermal systems if challenges such as stability, high prices, and environmental impact are overcome.

10.8   APPLICATIONS

Nanofluidics is a widely explored domain under the umbrella of nanotechnology. Over the years, nanofluid research has grown from measuring and modelling several nanofluids' thermophysical properties to exploring the nanofluids' performance in numerous engineering applications, such as industrial applications, nuclear reactors, transportation, electronics, as well as biomedicine and food. In this section, we discuss prominent areas of applications for which nanofluids are actively explored and developed.

For decades, research and development activities on improving the heat transfer capabilities of heat exchangers have played a pre-eminent role in providing solutions to various thermal problems in various engineering applications. For instance, radiator (a heat exchanger) is a critical vehicle component that prevents the engine from overheating. An improved heating rate by modifications in geometries has enabled a significant reduction in the size and weight of the radiator, thereby reducing fuel consumption and improving efficiency. Similarly, miniaturisation and high-power density of modern electronic devices required thermal solutions for heat dissipation to realise higher performances, which was effectively addressed using microchannel heat sinks (heat exchangers) with cooling fluids. Heat exchangers are critical components in chemical and processing industries, as they maintain diverse components of machinery, water, chemical, and oils in a secure operating temperature range. Moreover, the heat transfer process in solar energy systems is critical for achieving the better performance of these systems with compact designs. In solar devices, efficient heat transfer enables more power generation and has fuel conservation features, by improving the energy utilisation efficiency and reducing the heat transfer time thus establishing a strong interlink between energy generation and its conversion.

10.8.1 Automotive

Thermal management in automotive engines is a critical issue affecting their performance and operability in harsh conditions. The heat exchanger used in vehicles is referred to as a radiator, whose cooling capabilities are directly related to the engine performance and fuel consumption. With rising heat generation due to higher engine power and exhaust gas recirculation, higher cooling needs are considered critical design challenges in engine cooling. As with existing cooling technology, it requires larger radiators with increased frontal areas, thereby increasing fuel consumption and aerodynamic drag. Stable nanofluids mainly address these limitations by enhancing the thermal properties of coolants and oils [79]. As demonstrated in many studies, a mixture of water and ethylene glycol is a standard engine coolant with poor heat transfer performance that can be significantly improved by NPs addition. However, problems such as surface erosion, dispersibility and agglomeration are routinely highlighted as major impediments in such studies. Thus, adapting nanofluids-based engine cooling, albeit capable of dissipating heat more efficiently requires optimal design. Several (TiO_2, Al_2O_3, Cu/graphene, and CeO_2) nanofluids as automobile engine lubricants showed a reduced friction and wear rate coefficient with enhanced brake thermal efficiency during experimental studies [79]. Also, Cu-based nanofluids showed higher boiling temperatures, viscosity, and conductivities as brake fluid, which can be developed to prevent the undesirable thermal degradation of the brake system during braking [80]. Thus, nanofluids' superior thermophysical, rheological, and tribological properties have been shown to improve operational capabilities by enhancing heat dissipation. Automotives with a better thermal design using nanofluids will offer increased power output, better mileage, and fewer emissions if associated challenges are overcome soon.

10.8.2 Electronics Cooling

Commercial liquid cooling kits using (CuO, Al_2O_3, and water) nanofluids are emerging products that lower the processor operating temperatures by convective cooling. A typical cooling system of CPUs contains an evaporator and condenser sections interconnected by a tube filled with a nanofluid. The growing acceptability of nanofluids as HTFs in microchannels is expected to enhance heat dissipation efficiency further [81]. Interestingly, nanofluids comprising nano-encapsulated PCM are emerging HTFs for electronic cooling and thermal energy storage, as they inherit a higher heat capacity and display a high latent heat of absorption. High particle loading in pre-existing nanofluids may offer high convective cooling, but it can also reduce hydraulic performance and cause significant pressure drop issues, which in turn increases power consumption. Designing heat sinks with flat and corrugated surfaces will enhance the turbulent intensity and mixing of fluid flow to provide a higher heat transfer rate. Ongoing research aims to integrate nanofluids in developing next-generation cooling devices for dissipating heat in ultrahigh-heat-flux electronic systems [81].

10.8.3 Nuclear Reactors

The feasibility assessment of nanofluids in nuclear power plant systems, such as pressurised water reactor (PWR) primary coolants, standby safety systems, accelerator targets, plasma diverters, etc., for heat exchange has shown excellent prospects in preliminary experiments [82]. For instance, in PWR, the obstructed critical heat flux (CHF) between the fuel's rods and the water due to vapour bubbles can be prevented with CHF enhancement using alumina-based nanofluids. Thus, the boiling of nanofluids shows an enhanced surface without altering the pool-boiling characteristics of water. Also, for faster cooling of overheated surfaces in critical systems and thermal control of plasma diverters and accelerator targets, nanofluids display better performances over traditional HTFs. However, in boiling water reactors, the NPs carried over to the turbine and condenser may enhance erosion and fouling, limiting their use. Moreover, fewer studies on nanofluid's thermophysical performance, particularly in prototypical reactor conditions with an obscure understanding of nanofluid's compatibility with the reactor materials, require more validation.

10.8.4 Solar Thermal Systems Applications

Solar energy conversion into thermal energy involves efficient heat transfer (convective, conductive, and radiative) between the mediums, which can be augmented using nanofluids. For solar thermal energy absorption, water, oil, and ethylene glycol are commonly HTFs. In several types of solar collectors, such as flat-plate, evacuated tube, cylindrical, conical, parabolic, and triangular shape solar collectors, the employability of nanofluids is actively studied [83]. It was found that for maximum efficiency, a lower concentration of NPs in a base fluid is optimal for most solar collectors excluding flat-plate collectors, where exergy efficiency increases with volume concentration. The expanding applicability of nanofluids holds excellent prospects in solar thermal systems, such as photovoltaic/thermal systems, solar water heaters, solar-geothermal combined cooling heating and power system, evaporative cooling for greenhouses, and water desalination, etc., for improving energy conversion efficiency.

10.8.5   Other Emerging Applications

Employing nanofluids in closed-loop cooling cycles by replacing the pre-existing HTFs, i.e. in most cases, water holds excellent prospects in energy savings and resulting in emissions reductions, particularly in chemical and process industries. In the extraction of geothermal power via drilling, the working equipments are often subjected to high friction and a high-temperature environment, wherein "fluid superconductor" nanofluids can provide rapid cooling for the machinery [84]. Also, by improved cooling and lubrication using nanofluids, the operability of sensors and electronics in harsh and higher-temperature surroundings can be significantly improved, enabling capabilities to access deeper, hotter regions in high-grade and economically acceptable temperatures in lower-grade formations. In machining industries, (SiO_2, MoS_2, Cu, Al_2O_3, and diamond) nanofluids as cutting fluid are widely demonstrated in turning, milling, drilling, and grinding processes for enhancing the cooling efficiency and lubricating the tool-workpiece interface [85]. Compared to traditional oil-based lubricants and dry machining, nanofluid provides better surface finish and cooling efficiency during machining.

10.9   CHALLENGES AND SUSTAINABILITY ASSESSMENT

High thermal conductivity and low viscosity are crucial attributes for selecting nanofluids in heat pipes. However, thermal conductivity enhancement is invariably observed at the expense of viscosity. The major bottlenecks for the applicability of nanofluids as HTFs shown in Figure 10.6 includes the absence of standardisation resulting in inconsistent data and conclusions on the nature and effects of particle dispersion. For instance, different dimensions of a test setup with varied concentrations and nanofluids type often skew any comparison of heat transfer and flow properties. The technical specifications, such as evaluation standards, filling rate, and inclination angle are yet to be formulated and adapted, impeding scalable application at the industry level. Although the round-robin test validates several observations, their adaptation in experimental studies remains critical for the intercomparison of several nanofluid alternatives and their suitability for various types of heat pipes under different application scenarios.

Among reported studies on thermal conductivity, investigations on convective heat transfer of nanofluids are few. Also, agglomeration tendency and sedimentation of NPs, usually increase during operation with time. Even the time-dependent evaluation of heat pipes charged with nanofluids is rarely considered. After use in the heat exchanger circuits, the manufactured nanofluids will get discharged to air, soils, and water systems, triggering environmental concerns. It has impeded the significant scalability of well-established nanofluids, which necessitates prior studies into nanofluids' physicochemical properties, behaviour, and toxicity under realistic conditions such as the natural aquatic environment. Moreover, the cost-effectiveness of formulating nanofluids and pumping power requirements further requires evaluation for assessing nanofluids' scalability and broader applicability.

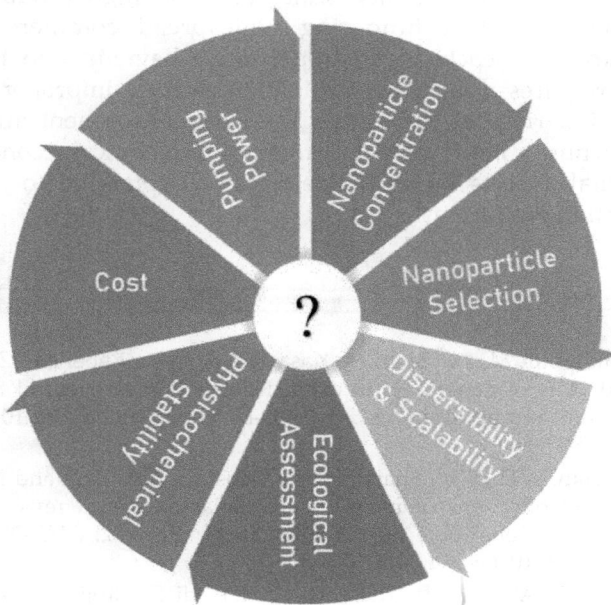

Figure 10.6   Challenges and sustainability aspects of nanofluids in heat transfer applications.

10.10 CONCLUSION AND OUTLOOK

The advent of nanoscience and nanotechnology and its rapid expansion over the past two decades have rejuvenated newer solutions to rapidly emerging heat transfer problems by employing nanofluids. Nanofluids, an engineered nanoscale colloidal solution, has attracted great attention from investigators for its superior thermal properties and has found wider acceptability in many applications, despite the ongoing challenges for addressing stability problems and inconsistencies within measurement techniques/reports and ecological implications. NPs of metals and their oxides, ceramics, and carbon composites were mainly utilised to formulate homogeneous nanofluids with the desired concentration. The significance of the complex interaction between NPs and base fluids in determining thermal transport based on the existing mechanism of heat transfer with inputs from theoretical and experimental research is yet to reach conformity in explaining the anomalous enhancement. However, the synergy within an integrated experimental and theoretical approach is being actively applied to guide the development of nanofluids with maximum heat transfer. Nanofluids' enhanced heat transfer capabilities make their use in heat exchangers an exciting option, leading to better system performance and abilities in energy efficiency. Conversely, nanofluid's stability, scalability, and high production cost are detrimental factors hindering widespread commercialisation. Moreover, the physicochemical properties, behaviour, and toxicity of nanofluids requires a thorough evaluation as their improper handling poses potential risks of disrupting the soil environment and aquatic ecosystem. Thus, a more careful evaluation of technical, economic, and environmental aspects may provide wider acceptability to nanofluids for large-scale applications.

REFERENCES

[1] Sheikhpour, M., M. Arabi, A. Kasaeian, A. Rokn Rabei and Z. Taherian. 2020. Role of nanofluids in drug delivery and biomedical technology: methods and applications. NSA 13: 47–59. https://doi.org/10.2147/NSA.S260374.

[2] Elsheikh, A.H., S.W. Sharshir, M.E. Mostafa, F.A. Essa and M.K.A. Ali. 2018. Applications of nanofluids in solar energy: a review of recent advances. Renewable Sustainable Energy Rev. 82: 3483–3502. https://doi.org/10.1016/j.rser.2017.10.108.

[3] Xiong, Q., A. Hajjar, B. Alshuraiaan, M. Izadi, S. Altnji and S.A. Shehzad. 2021. State-of-the-art review of nanofluids in solar collectors: a review based on the type of the dispersed nanoparticles. J. Cleaner Prod. 310: 127528. https://doi.org/10.1016/j.jclepro.2021.127528.

[4] Rasheed, A.K., M. Khalid, W. Rashmi, T.C.S.M. Gupta and A. Chan. 2016. Graphene based nanofluids and nanolubricants—review of recent developments. Renewable Sustainable Energy Rev. 63: 346–362. https://doi.org/10.1016/j.rser.2016.04.072.

[5] Chamkha, A.J., M. Molana, A. Rahnama and F. Ghadami. 2018. On the nanofluids applications in microchannels: a comprehensive review. Powder Technol. 332: 287–322. https://doi.org/10.1016/j.powtec.2018.03.044.

[6] Sheremet, M.A. 2021. Applications of nanofluids. Nanomaterials 11: 1716. https://doi.org/10.3390/nano11071716.

[7] Choi, S.U.S. 2009. Nanofluids: from vision to reality through research. J. Heat Transfer. 131: 033106. https://doi.org/10.1115/1.3056479.

[8] Choi, S.U.S. and J. Eastman. 1995. Enhancing thermal conductivity of fluids with nanoparticles, developments and applications of non-newtonian flows. ASME, New York. 38.

[9] Devendiran, D.K. and V.A. Amirtham. 2016. A review on preparation, characterization, properties and applications of nanofluids. Renewable Sustainable Energy Rev. 60: 21–40. https://doi.org/10.1016/j.rser.2016.01.055.

[10] Souza, R.R., I.M. Gonçalves, R.O. Rodrigues, G. Minas, J.M. Miranda, A.L.N. Moreira, et al. 2022. Recent advances on the thermal properties and applications of nanofluids: from nanomedicine to renewable energies. Appl. Therm. Eng. 201: 117725. https://doi.org/10.1016/j.applthermaleng.2021.117725.

[11] Wu, D., H. Zhu, L. Wang and L. Liu. 2009. Critical issues in nanofluids preparation, characterization and thermal conductivity. Curr. Nanosci. 5: 103–112. https://doi.org/10.2174/157341309787314548.

[12] Okonkwo, E.C., I. Wole-Osho, I.W. Almanassra, Y.M. Abdullatif and T. Al-Ansari. 2021. An updated review of nanofluids in various heat transfer devices. J. Therm. Anal. Calorim. 145: 2817–2872. https://doi.org/10.1007/s10973-020-09760-2.

[13] Adun, H., I. Wole-Osho, E.C. Okonkwo, D. Kavaz and M. Dagbasi. 2021. A critical review of specific heat capacity of hybrid nanofluids for thermal energy applications. J. Mol. Liq. 340: 116890. https://doi.org/10.1016/j.molliq.2021.116890.

[14] Awais, M., A.A. Bhuiyan, S. Salehin, M.M. Ehsan, B. Khan and Md.H. Rahman. 2021. Synthesis, heat transport mechanisms and thermophysical properties of nanofluids: A critical overview, Int. J. Thermofluids 10: 100086. https://doi.org/10.1016/j.ijft.2021.100086.

[15] Keblinski, P., S.R. Phillpot, S.U.S. Choi and J.A. Eastman. 2002. Mechanisms of heat flow in suspensions of nano-sized particles (nanofluids). Int. J. Heat Mass Transfer. 45: 855–863. https://doi.org/10.1016/S0017-9310(01)00175-2.

[16] Pinto, R.V. and F.A.S. Fiorelli. 2016. Review of the mechanisms responsible for heat transfer enhancement using nanofluids. Appl. Therm. Eng. 108: 720–739. https://doi.org/10.1016/j.applthermaleng.2016.07.147.

[17] Wang, L. and J. Fan. 2010. Nanofluids research: key issues. Nanoscale Res. Lett. 5: 1241–1252. https://doi.org/10.1007/s11671-010-9638-6.

[18] Buongiorno, J., D.C. Venerus, N. Prabhat, T. McKrell, J. Townsend, R. Christianson, et al. 2009. A benchmark study on the thermal conductivity of nanofluids. J. Appl. Phys. 106: 094312. https://doi.org/10.1063/1.3245330.

[19] Lee, J.-H., S.-H. Lee, C. Choi, S. Jang and S. Choi. 2010. A review of thermal conductivity data, mechanisms and models for nanofluids. Int. J. Micro-Nano Scale Transp. 1: 269–322. https://doi.org/10.1260/1759-3093.1.4.269.

[20] Lee, W.-H., C.-K. Rhee, J. Koo, J. Lee, S.P. Jang, S.U. Choi, et al. 2011. Round-robin test on thermal conductivity measurement of ZnO nanofluids and comparison of experimental results with theoretical bounds. Nanoscale Res. Lett. 6: 258. https://doi.org/10.1186/1556-276X-6-258.

[21] Maxwell, J.C.. 1881. A Treatise on Electricity and Magnetism: Pt. III. Magnetism. Pt. IV. Electromagnetism. Clarendon Press.

[22] Angayarkanni, S.A. and J. Philip. 2015. Review on thermal properties of nanofluids: recent developments. Adv. Colloid Interface Sci. 225: 146–176. https://doi.org/10.1016/j.cis.2015.08.014.

[23] Philip, J. and P.D. Shima. 2012. Thermal properties of nanofluids, Adv. Colloid Interface Sci. 183–184: 30–45. https://doi.org/10.1016/j.cis.2012.08.001.

[24] Murshed, S.M.S., K.C. Leong and C. Yang. 2008. Thermophysical and electrokinetic properties of nanofluids—a critical review. Appl. Therm. Eng. 28: 2109–2125. https://doi.org/10.1016/j.applthermaleng.2008.01.005.

[25] Bruggeman, D.A.G. 1935. Berechnung verschiedener physikalischer Konstanten von heterogenen Substanzen. I. Dielektrizitätskonstanten und Leitfähigkeiten der Mischkörper aus isotropen Substanzen. Ann. Phys. 416: 636–664. https://doi.org/10.1002/andp.19354160705.

[26] Hamilton, R.L. and O.K. Crosser. 1962. Thermal conductivity of heterogeneous two-component systems. Ind. Eng. Chem. Fund. 1: 187–191. https://doi.org/10.1021/i160003a005.

[27] Das, S.K., S.U.S. Choi, W. Yu and T. Pradeep. 2007. Nanofluids. John Wiley & Sons, Inc., Hoboken, NJ, USA. https://doi.org/10.1002/9780470180693.

[28] Xuan, Y. and W. Roetzel. 2000. Conceptions for heat transfer correlation of nanofluids. Int. J. Heat Mass Transfer. 43: 3701–3707. https://doi.org/10.1016/S0017-9310(99)00369-5.

[29] Bhattacharya, P., S.K. Saha, A. Yadav, P.E. Phelan and R.S. Prasher. 2004. Brownian dynamics simulation to determine the effective thermal conductivity of nanofluids. J. Appl. Phys. 95: 6492–6494. https://doi.org/10.1063/1.1736319.

[30] Jang, S.P. and S.U.S. Choi. 2004. Role of Brownian motion in the enhanced thermal conductivity of nanofluids. Appl. Phys. Lett. 84: 4316–4318. https://doi.org/10.1063/1.1756684.

[31] Prasher, R., P. Bhattacharya and P.E. Phelan. 2006. Brownian-Motion-Based Convective-Conductive Model for the Effective Thermal Conductivity of Nanofluids. J. Heat Transfer 128: 588–595. https://doi.org/10.1115/1.2188509.

[32] Shima, P.D., J. Philip and B. Raj. 2009. Role of microconvection induced by Brownian motion of nanoparticles in the enhanced thermal conductivity of stable nanofluids. Appl. Phys. Lett. 94: 223101. https://doi.org/10.1063/1.3147855.

[33] Prasher, R., W. Evans, P. Meakin, J. Fish, P. Phelan and P. Keblinski. 2006. Effect of aggregation on thermal conduction in colloidal nanofluids. Appl. Phys. Lett. 89: 143119. https://doi.org/10.1063/1.2360229.

[34] Xuan, Y., Q. Li and W. Hu. 2003. Aggregation structure and thermal conductivity of nanofluids. AIChE J. 49: 1038–1043. https://doi.org/10.1002/aic.690490420.

[35] Xue, L., P. Keblinski, S.R. Phillpot, S.U.-S. Choi and J.A. Eastman. 2004. Effect of liquid layering at the liquid–solid interface on thermal transport. Int. J. Heat Mass Transfer. 47: 4277–4284. https://doi.org/10.1016/j.ijheatmasstransfer.2004.05.016.

[36] Talbot, L., R.K. Cheng, R.W. Schefer and D.R. Willis. 1980. Thermophoresis of particles in a heated boundary layer. J. Fluid Mech. 101: 737–758. https://doi.org/10.1017/S0022112080001905.

[37] Piazza, R. and A. Parola. 2008. Thermophoresis in colloidal suspensions. J. Phys.: Condens. Matter. 20: 153102. https://doi.org/10.1088/0953-8984/20/15/153102.

[38] Xu, M. and L. Wang. 2002. Thermal oscillation and resonance in dual-phase-lagging heat conduction. Int. J. Heat Mass Transfer. 45: 1055–1061. https://doi.org/10.1016/S0017-9310(01)00199-5.

[39] Xu, M. and L. Wang. 2005. Dual-phase-lagging heat conduction based on Boltzmann transport equation. Int. J. Heat Mass Transfer. 48: 5616–5624. https://doi.org/10.1016/j.ijheatmasstransfer.2005.05.040.

[40] Wasp, E.J., J.P. Kenny and R.L. Gandhi. 1977. Solid--liquid flow: slurry pipeline transportation. [Pumps, valves, mechanical equipment, economics]. Ser. Bulk Mater. Handl. (United States). 1.

[41] Davis, R.H. 1986. The effective thermal conductivity of a composite material with spherical inclusions. Int. J. Thermophys. 7: 609–620. https://doi.org/10.1007/BF00502394.

[42] Lu, S. and H. Lin. 1996. Effective conductivity of composites containing aligned spheroidal inclusions of finite conductivity. J. Appl. Phys. 79: 6761–6769. https://doi.org/10.1063/1.361498.

[43] Jeffrey, D.J. 1973. Conduction through a random suspension of spheres. Proc. R. Soc. Lond. A 335: 355–367. http://doi.org/10.1098/rspa.1973.0130.

[44] Xie, H., M. Fujii and X. Zhang. 2005. Effect of interfacial nanolayer on the effective thermal conductivity of nanoparticle-fluid mixture. Int. J. Heat Mass Transfer. 48: 2926–2932. https://doi.org/10.1016/j.ijheatmasstransfer.2004.10.040.

[45]   Yu, W. and S.U.S. Choi. 2003. The role of interfacial layers in the enhanced thermal conductivity of nanofluids: a renovated maxwell model. J. Nanopart. Res. 5: 167–171. https://doi.org/10.1023/A:1024438603801.

[46]   Koo, J. and C. Kleinstreuer. 2004. A new thermal conductivity model for nanofluids. J. Nanopart Res. 6: 577–588. https://doi.org/10.1007/s11051-004-3170-5.

[47]   Koo, J. and C. Kleinstreuer. 2005. Laminar nanofluid flow in microheatsinks. Int. J. Heat Mass Transfer. 48: 2652–2661. https://doi.org/10.1016/j.ijheatmasstransfer.2005.01.029.

[48]   Chon, C.H., K.D. Kihm, S.P. Lee and S.U.S. Choi. 2005. Empirical correlation finding the role of temperature and particle size for nanofluid (Al_2O_3) thermal conductivity enhancement. Appl. Phys. Lett. 87: 153107. https://doi.org/10.1063/1.2093936.

[49]   Hwang, Y., J.K. Lee, C.H. Lee, Y.M. Jung, S.I. Cheong, C.G. Lee, et al. 2007. Stability and thermal conductivity characteristics of nanofluids. Thermochim. Acta 455: 70–74. https://doi.org/10.1016/j.tca.2006.11.036.

[50]   Che Sidik, N.A., M. Mahmud Jamil, W.M.A. Aziz Japar and I. Muhammad Adamu. 2017. A review on preparation methods, stability and applications of hybrid nanofluids. Renewable Sustainable Energy Rev. 80: 1112–1122. https://doi.org/10.1016/j.rser.2017.05.221.

[51]   Suganthi, K.S. and K.S. Rajan. 2017. Metal oxide nanofluids: review of formulation, thermo-physical properties, mechanisms, and heat transfer performance. Renewable Sustainable Energy Rev. 76: 226–255. https://doi.org/10.1016/j.rser.2017.03.043.

[52]   Patel, H.E., T. Sundararajan and S.K. Das. 2010. An experimental investigation into the thermal conductivity enhancement in oxide and metallic nanofluids. J. Nanopart Res. 12: 1015–1031. https://doi.org/10.1007/s11051-009-9658-2.

[53]   Das, S.K. and S.U.S. Choi. 2009. A review of heat transfer in nanofluids. Adv. Heat Transfer 41: 81–197. https://doi.org/10.1016/S0065-2717(08)41002-X.

[54]   Huminic, G., A. Huminic, F. Dumitrache, C. Fleacă and I. Morjan. 2020. Study of the thermal conductivity of hybrid nanofluids: recent research and experimental study. Powder Technol. 367: 347–357. https://doi.org/10.1016/j.powtec.2020.03.052.

[55]   Sadeghinezhad, E., M. Mehrali, R. Saidur, M. Mehrali, S. Tahan Latibari, A.R. Akhiani, et al. 2016. A comprehensive review on graphene nanofluids: recent research, development and applications. Energy Convers. Manage. 111: 466–487. https://doi.org/10.1016/j.enconman.2016.01.004.

[56]   Khadangi Mahrood, M.R., S.G. Etemad and R. Bagheri. 2011. Free convection heat transfer of non-Newtonian nanofluids under constant heat flux condition. Int. Commun. Heat Mass Transfer 38: 1449–1454. https://doi.org/10.1016/j.icheatmasstransfer.2011.08.012.

[57]   Hojjat, M., S.Gh. Etemad, R. Bagheri and J. Thibault. 2011. Convective heat transfer of non-Newtonian nanofluids through a uniformly heated

circular tube. Int. J. Therm. Sci. 50: 525–531. https://doi.org/10.1016/j.ijthermalsci.2010.11.006.

[58] Anoop, K.B., T. Sundararajan and S.K. Das. 2009. Effect of particle size on the convective heat transfer in nanofluid in the developing region. Int. J. Heat Mass Transfer 52: 2189–2195. https://doi.org/10.1016/j.ijheatmasstransfer.2007.11.063.

[59] Ding, Y., H. Alias, D. Wen and R.A. Williams. 2006. Heat transfer of aqueous suspensions of carbon nanotubes (CNT nanofluids). Int. J. Heat Mass Transfer 49: 240–250. https://doi.org/10.1016/j.ijheatmasstransfer.2005.07.009.

[60] Ferrouillat, S., A. Bontemps, O. Poncelet, O. Soriano and J.-A. Gruss. 2013. Influence of nanoparticle shape factor on convective heat transfer and energetic performance of water-based SiO_2 and ZnO nanofluids. Appl. Therm. Eng. 51: 839–851. https://doi.org/10.1016/j.applthermaleng.2012.10.020.

[61] Timofeeva, E.V., J.L. Routbort and D. Singh. 2009. Particle shape effects on thermophysical properties of alumina nanofluids. J. Appl. Phys. 106: 014304. https://doi.org/10.1063/1.3155999.

[62] Lee, S., S.U.-S. Choi, S. Li and J.A. Eastman. 1999. Measuring thermal conductivity of fluids containing oxide nanoparticles. J. Heat Transfer 121: 280–289. https://doi.org/10.1115/1.2825978.

[63] Das, S.K., N. Putra, P. Thiesen and W. Roetzel. 2003. Temperature dependence of thermal conductivity enhancement for nanofluids. J. Heat Transfer. 125: 567–574. https://doi.org/10.1115/1.1571080.

[64] Yang, B. and Z.H. Han. 2006. Temperature-dependent thermal conductivity of nanorod-based nanofluids, Appl. Phys. Lett. 89: 083111. https://doi.org/10.1063/1.2338424.

[65] Wu, J.M. and J. Zhao. 2013. A review of nanofluid heat transfer and critical heat flux enhancement—research gap to engineering application, Prog. Nucl. Energy 66: 13–24. https://doi.org/10.1016/j.pnucene.2013.03.009.

[66] Pordanjani, A.H., S. Aghakhani, M. Afrand, M. Sharifpur, J.P. Meyer, H. Xu, et al. 2021. Nanofluids: physical phenomena, applications in thermal systems and the environment effects—a critical review. J. Cleaner Prod. 320: 128573. https://doi.org/10.1016/j.jclepro.2021.128573.

[67] Keblinski, P., J.A. Eastman and D.G. Cahill. 2005. Nanofluids for thermal transport, Materials Today 8: 36–44. https://doi.org/10.1016/S1369-7021(05)70936-6.

[68] Paul, G., M. Chopkar, I. Manna and P.K. Das. 2010. Techniques for measuring the thermal conductivity of nanofluids: a review. Renewable Sustainable Energy Rev. 14: 1913–1924. https://doi.org/10.1016/j.rser.2010.03.017.

[69] Souza, R.R., V. Faustino, I.M. Gonçalves, A.S. Moita, M. Bañobre-López and R. Lima. 2022. A review of the advances and challenges in measuring the thermal conductivity of nanofluids. Nanomaterials 12: 2526. https://doi.org/10.3390/nano12152526.

[70] Pryazhnikov, M.I., A.V. Minakov, V.Ya. Rudyak and D.V. Guzei. 2017. Thermal conductivity measurements of nanofluids. Int. J. Heat Mass Transfer. 104 (2017) 1275–1282. https://doi.org/10.1016/j.ijheatmass transfer.2016.09.080.

[71] Pak, B.C. and Y.I. Cho. 1998. Hydrodynamic and heat transfer study of dispersed fluids with submicron metallic oxide particles. Exp. Heat Transfer. 11: 151–170. https://doi.org/10.1080/08916159808946559.

[72] Pandya, N.S., H. Shah, M. Molana and A.K. Tiwari. 2020. Heat transfer enhancement with nanofluids in plate heat exchangers: a comprehensive review. Eur. J. Mech. B. Fluids 81: 173–190. https://doi.org/10.1016/j. euromechflu.2020.02.004.

[73] Bretado-de los Rios, M.S., C.I. Rivera-Solorio and K.D.P. Nigam. 2021. An overview of sustainability of heat exchangers and solar thermal applications with nanofluids: a review. Renewable Sustainable Energy Rev. 142: 110855. https://doi.org/10.1016/j.rser.2021.110855.

[74] Sivashanmugam, P. 2012. Application of nanofluids in heat transfer. *In:* S.N. Kazi (ed.). An Overview of Heat Transfer Phenomena. InTech. https://doi.org/10.5772/52496.

[75] He, Y., Y. Jin, H. Chen, Y. Ding, D. Cang and H. Lu. 2007. Heat transfer and flow behaviour of aqueous suspensions of TiO_2 nanoparticles (nanofluids) flowing upward through a vertical pipe, Int. J. Heat Mass Transfer. 50: 2272–2281. https://doi.org/10.1016/j. ijheatmasstransfer.2006.10.024.

[76] Fotukian, S.M. and M. Nasr Esfahany. 2010. Experimental investigation of turbulent convective heat transfer of dilute γ-Al_2O_3/water nanofluid inside a circular tube. Int. J. Heat Fluid Flow. 31: 606–612. https://doi. org/10.1016/j.ijheatfluidflow.2010.02.020.

[77] Fotukian, S.M. and M. Nasr Esfahany. 2010. Experimental study of turbulent convective heat transfer and pressure drop of dilute CuO/ water nanofluid inside a circular tube. Int. Commun. Heat Mass Transfer 37: 214–219. https://doi.org/10.1016/j.icheatmasstransfer.2009.10.003.

[78] Corcione, M., M. Cianfrini and A. Quintino. 2012. Heat transfer of nanofluids in turbulent pipe flow. Int. J. Therm. Sci. 56: 58–69. https:// doi.org/10.1016/j.ijthermalsci.2012.01.009.

[79] Patel, J., A. Soni, D.P. Barai and B.A. Bhanvase. 2023. A minireview on nanofluids for automotive applications: current status and future perspectives. Appl. Therm. Eng. 219: 119428. https://doi.org/10.1016/j. applthermaleng.2022.119428.

[80] Kao, M.J., C.H. Lo, T.T. Tsung, Y.Y. Wu, C.S. Jwo and H.M. Lin. 2007. Copper-oxide brake nanofluid manufactured using arc-submerged nanoparticle synthesis system. J. Alloys Compd. 434–435: 672–674. https://doi.org/10.1016/j.jallcom.2006.08.305.

[81] Bahiraei, M. and S. Heshmatian. 2018. Electronics cooling with nanofluids: a critical review, energy convers. Manage. 172: 438–456. https://doi.org/10.1016/j.enconman.2018.07.047.

[82] Buongiorno, J., L.-W. Hu, S.J. Kim, R. Hannink, B. Truong and E. Forrest. 2008. Nanofluids for enhanced economics and safety of nuclear reactors: an evaluation of the potential features, issues, and research gaps. Nuclear Techno. 162: 80–91. https://doi.org/10.13182/NT08-A3934.

[83] Wahab, A., A. Hassan, M.A. Qasim, H.M. Ali, H. Babar and M.U. Sajid. 2019. Solar energy systems—potential of nanofluids. J. Mol. Liq. 289: 111049. https://doi.org/10.1016/j.molliq.2019.111049.

[84] Soltani, M., F. Moradi Kashkooli, M. Alian Fini, D. Gharapetian, J. Nathwani and M.B. Dusseault. 2022. A review of nanotechnology fluid applications in geothermal energy systems. Renewable Sustainable Energy Rev. 167: 112729. https://doi.org/10.1016/j.rser.2022.112729.

[85] Hemmat Esfe, M., M. Bahiraei and A. Mir. 2020. Application of conventional and hybrid nanofluids in different machining processes: a critical review. Adv. Colloid Interface Sci. 282: 102199. https://doi.org/10.1016/j.cis.2020.102199.

Index

For Product Safety Concerns and Information please contact our EU
representative GPSR@taylorandfrancis.com
Taylor & Francis Verlag GmbH, Kaufingerstraße 24, 80331 München, Germany